PLANT TISSUE CULTURE: BASIC AND APPLIED

Plant Tissue Culture: Basic and Applied

Editor:

Sarika Garg

Delve Publishing

www.delvepublishing.com

Plant Tissue Culture: Basic and Applied

Editor: **Sarika Garg**

708 3rd Avenue, 6th Floor
New York, NY 10017
United States of America
www.delvepublishing.com

ISBN: 978-1-68095-763-1

Library of Congress Control Number: 2016951519

About the Editor

Sarika Garg

Ph.D. Montreal, Canada

Sarika Garg procured her doctoral degree at the Max Planck Unit For Structural Molecular Biology, University of Hamburg, Germany in 2011. This was followed by Postdoctoral Scientist position at University of Saskatchewan, Canada. Later, she moved to CRCHUM, University of Montreal for her second Postdoc. She has diverse research interests and has contributed immensely in many research fields viz. Enzyme Technology, Environmental Science, and Neuroscience. She has numerous research papers and review articles to her credit. She has coauthored two books 'Enzymes & Enzyme Technology'; and 'Biochemical Tests: Principles and Protocols'. She has also contributed several chapters in various books including a chapter on Regulation of Gene Expression in Encyclopedia of Molecular Cell Biology and Molecular Medicine. She is a recipient of 2012 SHRF Postdoctoral Fellowship Award and 2012 Laura E. Chapman Research Award. She is currently working as a Career Counselor in her very own venture 'higherstudiescounseling.com', Montreal, Canada and as a freelancer science writer.

Contents

List of Contributors

Kumar A (2016)
School of Biotechnology, Devi Ahilya University, Indore, India

Iyyakkannu Sivanesan
Department of Molecular Biotechnology, Konkuk University, Seoul, South Korea

Se Won Park
Department of Molecular Biotechnology, Konkuk University, Seoul, South Korea

Katarzyna Mac´kowska
Institute of Plant Biology and Biotechnology, Faculty of Horticulture, University of Agriculture in Krakow, Al. 29 Listopada 54, 31-425 Krako´w, Poland

Agata Jarosz
Institute of Plant Biology and Biotechnology, Faculty of Horticulture, University of Agriculture in Krakow, Al. 29 Listopada 54, 31-425 Krako´w, Poland

Ewa Grzebelus
Institute of Plant Biology and Biotechnology, Faculty of Horticulture, University of Agriculture in Krakow, Al. 29 Listopada 54, 31-425 Krako´w, Poland

Ewa Grzebelus Maria
Department of Genetics, Plant Breeding and Seed Science, Faculty of Horticulture, University of Agriculture in Krakow, Al. 29 Listopada 54, 31-425 Krako´w, Poland

Kruk Alicja
Department of Genetics, Plant Breeding and Seed Science, Faculty of Horticulture, University of Agriculture in Krakow, Al. 29 Listopada 54, 31-425 Krako´w, Poland

Macko-Podgo´rni
Department of Genetics, Plant Breeding and Seed Science, Faculty of Horticulture, University of Agriculture in Krakow, Al. 29 Listopada 54, 31-425 Krako´w, Poland

Dariusz Grzebelus
Department of Genetics, Plant Breeding and Seed Science, Faculty of Horticulture, University of Agriculture in Krakow, Al. 29 Listopada 54, 31-425 Krako´w, Poland

Aneta Gerszberg ,Katarzyna
Department of Genetics, Plant Molecular Biology and Biotechnology, University of Lodz, Banacha Street 12/16, 90-237 Lodz, Poland

Hnatuszko-Konka
Department of Genetics, Plant Molecular Biology and Biotechnology, University of Lodz, Banacha Street 12/16, 90-237 Lodz, Poland

Tomasz Kowalczyk,
Department of Genetics, Plant Molecular Biology and Biotechnology, University of Lodz, Banacha Street 12/16, 90-237 Lodz, Poland

Andrzej K. Kononowicz
Department of Genetics, Plant Molecular Biology and Biotechnology, University of Lodz, Banacha Street 12/16, 90-237 Lodz, Poland

J. Machczyn´ska R. Orłowska P. T. Bednarek
Department of Plant Physiology and Biochemistry, Plant Breeding and Acclimatization Institute, National Research Institute, 05-870 Błonie, Radziko´w, Poland

D. R. Man´kowski
Department of Seed Science and Technology, Plant Breeding and Acclimatization Institute, National Research Institute, 05-870 Błonie, Radziko´w, Poland

J. Zimny
Department of Plant Biotechnology and Cytogenetics, Plant Breeding and Acclimatization Institute, National Research Institute, 05-870 Błonie, Radziko´w, Poland

Roberto Bobadilla Landey
UMR IPME, CIRAD, 911 Avenue Agropolis, BP 64501, 34394 Montpellier, France

Alberto Cenci
Bioversity, Montpellier, France

Romain Guyot
UMR IPME, IRD, 911 Avenue Agropolis, BP 64501, 34394 Montpellier, France

Benoıˆt Bertrand
UMR IPME, CIRAD, 911 Avenue Agropolis, BP 64501, 34394 Montpellier, France

Fre´de´ric Georget
UMR IPME, CIRAD, 911 Avenue Agropolis, BP 64501, 34394 Montpellier, France

Eveline Dechamp
UMR IPME, CIRAD, 911 Avenue Agropolis, BP 64501, 34394 Montpellier, France

Juan-Carlos Herrera
Centro Nacional de Investigaciones de Cafe´, CENICAFE, Manizales, Caldas, Colombia

Jamel Aribi
UMR IPME, IRD, 911 Avenue Agropolis, BP 64501, 34394 Montpellier, France

Philippe Lashermes
UMR DIADE, IRD, 911 Avenue Agropolis, BP 64501, 34394 Montpellier, France

Herve´ Etienne
UMR IPME, CIRAD, 911 Avenue Agropolis, BP 64501, 34394 Montpellier, France

Ewa Skała
Department of Biology and Pharmaceutical Botany, Medical University of Ło´dz´, Muszyn´skiego 1, 90-151 Ło´dz´, Poland

Renata Gra˛bkowska
Department of Biology and Pharmaceutical Botany, Medical University of Ło´dz´, Muszyn´skiego 1, 90-151 Ło´dz´, Poland

Przemysław Sitarek
Department of Biology and Pharmaceutical Botany, Medical University of Ło´dz´, Muszyn´skiego 1, 90-151 Ło´dz´, Poland

Łukasz Kuz´ma
Department of Biology and Pharmaceutical Botany, Medical University of Ło´dz´, Muszyn´skiego 1, 90-151 Ło´dz´, Poland

Andrzej Błauz_
Cytometry Laboratory, Department of Molecular Biophysics, University of Lo´dz´, ul. Pomorska 141/143, 90-236 Ło´dz´, Poland

Halina Wysokin´ska
.Department of Biology and Pharmaceutical Botany, Medical University of Ło´dz´, Muszyn´skiego 1, 90-151 Ło´dz´, Poland

Inga Kwiecien´
Chair and Department of Pharmaceutical Botany, Jagiellonian University, Collegium Medicum, Medyczna Street 9, 30-688 Krako´w, Poland

Aleksandra Szydłowska
Chair and Department of Pharmaceutical Botany, Jagiellonian University, Collegium Medicum, Medyczna Street 9, 30-688 Krako´w, Poland

Beata Kawka
Chair and Department of Pharmaceutical Botany, Jagiellonian University, Collegium Medicum, Medyczna Street 9, 30-688 Krako´w, Poland

Ludger Beerhues
Institut fu¨r Pharmazeutische Biologie, Technische Universita¨t Braunschweig, Mendelssohnstrasse 1, 38106 Braunschweig, Germany

Halina Ekiert
Chair and Department of Pharmaceutical Botany, Jagiellonian University, Collegium Medicum, Medyczna Street 9, 30-688 Krako´w, Poland

Izabela Grzegorczyk-Karolak
Department of Biology and Pharmaceutical Botany, Medical University of Ło´dz´, ul. Muszyn´skiego 1, 90-151 Lodz, Poland

Łukasz Kuz´ma
Department of Biology and Pharmaceutical Botany, Medical University of Ło´dz´, ul. Muszyn´skiego 1, 90-151 Lodz, Poland

Halina Wysokin´ska
Department of Biology and Pharmaceutical Botany, Medical University of Ło´dz´, ul. Muszyn´skiego 1, 90-151 Lodz, Poland

A. Kokotkiewicz M. Luczkiewicz
Department of Pharmacognosy, Faculty of Pharmacy, Medical University of Gdansk, al. gen. J. Hallera 107, 80-416 Gdan´sk, Poland

A. Bucinski
Department of Biopharmacy, Faculty of Pharmacy, Ludwik Rydygier Collegium Medicum in Bydgoszcz, Nicolaus Copernicus University in Torun, ul. dr A. Jurasza 2, 85-089 Bydgoszcz, Poland

Danuta Raj
Department of Pharmacognosy, Faculty of Pharmacy, Wroclaw Medical University, ul. Borowska 211 A, 50-556 Wrocław, Poland

Adam Kokotkiewicz
Department of Pharmacognosy, Faculty of Pharmacy, Medical University of Gdansk, al. gen. J. Hallera 107, 80-416 Gdan´sk, Poland

Andrzej Drys
Department of Physical Chemistry, Faculty of Pharmacy, Wroclaw Medical University, ul. Borowska 211 A, 50-556 Wrocław, Poland

Maria Luczkiewicz
Department of Pharmacognosy, Faculty of Pharmacy, Medical University of Gdansk, al. gen. J. Hallera 107, 80-416 Gdan´sk, Poland

Karolina Tomiczak
Department of Experimental Plant Biology, Polish Academy of Sciences Botanical Garden, Center for Biological Diversity Conservation in Powsin, Prawdziwka 2, 02-973 Warsaw, Poland

Elwira Sliwinska
Laboratory of Molecular Biology and Cytometry, Department of Plant Genetics, Physiology and Biotechnology, UTP University of Science and Technology, Kaliskiego Ave. 7, 85-789 Bydgoszcz, Poland

Jan J. Rybczyn´ski
Department of Experimental Plant Biology, Polish Academy of Sciences Botanical Garden, Center for Biological Diversity Conservation in Powsin, Prawdziwka 2, 02-973 Warsaw, Poland

Munguatosha Ngomuo
School of Life Sciences, Nelson Mandela African Institute of Science and Technology, Arusha, Tanzania

Emerald Mneney
Mikocheni Agricultural Research Institute, Dar es Salaam, Tanzania

Patrick A. Ndakidemi
School of Life Sciences, Nelson Mandela African Institute of Science and Technology, Arusha, Tanzania

Ciddi Veeresham
University college of Pharmaceutical Sciences, Kakatiya University, Warangal, Andhra Pradesh, India

Praveena Chitti
University college of Pharmaceutical Sciences, Kakatiya University, Warangal, Andhra Pradesh, India

L Chalak
Lebanese Agricultural Research Institute, Tal Amara, P.O. Box 287, Zahlé, Lebanon
Lebanese University, Faculty of Agricultural Sciences, Dekwaneh, Beirut, Lebanon

A Elbitar
Lebanese Agricultural Research Institute, Tal Amara, P.O. Box 287, Zahlé, Lebanon

N Mourad
Lebanese Agricultural Research Institute, Tal Amara, P.O. Box 287, Zahlé, Lebanon

C Mortada
Lebanese Agricultural Research Institute, Tal Amara, P.O. Box 287, Zahlé, Lebanon

E Choueiri
Lebanese Agricultural Research Institute, Tal Amara, P.O. Box 287, Zahlé, Lebanon

Hatıra Taşkın
Department of Plant Production and Technologies, Faculty of Agricultural Sciences and Technologies, University of Niğde, 51240 Niğde, Turkey

Gökhan Baktemur
Department of Biology, Faculty of Arts and Science, University of Osmaniye Korkut Ata, 80000 Osmaniye, Turkey

Mehmet Kurul,
Roche Diagnostics, 34394 Istanbul, Turkey

Saadet Büyükalaca
Department of Horticulture, Faculty of Agriculture, University of Çukurova, 01330 Adana, Turkey

Preface

Plant tissue culture (PTC) is the regeneration of a whole plant from plant cells, tissues or organs in a suitable nutrient medium under aseptic conditions. This plant sciences methodology is much useful approach for natural resources in an era where these resources are largely based on a number of economical and environmental factors. Regenerated plants are favorable for boosting the quality of human life. This book brings together a wide range of published research studies under this central theme and highlights the large number of commercial plants whose growth and development have been upgraded using tissue culture techniques.

The first chapter reviews the history and basics of PTC. The main steps and various techniques of PTC are briefly discussed. In Chapter 2, the role of silicon in PTC has been reviewed. The plant regeneration from leaf protoplast cultures of Daucus genus to facilitate their use in somatic hybridization with cultivated carrot; and carrot plants regeneration with improved tolerance to black rot are discussed in Chapters 3 and 4, respectively. The next chapter reviews the efficient methods for in vitro generation of different tomato varieties. The production of virus free plants in garlic; the alterations in DNA methylation pattern in tissue cultured triticale; and production of bacteria free grapevines using tissue culture techniques are described in Chapters 6, 7 and 8, respectively. Production of banana using tissue culture techniques is discussed in the following two chapters. In Chapter 11, discussion is on the genetic and epigenetic changes associated with somaclonal variation in a coffee species. Applications of PTC technology for the regeneration of numerous medicinal plants and production of secondary metabolites/ bioactive compounds have been discussed in length in Chapters 12-18.

I hope that this book will prove to be a convenient and beneficial source of information for both who are well versed in PTC as well as the beginners.

Editor
Sarika Garg, Ph.D.
Montreal, Canada

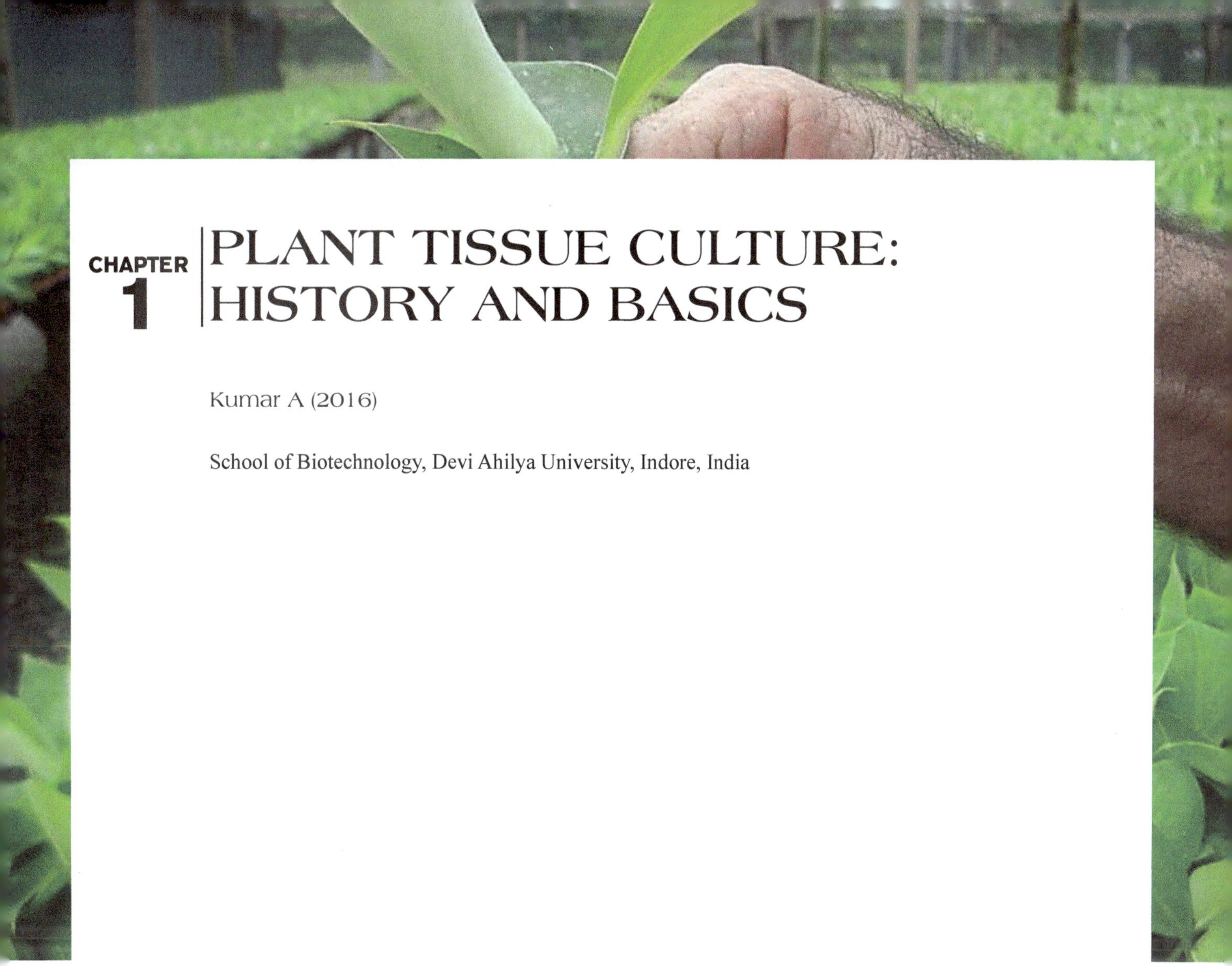

CHAPTER 1 PLANT TISSUE CULTURE: HISTORY AND BASICS

Kumar A (2016)

School of Biotechnology, Devi Ahilya University, Indore, India

HISTORY OF PLANT TISSUE CULTURE

The plant tissue culture takes its roots from the cell theory given by Schleiden (1838) and Schwann (1839). They proposed that cell is the basic unit of life and is capable to reproduce. In 1902, a German physiologist, Gottlieb Haberlandt gave birth to the science of plant tissue culture. The science of plant tissue culture flourished with the discovery of plant growth regulators and many new instruments. Initially complex medium was being employed since the nutrient requirements of plants were not known. With the latest advancements in the field of plant tissue culture now it is possible to generate the entire plant starting from anything might be a single cell, pollen, anther, egg or leaf etc. Work of many eminent scientists has led to the constant development of the science of plant tissue culture.

The idea of cell theory was tested by several scientists but the work of Vöchting (1878) on callus formation and limitation of plant segments to divide was important. He suggested 'polarity' as a characteristic feature guiding the development of plant fragments. In 1902, G. Haberlandt for the first time started work in the field of Plant tissue culture. He is undoubtedly called the father of plant tissue culture because of his contributions, especially the postulates given regarding plant tissue culture. He isolated single cells from the leaves of carrot and tried to culture them in a medium containing Knop's solution along with sucrose, asparagine and peptone. He reported that the cells remained alive for a month in the medium. He also observed that the cells grew in size, changed shape, thickening of the cell wall occurred, but none of them underwent division. Haberlandt failed in inducing successful divisions in the cells due to lack of the knowledge of plant hormones at that time. The most important part of his results was some predictions, made by him. Though these were just predictions having no experimental proof, later these were confirmed through various experiments. These predictions are now the facts of plant tissue culture. The followings are the predictions given by Haberlandt:

- The growth of a cell *in vivo* stops, once it acquires the features needed to meet the requirement of the whole plant.
- He perceived the concept of plant hormones at that time. He felt that some compounds called as growth enzymes are present in the plant system. These are released from specific cells of the plant and stimulate growth and development in other cells.
- Based on the observation that, in nature growth of the pollen tube induces division of the female gametophyte, he suggested that vegetative cells should be cultured along with pollen tubes, which would induce the cells to divide.
- He also suggested the addition of vegetative apices in the medium used for culture of plant cells.
- He said that addition of embryo sac fluid in tissue culture medium might also be useful.

Citation: Kumar A (2016). Plant Tissue Culture: History and Basics.

- He predicted at that time itself that one could successfully cultivate artificial embryos from vegetative cells.

All the predictions made by Haberlandt are true and are practiced today also. Coconut milk (embryo sac fluid) is a rich source of cytokinins and is added in the culture medium. It stimulates shoot apices known to produce an important category of plant growth regulators, the auxins.

In 1904, Hannig tried to culture nearly immature embryos of various crucifers (*Raphanus sativus, R. landra*, etc) in a culture medium containing mineral salts and sugar solution. He successfully raised transplantable seedling from the embryos cultured in a medium. After that many workers cultured embryos especially of species which failed to germinate because the embryos got aborted. The discovery by Laibach (1925) displayed successful elevation of the hybrid plants. He successfully cultured embryos from the seeds of inter specific cross of *Linum perenne* and *Linum austrianum*. The hybrid of this cross did not occur in nature because the seeds formed after crossing get shriveled, become light and incapable of germination. Two important discoveries in mid 1930s gave a big boost to the development of the science of plant tissue culture. One was the discovery of auxins by Kogl (1934) and their role as plant growth regulators; and the other was the recognition of the importance of Vitamin B complex in plant growth. The latter was also discovered in 1934 by Thimmen, who found a way to control growth of plants, tissues and cells.

The first true plant tissue cultures were obtained by Gautheret from cambial tissue from *Salix capraea*, *Robinia pseudoacacia*, *Populus nigra* and other trees using agar-solidified medium of Knop's solution, glucose and cysteine hydrochloride under aseptic condition. He observed that the cells were able to proliferate for a few months only. He added vitamin B - complex and IAA in the medium, which enhanced the growth of the cells tremendously. These studies were continued and in 1939 Gautheret, Nobecourt and White published their independent work on the successful cultivation for prolonged periods of cambial tissues of carrot root, carrot and tobacco, respectively. White was also successful in achieving indefinite divisions of tomato cells by using Knop's solution, sucrose and yeast extract. In later experiments he replaced yeast extract by pyridoxine, thiamine, and nicotinic acid, which gave a way towards the formulation of a pure synthetic medium. Later, he formulated a medium most popularly known as White's medium. This was the first chemically defined medium for use in plant tissue culture, which is till date amongst the most commonly used media. Gautheret was also successful in plant regeneration without the intermediary formation of callus. He for the first time was successful in achieving unlimited divisions in plant cells. He maintained the continuous growing cultures.

1940s, and 1950's proved to be an exciting time for the development of new techniques and the improvement of those already available. In 1941, Van Overbeek, Conklin and Blakeslee for the first time showed the stimulatory effect of coconut milk (coconut water) on embryo development in callus cultures of *Dhatura* species. Stem tip cultures yielded success when Ernest Ball (1946) devised a method to identify the exact part of shoot meristem that gives rise to whole plant. In 1950, cell division factor was discovered by Folke and Skoog. They also discovered zeatin from maize, which led to the discovery of cytokinins as the second category of plant growth hormones. In 1955, Miller and co-workers separated the first cytokinin from the DNA of Herring sperm and named it cytokinin. This enabled the initiation of callus cultures from differentiated tissues. In 1957, Skoog and Miller gave the concept of hormonal control of organogenesis in callus culture of tobacco. They postulated that differentiation of roots and shoots is dependent on the relative auxin and cytokinin ratio in the culture medium. They showed that the process of organ formation could be controlled by changing the relative concentration of auxin and cytokinin in the culture. They also observed that a high concentration of auxin promoted root formation while a high cytokinin concentration promoted development of shoots. First report of somatic embryogenesis came from the work of Reinert (Germany) and Steward (USA) in 1958-59.

A revolution occurred in the field of industrial production of economically important plants by the work of G. Morel in 1960. He was amongst those who started tissue culture of monocots viz. gladiolus, iris, lily etc. He practiced the addition of yeast extract and coconut milk in tissue culture media. He was successful in raising the first virus free plant. He used shoot tip culture technique for raising virus free plants and in this series prepared virus free plants of many species like dahlia, potato etc. He also used shoot tip culture technique for the rapid propagation of plants. He worked on orchids and became successful in developing a technique for raising 4 million genetically identical plants in a single year from a single bud of orchid. Orchids generally propagate through seeds in nature. But this method of propagation renders genetic variations in orchids which is not preferred in horticulture industry. This technique developed by Morel brought a revolution in the orchid industry. It also opened new avenues for the large scale industrial propagation of economically important plants.

Many techniques of cell and protoplast cultures were developed by E.C. Cocking. He for the first time used enzymatic method for the isolation of protoplast from *Lycopersicon esculentum.* The improved techniques of protoplast culture developed by him opened a new method of producing improved crop varieties. In 1962, another widely used chemically defined medium was given by T. Murashige, popularly known as MS (Murashige and Skoog) medium. He prepared a medium by increasing the concentration of salts twenty-five times higher than Knops. This media enhanced the growth of tobacco tissues by five times. Even today MS medium has immense commercial application in tissue culture. In 1963, an important contribution was from Panchanan Maheshwari who laid the foundation of experimental methods for the study of embryogenesis with the help of plant tissue culture. He was actively engaged in the *in vitro* culture of reproductive parts of flowering plants. A revolution occurred in the field of plant tissue culture by the regeneration of haploid plants using culture of anthers and isolated microspores, by Guha and Maheshwari during 1964-1966. They provided new dimensions to the development of the technique of anther and pollen culture. They also worked on the modification of crop plants by protoplast culture and gene transfer (wheat, moong, rice etc.).

Kameya and Hinata showed callus formation in isolated pollen cultures of *Brassica* species in 1970. Since then technique of pollen culture has been considerably improved. In

1971, Morel was successful in the transfer of bacterial genetic material in the crown gall cells based on his observation that crown gall cells acquired the new trait for the synthesis of some opines and amino acids. First somatic hybrid by fusing the protoplasts of *Nicotiana langsdorfii* and *N. glauca* was obtained by Carlson in 1972. Cocking and Bhojwani in 1972 were successful in the isolation of protoplasts for the first time from pollen grains. They were successful in isolating protoplasts from pollen grains and pollen mother cells. They used protoplasts for understanding tobacco mosaic virus infection and multiplication in plant cells. In 1973, Nag and Street examined the cell suspension of *Daucus carota* and found the subsequent recovery on freezing the experimental material at -196°C. It was the first time when they reported regeneration of plants from frozen cells of carrot stored at a temperature of -196° C. Later, Nitsch and coworkers became successful in raising haploid plants from the culture of isolated microspores of *Nicotiana tobaccum* and *Dhatura innoxa*. In 1978, the first intergeneric hybrid between potato and tomato was obtained by fusing protoplasts of the two plants by Melchers, which became popular by the name pomato. But this intergeneric hybrid was sterile. In 1979, Brodelius and co-workers developed the technique of immobilization of plant cells so that the biomass could be utilized for longer periods without substantial loss in the yield of the bioactive compound which was produced from it.

Indra Kumar Vasil made many contributions towards the development of the technique of somatic embryogenesis in 1980. He worked for raising somatic hybrids of milet, maize and wheat. He used protoplasts isolated from immature zygotic embryos for genetic manipulation. He gave immense contribution in the field of raising genetically engineered crop plants. He used the microprojectile bombardment method for obtaining herbicide resistant varieties of crops like wheat, milet, maize, sugarcane, etc. In 1981, term somaclonal variation was introduced by Larkin and Scowcroft. In 1982, Zimmermann was able to fuse the protoplast using electric stimuli; and Krens and coworkers discovered that protoplasts were able to incorporate naked DNA. In 1984, development of the genetic fingerprinting technique for identifying individuals by analyzing Polymorphism at DNA sequence level was done by Alec Jeffreys and transformation of plant cells with plasmid DNA was done by Paszkowski and co workers. An outstanding research was done in 1987 by Barton and his coworkers. They isolated Bt gene from bacterium *Bacillus thuringiensis*. A method for development of biolistic gene transfer method for plant transformation was also developed in 1987. Till now, many outstanding techniques have been developed for Plant tissue culture.

BASICS OF PLANT TISSUE CULTURE

Plant tissue culture is the technique for in vitro multiplication of plants using a small portion of a tissue called as explants using synthetic medium under controlled conditions in an aseptic environment outside an intact plant under defined chemical and physical conditions. It includes culture of individual or groups of plant cells and even whole organs, and has contributed in the understanding of both fundamental and applied science. The plant tissue culture is based on maintaining plant cells in aseptic conditions on a suitable nutrient medium. Tissue culture supports the growth and/or development of isolated plant cells, tissues or organs on an artificial, nutritive medium. Plant tissue cultures are axenic, which means that there is no other life form present in the culture medium except the plant tissue of interest. Plant tissues are maintained on the culture medium for a specified period of time and may be transferred to fresh medium periodically, or to a different medium to alter the path of development. The culture can be sustained as a mass of undifferentiated cells called callus, for an extended period of time, or regenerated into whole plants.

The process of designing the strategy for culturing cells from a plant for the first time is a matter of trial and error. It is a fascinating and useful tool, which allows the rapid production of many genetically identical plants, called clones, using relatively small amounts of space, supplies and time. Basically the technique involves use of a piece of a plant called as explant (such as a stem tip, node, meristem, embryo, or even a seed) and placing it on a sterile, (usually gel-based) nutrient medium for multiplication. The formulation of the growth medium depends on whether to produce undifferentiated mass of cells called callus, or to multiply the number of plantlets, grow roots, or multiply embryos for artificial seeds.

The major steps in plant tissue culture are outlined below:

Selecting Plant Sources

The task of selecting a species of a particular plant for use in tissue culture is quite cumbersome because some species or their clones grow easily in artificial medium while others may not. It has been observed that some species respond reluctantly to culture, some do not respond at all, and there are many plants, which could never have been tried.

Different parts of the plants are used as explant for tissue culture. Depending on the plant species, a choice of an explant can be made from a wide variety of tissues or plant parts. . For example: shoot tip can be a good choice for micropropagation.

Choosing a Growth Medium

The success of the plant tissue culture depends on the choice of the medium. Price, convenience, type of plant, type of response required and the purpose of micropropagation all are considered during the selection of the medium for culture. The plant tissue culture media should contain the same nutrients as required by the whole plant. Plants in nature can synthesize their own food material. However, plants grown *in vitro* are mainly heterotrophic i.e. they cannot synthesize their own food. The media used may be solid (solid medium) or liquid (liquid medium) in nature. The selection of solid or liquid medium is dependent on the better response of a culture.

Media Preparation and Sterilization

Tissue culture media are nutritive solutions prepared in both liquid and semi-solid state. The medium is gelled by the addition of agar, a heteropolysaccharide derived from some varieties of red algae. This gelled agar supports the plant tissue and saves it from submerging in the medium. The agar also allows the diffusion of medium ingredients into the plant tissues.

There are many different formulations of media, and the selection depends largely upon the plant species and the cell, tissue or organ selected for culture. One of the most widely employed medium is the medium developed by Toshio Murashige and Folke Skoog (Murashige, T. and F. Skoog. 1962. Physiologia Plantarum 15:473-497). The Murashige and Skoog medium is so well known to the scientific community that it is often abbreviated simply as the MS medium. Other important media include White's medium, Gamborg's B-5 medium, McCown's Woody Plant medium to name a few.

The nutritive medium usually contains macronutrients, micronutrients, vitamins, carbon source, plant growth regulators. Macronutrients such as N, P, K, Mg, S, Ca are required in large concentrations. There should be at least 25-60 mM of inorganic nitrogen for satisfactory plant cell growth. Potassium is required for the cell growth of most plant species. Most media contain K in the form of nitrate chloride salts at concentrations ranging between 20 and 30 mM. The optimum concentrations of P, Mg, S and Ca range from 1-3 mM if other requirements for cell growth are provided. Micronutrients also called as trace elements such as Fe, Zn, B, I, Mn, Mo, Co, Cu are required in low concentrations. Iron is usually the most critical of all the micronutrients. Cobalt (Co) and iodine (I) may be added to certain media, but their requirements for cell growth has not been precisely established. Sodium (Na) and chlorine (Cl) are also used in some media, in spite of reports that they are not essential for growth. Copper and cobalt are added to culture media at concentrations of 0.1μM, iron and molybdenum at 1μM, iodine at 5μM, zinc at 5-30 μM, manganese at 20-90 μM and boron at 25-100 μM. Besides these nutrients, vitamins (thiamine, nicotinic acid, riboflavin etc.), sugars (sucrose and glucose are common) and sugar alcohols (myo-inositol) are also required in media. A medium may also include amino acids, nucleic acid bases, or other organic molecules. Incorporation of amino acids like glycine depends on the type of response required. Besides the regularly used components, some plants require incorporation of extra components in the medium. The common examples are complex constituents viz. casein hydrolyzate, yeast extract, coconut milk, peptone, etc.

Phytohormones or plant growth regulators are added to the basal medium to stimulate the growth and development of the explant in a desired manner. If the medium has no phytohormones (plant growth regulators), it is referred to as a basal medium. There are five classes of phytohormones: auxins, cytokinins, gibberellins, abscisic acid and ethylene. Auxins and cytokinins are most commonly employed in tissue culture media. In general, auxins promote root initiation, although they may inhibit root elongation and subsequent development. Cytokinins tend to promote the development of vegetative buds into mature shoots. Different combinations and concentrations of these hormones are to be added in the media depending on the plant species and type of the response.

Auxins- In nature, auxins are involved in the elongation of stem and internodes, apical dominance, abscission etc. In plant tissue culture auxins are known to promote cell division and root differentiation. 3-Indole butyric acid (IBA) and indole acetic acid (IAA) are commonly used auxins for rooting in the *in vitro* developed shoots. Similarly 2,4 dichlorophenoxy acetic acid (2,4-D) is used for induction of callus and somatic embryogenesis. A stock solution of the auxin is prepared in a concentration of 1 to 10 mmol/litre by dissolving it initially in minimal quantity of ethanol or dilute NaOH and then adjusting the desired volume using double distilled water.

Cytokinins- These hormones are reported to cause cell division, modification of apical dominance, shoot differentiation, etc. in nature. In plant tissue culture also they promote cell division, differentiation of shoots from callus and organs. A stock solution similar in concentration to that of auxins is prepared by dissolving cytokinin in minimal quantity of dilute NaOH or dilute HCl, and then adjusting the volume to the desired volume using double distilled water.

Gibberellins - These hormones are reported to cause stem elongation, germination, dormancy, flowering, sex expression, enzyme induction, and leaf and fruit senescence. Gibberellins comprise more than twenty compounds, of which GA3 is the most frequently used gibberellin. Abscisic acid is a strong antagonist of Gibberellins action.

Abscisic acid- Abscisic acid (ABA) is often referred to as inhibitory rather than stimulatory hormone. Abscisic acid is involved in several plant functions. It is involved in the closure of stomata, bud and seed dormancy and is known to inhibit other hormonal actions. The plant hormone abscisic acid (ABA) plays a major role in mediating the adaptation of the plant to stress.

Ethylene- Ethylene is a gaseous plant hormone. It is also called as fruit ripening gas. Ethylene effects include fruit ripening, loss of chlorophyll, abortion of plant parts, stem shortening, abscission of plant parts, and epinasty (bending of stems). Ethylene also plays a role in response to water stress, chilling and mechanical injury.

There is always a problem of aeration associated with the suspension cultures. Due to improper aeration, usually the cells die out only after a few cycles of division. Therefore, in most of the cases, in plant tissue culture solid media are preferred. For the purpose, a suitable gelling agent is added in the medium along with sucrose. The most commonly used gelling agent is agar- agar which is obtained from a seaweed Gelidium amansii. Besides agar, other gelling agents in use are agarose and phytagel.

Steps of Media Preparation

1. 5 x or 10 x stock solutions of the different salts are prepared in clean glasswares using double distilled water and are stored in deep freezers. For example if one wishes to prepare MS medium, 10 x stocks of all the 7 solutions should be prepared.
2. At the time of requirement, these stocks are diluted to the required concentration and mixed in the given volume. Sucrose in the range of 2-4% and the desired growth hormones are added separately at the time of medium preparation itself.
3. The medium can be prepared in liquid or solid form depending whether one needs suspension culture or culture on solidified medium.
4. For preparation of solidified medium, as written above, a suitable gelling agent like agar-agar is added in the medium in the range of 0.6-1%. For solid medium, two thirds of the final volume of double distilled water is taken. To it, weighed amount of agar is added and dissolved by warming the solution.
5. Afterwards, sucrose is dissolved and other stock solutions are added and mixed. The pH of the medium

is adjusted between 5.0-6.0 (as per requirement) with the help of 0.1 N NaOH and 0.1 N HCl. If the pH is kept higher than 6.0, the media becomes too hard and if below 5.0, problem occurs in the setting of the gelling agent.

6. The final volume is made up with double distilled water. The medium is dispensed in conical flasks upto 1/3rd of their capacity and mouth is plugged with the help of cotton plugs prepared of non-absorbent cotton wrapped in cheese cloth.

7. The plugs are wrapped with plastic coated brown paper or aluminum foil to prevent them from wetting during autoclaving. The media is then autoclaved at 121°C for 20 min. The sterile media can then be used directly or even can be stored at 4°C and used when required.

Some components used in the medium are heat labile. If any of these components is required in the medium, the medium minus that compound is first autoclaved. The compound is added afterwards in the cooled medium by filter sterilization. For the purpose pre-sterilized filters of pore diameter 0.22 - 0.45 µm are used with the help of a syringe. This whole process is carried out in the laminar airflow cabinet. The most suitable method for medium selection is to select three different media of different mineral strengths and media are prepared using different combinations of growth hormones suitable for the desired response with all the three. The work is continued with the medium showing most promising results.

Sterilization- Plant tissue culture media are enriched in nutrients and therefore, can harbor a number of microbes. If the media gets contaminated with any microbe, the microbe(s) grows rapidly and destroys the whole plant material in culture. The most common contaminations are bacterial and fungal contaminations. Contamination may be introduced at any level during the entire course of culture technique, starting from medium preparation till the end of culture period. The main sources of contamination are – the culture vessel itself, the medium, the researcher, media preparation room, working place, transfer area, inoculation, culture room, observation area etc. Proper precautions should be taken during each step of the experiment so that the culture remains free of contamination. It should be kept in mind that despite all proper measures, there are always chances of contamination of the cultures. Therefore, it is always suggested to culture the material in multiple sets of 10, 15 etc, according to the convenience. If few cultures are contaminated, one would always have at least ample number of few uncontaminated cultures for pursuing further work.

To generate contamination free cultures, it is necessary to work under perfectly sterile conditions, especially during inoculation. The culture room and the transfer area should be perfectly sterile. Prior to inoculation, the medium and the vessels to be used for culturing should also be sterile. Thus, to avoid contamination, measures are to be taken which include sterilization at different stages as discussed below:

a) **Sterilization of the Medium** - Except a few, all the components of plant tissue culture media are stable towards heat. Benefited by this property, autoclaving proves to be the most promising method of sterilization of the medium. For the purpose, medium is prepared in perfectly clean and dry glassware. The medium components are dissolved in double distilled water and are dispensed in suitable glasswares. There are two ways, either the medium may be dispensed directly in the vessel to be used for culture and autoclaved, or autoclaving is done in a suitable glassware and afterwards, the medium is dispensed in clean sterile glasswares which are to be used for culture. Irrespective of the approach, the medium is poured in the desired vessel up to one third of its capacity and the mouth is closed with the help of a stopper. The stopper is usually a cotton plug wrapped in cheesecloth. However, other types of plugs like stainless steel plugs can also be used. After closing the mouth of the vessel with the plug it is generally covered by the aluminum foil or paper to avoid it from wetting during autoclave, tied with a piece of thread, arranged in wire baskets and are placed in an autoclave. The conditions for sterilization are 1.06-Kg cm^{-2} (121°C) for 15 – 40 min.

After sterilization at such a high temperature and pressure, still some microorganisms like *Bacillus* may survive. To check it, the autoclaved medium is stored for few days before use so that the improperly sterilized vessels may be discarded. Some constituents of the medium are thermolabile that is they are degraded when subjected to a high temperature. The compounds in this category include some growth regulators (thiadiazuran), urea, certain vitamins, antibiotics etc. Such compounds if required in the media are filter sterilized in the laminar airflow cabinet. The medium is cooled to 40°C and to it, the filter sterilized compound is added under sterile conditions in the laminar airflow cabinet. The filters used are usually bacteria proof membrane filters of pore size between 0.22 - 0.45 µm. A large number of different types of filters are available commercially. These can be easily fitted either on a syringe or filter holder available commercially. Some companies provide disposable filters while others can be reused after autoclaving.

b) **Sterilization of the Glasswares-** Glasswares can be sterilized by autoclaving or by dry heat sterilization in a hot air oven. For autoclaving, clean glasswares plugged and wrapped in paper are autoclaved at 1.06-Kg cm^{-2} (121°C) for 15 min. For dry heat sterilization, the glasswares are heated in a hot air oven at 160 - 180°C for 3 hours. After the required duration of 3 hours, the oven is switched off and the glasswares are allowed to sufficiently cool down and afterwards removed from the oven. These glasswares are now transferred to the transfer area for use, which is pre-sterilized. Autoclaving is preferred over dry heat sterilization because the latter has many drawbacks. For example the circulation of hot air is poor in the oven and penetration of air into the glassware is also slow thus providing improper sterilization.

Plasticwares are usually sterilized by wet heat. A large number of such plasticwares of different materials like polypropylene, polymethyl pentene, polyallone, teflon etc are available. Some plasticwares like those made of teflon can be autoclaved. Now a days, large number of different types of autoclavable plasticwares are available. Besides this, pre-sterilized plasticwares which are available commercially can also be used which are discarded post-work. The sterilized plasticwares and glasswares should always be wiped on their outer surface before placing in the laminar airflow cabinet.

c) **Sterilization of the instruments** - The basic instruments required in the experiments of plant tissue culture includes forceps of different sizes, scalpels, scissors, spatula, inoculation needles, etc. All these items should be perfectly sterile at the time of use as these are in direct contact with the material to be inoculated. For the purpose, the most common and widely used method is dipping the instrument in 95% alcohol and then flaming. The instruments should be kept immersed in alcohol and the alcohol should be changed regularly as few microorganisms may grow in it.

Sterilization by 95% alcohol too is not completely efficient in killing the microorganisms. Now adays, sophisticated instruments which provide 100% sterility are in use. The most common ones are the glass bead sterilizer and infra red sterilizer. Both the instruments provide a very high temperature. The instruments to be sterilized are kept in these instruments before use. The glass bead sterilizer contains a high watt element. The beads are heated in a brass crucible with the help of this element. A temperature of about 250°C is reached in it in 15-20 min and the instruments get sterilized in just 5-7 seconds.

Similarly, the infra red sterilizer provides a temperature as high as 700°C which can sterilize the instruments in just 2-5 seconds. These sterilizers are always kept in the laminar airflow cabinet and are switched on before starting the work.

d) **Sterilization of the Plant Material** - The plant material is the most prone source of contamination and should be well sterilized before inoculation and sub-culturing. Plant material contains a number of flora on its surface. It is the site of harbor of bacteria, fungi and their spores. This load of microorganisms is too large if the material is obtained from an open field. To avoid this contamination, usually the plant material is used from the seedlings obtained by germination of seeds in the laboratory itself, which requires less harsh treatments for sterilization. However, if this is not possible then the available material from the field too can be taken and a harsher treatment is given. The most common agent of sterilization of the plant material is sodium or calcium hypochlorite. A 0.3-0.6% solution of sodium hypochlorite is prepared and the plant material is dipped in it for 15-30 min. Besides hypochlorite, the second preferred sterilant of choice is 0.1 to 0.5% solution of mercuric chloride in which the plant is dipped for 5-20 min. The time for subjecting the material for sterilization depends on the tissue. For example harder tissues like nodal and internodal explants, leaf discs require more time as compared to pollen, anthers, flower buds etc., which are delicate and are damaged by harsh treatments. Another common treatment involves a 30-sec rinse in 70% alcohol prior to dip in the sterilant or addition of a few drops of a surfactant like Tween-20. Triton-X-100 has also been proved to give better results in some cases. Still another method of choice is the use of 1% cetavalon for 2 minutes prior the use of sterilant. The sterilant is taken in a screw capped tubes, plant material is added and is shaken well for the desired period of time. After sterilization, the plant material is washed 3-4 times with sterile distilled water to remove the last traces of the sterilant adhering to it and is then transferred in sterile petri plates.

e) **Sterilization of the Transfer Area** - After proper sterilization of the plant material, the next precautionary aspect is proper sterilization of the transfer area. Usually laminar airflow cabinet is used as the transfer area, which comprises of a constant blow of sterile air current. Prior to work, the cabinet is cleaned properly with sterile distilled water and afterwards with 70% alcohol. The UV-lights are provided in the cabinet which are switched on for some time prior to work. The air current is switched on a few minutes before the start of work. A Bunsen burner is lighted and all transfer operations are carried out in the vicinity of the flame. The transfer area is usually separate, close to the culture room and entry of the people is restricted in it. It is regularly cleaned.

f) **Sterilization of the Culture Room** - Proper sterilization of the culture room is the most important aspect. For the purpose, regular cleaning of the room is required. Besides, the culture room should be sterilized at regular intervals by fumigation with formaldehyde or with the help of ultraviolet (UV) lights. The walls of the culture room should be waterproof for proper cleaning by the detergent. There should be no dust deposition in the room and it should be completely cut off from the direct air of the surroundings. The proper precautions mentioned here at all the steps and practices render all the cultures free of contamination.

Cleaning, Storing and Manipulating Explants

The tissue or organ removed from a plant and placed into culture is called an explant. Before the explant is placed into a sterile culture medium, any contaminating organism(s) associated with it must be removed or killed. The process of cleaning the explant is referred to as surface sterilization or surface disinfection. When the explant has been surface sterilized, it is usually removed from the sterilizing solution and rinsed several times in sterile, distilled water. This last step is performed inside a laminar flow hood to maintain the axenic condition of the explant and to prevent the re-introduction of contaminating microbe(s). The explant is then ready to be trimmed if necessary and transferred into a tissue culture medium.

Inoculation of the Explant into Medium

The explants after sterilization are transferred aseptically into a suitable medium using a laminar airflow cabinet. The medium may be liquid or semi-solid whose selection is based on the type of response required. The explants may be immersed in a liquid medium as in root culture, or may be positioned above the liquid medium, such that the lower surface is in contact with the medium.

Growth of the Explant

Once the explant has been transferred on the nutritive medium, the culture vessel containing the explant is placed in a controlled environment. The amount (quantity) of light, spectral quality, periodicity, temperature (range and periodicity) all vary with the plant tissue involved. These factors also depend on the nature of medium and the type of response required.

Plant tissue culture plays an important role in the production of ornamental and agricultural plants and in the manipulation of plants for raising improved crop varieties for agricultural purposes. Plant tissue culture is a multi- dimensional science that offers exciting prospects to future improvements in crop productivity. It has led to the development of a system for selecting pathogen or stress-resistant plant clones. It has also helped in obtaining, maintaining, and mass propagating these specific pathogen-free plants by meristem culture which has led to the creation of novel genetic combinations through somatic hybridization. In addition, mass propagation of plants is also an important application. It offers the potential to produce thousands, or even billions of plants per year.

VARIOUS TECHNIQUES OF PLANT TISSUE CULTURE

Plant tissue culture includes various techniques like *In vitro* propagation, Meristem culture and production of disease-free plants, Somatic embryogenesis, Somaclonal variation, *In vitro* selection, Protoplast culture and Somatic hybridization.

In Vitro Propagation

It includes regeneration of plants from any desired plant part. It also includes the technique of micropropagation which leads to large scale multiplication of plants in a short duration.

Micropropagation is the technique of reproduction of plants on an artificial medium in a controlled environment under sterile conditions. For example- orchids are propagated from seeds, mature and green pods, meristem tips, leaves and flowers. Other plants are propagated from stem cuttings, shoot buds, leaves, pollen, immature embryo and various other plant parts.

Advantages of Micropropagation

Endangered species are being proliferated and taken off the endangered list even though they are no longer found in their native habitat. In the beginning, only herbaceous plants were propagated by the technique of micropropagation, but now almost any plant species including woody species, ferns, palms and varieties of forest trees used for their hardwood and softwood in timber industry can be grown in the laboratory. The new aseptic propagation methods available are reliable and present a new tool in the plant propagation industry. Though the traditional methods of plant propagation, for instance through stem cutting, grafting, germination of seeds etc. are important and are still relevant in the nursery business, propagation with the help of tissue culture is more advantageous. The cost of production of these plants can be considerably lowered down due to the large-scale multiplication in a short duration, thus economical for any industry.

There are many advantages of micropropagation. The main advantages are rendering the plant material pathogen free by culturing the plant's apical meristem, increased yield of food grain crops, generation of trees with better form and faster growth, plants with disease resistance, etc. In addition to its biotechnological applications, micropropagation is used commercially for asexual propagation of plants. Using micropropagation, millions of new plants can be derived from a single plant.

The process of micropropagation may be divided into four stages:

a) Initiation Stage

An excised piece of plant tissue (called an explant) is (i) cut from the plant, (ii) disinfected (removal of surface contaminants), and (iii) placed on a medium. A medium typically contains mineral salts, sucrose, vitamins, plant growth regulators and a solidifying agent such as agar. The objective of this stage is to achieve an aseptic culture. An aseptic culture is one without contaminating bacteria or fungi.

b) Multiplication Stage

A growing explant can be induced to produce vegetative shoots by adding a cytokinin in the medium. A cytokinin is a plant growth regulator that promotes shoot formation from growing plant cells. When multiple roots are generated, these are taken out of the culture vessel, separated and are inoculated separately in a fresh medium for further multiplication.

c) Rooting or Preplant stage

Growing shoots can be induced to produce adventitious roots by adding an auxin in the medium. Auxins are plant growth regulators which at a high concentration promote root formation.

d) Acclimatization

Plants grown *in vitro* must be acclimatized to greenhouse conditions after planting out. This is because *in vitro* plants are far more delicate and require high humidity. The whole process is called as hardening of the plants. A growing, rooted shoot is removed from the tissue culture medium and is placed in paper cups containing a mixture of soil and vermiculite in the ratio of 1:2. The plants in these cups are covered with poly bags to prevent loss of moisture. These cups are placed in moist chambers containing high humidity for a couple of weeks. Afterwards, they are transferred to a poly house or green house for further hardening. During this period development of cuticle and stomata takes place, which were lacking in them initially. When this is done, the humidity must be gradually reduced over time since tissue-cultured plants are extremely susceptible to wilting.

Knowledge of organogenesis and developmental phenomena of many economically important species is still scarce. Commercial micropropagation is in its initial stage compared with traditional horticultural technologies, which have been conducted for hundreds of years. The lack of knowledge concerning plant tissue reactions to artificial *in vitro* conditions are the basic reasons for most of the problems encountered by the scientists during micropropagation.

Meristem Culture and Production of Disease-free Plants

Meristem culture was pioneered by Morel (1960) and it involves the removal of the meristem from the donor plant and its subsequent culture on a nutrient medium. The meristem is a group of actively dividing cells, about 0.1 mm in diameter. Endogenous contaminants do not easily invade the meristem, hence its culture results in the formation of disease-free plants. When combined with micropropagation techniques, large number of disease-free plants may be produced from meristematic explants. Meristem culture has been used successfully in the removal of viruses from many plants (example potato, sugarcane, strawberry) and is now used routinely for the eradication of many viral diseases from plant materials.

Somatic Embryogenesis

By this technique, cells of somatic (non-sexual) origin are cultured in an appropriate medium resulting in the formation of somatic embryos. Somatic embryos are formed either directly on the explant or more often indirectly in callus culture. The former technique is called as direct somatic embryogenesis and the latter is termed as indirect somatic embryogenesis. There are several advantages of recovery of plants from cells via somatic embryogenesis compared to micropropagation. Somatic embryos can be produced from cells growing in suspension, thereby making possible large scale production of plants using batch culture techniques which are cost effective. The multiplication rate is very high and, in some species such as carrot, celery, tomato, the embryos can be encapsulated and treated as artificial seeds.

The types and concentrations of auxin and cytokinin in the growth medium play a very important role in the production of somatic embryos. The whole process of somatic embryogenesis occurs in two major steps. First is induction of somatic embryos which requires a high auxin concentration and 2,4-Dichlorophenoxyacetic acid (2,4-D) gives the best results. The second is maturation of somatic embryos, which requires very less (negligible) concentration of auxins. In some plants there is no requirement of addition of auxins or cytokinins in the medium for generation of somatic embryos. Incorporation of ammonium ions along with nitrate ions in the culture medium enhances the yield of somatic embryos. For instance, in carrot 30 mM NH4Cl and 40 mM KNO3 enhance the yield of somatic embryos considerably. Addition of alanine, glutamine and an increased concentration of sugar in the medium also shows increased yield.

Somaclonal Variation

Although micropropagation results in large scale production of clones but a great deal of genetic variability has been observed in regenerated plants. This is seen especially if the cultures are established from explants that did not contain a pre-organized meristem, or if plant regeneration is from callus. This process is called as somaclonal variation and plants as somaclones. Somaclonal variation has led to the use of cell culture for mutagenesis and selection of genetic variants and for direct recovery of novel genotypes from cell cultures. It is useful in case of several crop plants where this technique has helped in the generation of agriculturally useful changes in the plants, for example increase in sugar yield in sugarcane, late blight resistance in potato, resistance to *Fusarium* in tomato, etc.

In Vitro Selection

Now a days, one of the most intensively studied areas of plant tissue culture is the concept of selecting disease, insect, or stress resistant plants through tissue culture. Significant improvements have been made in various species by selecting and propagating superior individuals from *in vitro* systems. Besides naturally occurring variability in the cultures, variability can be induced by chemical or physical agents known to cause mutations. *In vitro* selection usually involves subjecting a population of cells to a suitable selection pressure and recovering the variant cell lines which have developed resistance or tolerance towards stress. Next step is to regenerate the entire plant from such resistant cell lines. Current research in this area is being carried out to select salt tolerant lines, freeze resistant plants, herbicide resistant agronomic crops, and plants resistant to chemical molecules such as heavy metals (aluminium, manganese, etc.).

Protoplast Culture and Somatic Hybridization

Protoplasts are single cells which have been stripped off their cell walls mechanically or by enzymatic treatment. These protoplasts are able to produce the whole plant. Fusion of protoplasts is called as "somatic hybridization". The protoplasts from sexually incompatible species have the ability to fuse further extending the techniques of plant modifications through tissue culture to the limits of imagination. The potential use of somatic hybridization is to bring about novel combinations of genetic material. This has been demonstrated in the genera *Petunia* and *Nicotiana.*

There are many techniques of protoplast culture in liquid medium. The most important techniques are - Protoplasts are plated as thin film in petri plates; cultured in flasks of 100 ml capacity; plated in the form of 50 -100μl drops in petri plates which are then stored in a humid chamber. Protoplasts can also be cultured by embedding in agar or alginate beads or discs. These beads are then suspended in liquid medium for further growth.

Within 2 - 4 days of culture, the protoplasts lose their characteristic spherical shape and start regeneration of the cell wall. The time period required for cell wall regeneration depends on the plant species. For example, in case of *Solanaceae* the required duration is of 24 - 40 hours, for legumes it is 4 days and about 7 days for woody trees. Those protoplasts, which fail to synthesize cell wall, cannot undergo mitosis. Protoplasts require osmotic protection till cell wall formation. For the purpose, the media should contain a suitable osmoticum. The osmolarity is controlled around 500 - 600 mM by using mannitol or sorbitol. This osmolarity is reduced when the cells start to divide because a high concentration of solutes in the culture media may hinder the growth of cells. Ten days later when the cell wall develops fully, the osmolarity is decreased slowly by addition of fresh medium, which reduces the concentration of the osmoticum. For further cell division and callus generation, presence of a mature cell wall is necessary and a plating density between 0.1 - 10 % is required. Those cells, which are capable of division, undergo first division in

2 - 7 days. It takes about 2 - 3 weeks for the appearance of multicellular colonies. Next two weeks are required for the appearance of macroscopic colonies.

REFERENCES

1. Thorpe, T. A. (2006). History of plant tissue culture. Plant cell culture protocols, 9-32.
2. Gautheret, R. J. (1983). Plant tissue culture: A history. The botanical magazine= Shokubutsu-gaku-zasshi, 96(4), 393-410.
3. Gamborg, O. L. (2002). Plant tissue culture. biotechnology. milestones. In Vitro Cellular & Developmental Biology-Plant, 38(2), 84-92.
4. Gamborg, O. L., Murashige, T., Thorpe, T. A., & Vasil, I. K. (1976). Plant tissue culture media. In Vitro Cellular & Developmental Biology-Plant, 12(7), 473-478.
5. Saad, A. I., & Elshahed, A. M. (2012). Plant tissue culture media. INTECH Open Access Publisher.
6. Gaspar, T., Kevers, C., Penel, C., Greppin, H., Reid, D. M., & Thorpe, T. A. (1996). Plant hormones and plant growth regulators in plant tissue culture. In Vitro Cellular & Developmental Biology-Plant, 32(4), 272-289.
7. Davies, P. J. (2010). The plant hormones: their nature, occurrence, and functions. In Plant hormones (pp. 1-15). Springer Netherlands.
8. Pati, P. K., Rath, S. P., Sharma, M., Sood, A., & Ahuja, P. S. (2006). In vitro propagation of rose—a review. Biotechnology Advances, 24(1), 94-114.
9. Johnson, A. A., & Veilleux, R. E. (2010). Somatic hybridization and applications in plant breeding. Plant Breeding Reviews, Volume 20, 167-225.
10. Bridgen, M. P. (1994). A review of plant embryo culture. HortScience, 29(11), 1243-1246.
11. Azman, A. S., Mhiri, C., Grandbastien, M. A., & Tam, S. M. (2014). Transposable elements and the detection of somaclonal variation in plant tissue culture. Malaysian Applied Biology, 43(1), 1-12.

CHAPTER 2

THE ROLE OF SILICON IN PLANT TISSUE CULTURE

Iyyakkannu Sivanesan and Se Won Park

Department of Molecular Biotechnology, Konkuk University, Seoul, South Korea

ABSTRACT

Growth and morphogenesis of in vitro cultures of plant cells, tissues, and organs are greatly influenced by the composition of the culture medium. Mineral nutrients are necessary for the growth and development of plants. Several morpho-physiological disorders such as hooked leaves, hyperhydricity, fasciation, and shoot tip necrosis are often associated with the concentration of inorganic nutrient in the tissue culture medium. Silicon (Si) is the most abundant mineral element in the soil. The application of Si has been demonstrated to be beneficial for growth, development and yield of various plants and to alleviate various stresses including nutrient imbalance. Addition of Si to the tissue culture medium improves organogenesis, embryogenesis, growth traits, morphological, anatomical, and physiological characteristics of leaves, enhances tolerance to low temperature and salinity, protects cells and against metal toxicity, prevents oxidative phenolic browning and reduces the incidence of hyperhydricity in various plants. Therefore, Si possesses considerable potential for application in a wide range of plant tissue culture studies such as cryopreservation, organogenesis, micropropagation, somatic embryogenesis and secondary metabolites production.

INTRODUCTION

Plant tissue culture is a collection of experimental procedures for aseptic culture of isolated plant cells, tissues or organs on nutrient media under controlled environmental conditions. Growth and morphogenesis of in vitro cultures of plant cells, tissues and organs are greatly influenced by the composition of the culture medium. The composition of a culture medium has often been modified to stimulate the growth of particular plant material. In general, plant tissue culture medium composed of inorganic nutrients, organic supplements, carbon source, plant growth regulators and a solidifying agent. Mineral nutrients are necessary for growth and development of plants. The optimization of inorganic nutrients in the culture medium improves growth and morphogenesis of plant cells, tissues and organs in vitro. Several physiological disorders such as hooked leaves, hyperhydricity, fasciation and shoot tip necrosis are often associated with the concentration of inorganic nutrients in the culture medium (Reed et al., 2013).

Silicon (Si) is the most abundant mineral element in the soil (Epstein, 1999). Numerous studies have shown that Si treatment improves the growth and yield of various plants, particularly when they are subjected to both abiotic and biotic stresses (Ma, 2004). Several researchers have reviewed the role Si on plant tolerance to abiotic (Balakhnina and Borkowska, 2013; Zhu and Gong, 2014) and biotic stresses (Van Bockhaven et al., 2013). The availability of Si in hydroponic and substrate plant production system is restricted. Addition of Si to the nutrient

Citation: Sivanesan I and Park SW (2014) The role of silicon in plant tissue culture. Front. Plant Sci. 5:571. doi: 10.3389/fpls.2014.00571

solution or soilless substrate enhanced growth traits, yield and quality of several crops (Voogt and Sonneveld, 2001). The promoting effects of Si on plants might be due to increasing nutrient uptake and photosynthetic activity. Though Si is a ubiquitous contaminant, the use of silicon-free containers and double distilled water restricts its availability. Furthermore, Si has not been included in any commercial tissue culture media formulation. The inclusion of Si to the culture medium improved the morphogenetic potential of plant cells, tissues and organs. Several studies have shown that the inclusion of Si to the tissue culture medium enhances callus growth, shoot regeneration, and root induction and stimulates somatic embryogenesis, and improve morphological, anatomical and physiological characteristics of plantlets. In addition, Si treatment prolongs the longevity of calli and organs with a potential for plant regeneration. The inclusion of Si to the tissue culture medium also enhances tolerance to low temperature, metal toxicity and salinity. Si enhancing tolerance of plants to various stresses by altering activity of antioxidant enzymes, cation binding capacity of the cell walls, endogenous plant hormone level, increasing production of chitinase, glucanse, lignin, phenolics, and phytoalexins, nutrient uptake, improving strength of cell and plant, maintaining the structure of stomata, relative water content, and reducing uptake of heavy metals. This review concentrates the potential roles of Si in plant tissue culture.

ROLE OF SI IN PLANT TISSUE CULTURE

Organogenesis and Somatic Embryogenesis

Islam et al. (2005) investigated the effect of calcium silicate ($CaSiO_3$) on callus induction and plant regeneration from mature seed explants of rice 'Kalizira', 'Lucky', and 'Pajam'. The highest frequency of callus induction is achieved on Murashige and Skoog (MS) medium containing $CaSiO_3$. However, plant or root regeneration potential of rice calli is cultivar depended. Similarly, effects of Si on plant or root development depend on reed (Phragmites australis) genotype used for callus induction (Mathe et al., 2012). Addition of Si as sodium silicate (Na_2SiO_3) to the modified MS medium promotes the growth of calli obtained from stem nodal and root explants of P. australis while its effect on somatic embryogenesis is explants dependent: it stimulates embryogenesis of root calli, but it does not influence this process in stem nodal calli. Soares et al. (2011) evaluated the effect of Si source [potassium silicate (K_2SiO_3) and Na_2SiO_3] on shoot multiplication of Cattleya loddigesii. The highest number of shoots is observed on the modified Knudson C medium containing 5.0 mg L^{-1} K_2SiO_3. In Ajuga multiflora, addition of Si to MS medium containing 2iP and IAA, enhanced adventitious shoot regeneration (about threefold) by increasing the activity of antioxidant enzymes such as SOD, POD, APX, and CAT (Sivanesan and Jeong, 2014). In addition, the authors observed the Si accumulation in leaves of plants developed in the culture medium with Si, but not in plants developed in the medium without Si by wavelength dispersive X-ray analysis. These studies indicate that the effect of Si on morphogenetic potential ofin vitro plant cultures depends on plant species, genotype and concentration of Si in the culture medium. Still further studies are required to better understand the biochemical and molecular mechanism of Si on organogenesis and somatic embryogenesis.

Growth and Development

The application of Si has been reported to enhance the growth and development of various plants.Zhou (1995) observed silica bodies in leaf tissues of Phalaenopsis hybrid plantlets grown in Vacin and Went medium supplemented with $CaSiO_3$. Addition of $CaSiO_3$ also increased the leaf length.Soares et al. (2011) reported that the addition of 5.0 mg L^{-1} K_2SiO_3 and 20.0 mg L^{-1} Na_2SiO_3 to the modified Knudson C medium increased the number of roots and length of aerial part and root in seedlings of C. loddigesii. Subsequently the same group investigated the effect of GA_3 and Na_2SiO_3on growth and development of C. loddigesii (Soares et al., 2013). It was reported that the combination of GA_3 and Na_2SiO_3 increased the number of leaves and roots than GA_3 alone. The optimal concentration of Si varies within the same plant species and or genotype. The inclusion of $CaSiO_3$ at 0.5 and 2.0 mg L^{-1} to the MS medium stimulates the growth of native (Brassavolva perrinii) and hybrid (Laelia cattleya 'Culminant Tuilerie' ×L. cattleya 'Sons Atout Rotunda') ×Brassolaelia cattleya 'Startifire Moon Beach') orchid plants, respectively (Soares et al., 2012). Lim et al. (2012) also reported that the effect of Si (K_2SiO_3) on the growth traits of begonia 'Super Olympia Red' and 'Super Olympia Rose' and pansy 'Matrix White Blotch' and 'Matrix Yellow Blotch' are mainly dependent on the cultivars. Braga et al. (2009) investigated the effect of different Si sources such as $CaSiO_3$, K_2SiO_3, and Na_2SiO_3 on the growth and anatomical characteristics of strawberry 'Oso Grande' seedlings. The fresh and dry weight of seedlings increased in MS medium containing 1.0 g L^{-1} Na_2SiO_3. Seedlings of banana 'Maca' cultured in the medium supplemented with $CaSiO_3$ increased the chlorophyll content, whereas those cultured in the medium containing Na_2SiO_3 increased length, fresh and dry weight of shoots (Asmar et al., 2011).

The morphological and anatomical characteristics of in vitro-grown plantlets are different from the field-grown seedlings. Si inclusion to the rooting medium increased leaf tissue thickness and epicuticular wax deposition in banana (Asmar et al., 2013a) and strawberry (Braga et al., 2009) plantlets. Luz et al. (2012) reported that supplementation of $CaSiO_3$, K_2SiO_3, or Na_2SiO_3 to the rooting medium improved leaf anatomy of banana 'Maca' plantlets. The inclusion of $CaSiO_3$ to the culture medium also increased photosynthetic rate and chlorophyll content of banana plantlets (Asmar et al., 2013b). In strawberry, light and electron microscopic analysis showed deformation in chlorenchyma and the epidermis of leaves from plantlets grown in the culture medium devoid of Si (Soares et al., 2012). Recently, He et al. (2013) confirmed the deposition of Si within the cell walls of in vitro-cultured rice cells. Si improves the structural stability of cell walls during cell elongation and division and thereby maintained cell shape, which may be important for the function and survival of cells (Table 1).

Table 1: Role of Si in plant tissue culture

Plant species	Role of Si	Reference
Ajuga multiflora	Increased frequency of shoot regeneration	Sivanesan and Jeong (2014)
	Increased resistance to NaCl	
Begonia semperflorens	Increased growth, biomass and chlorophyll content	Lim et al. (2012)
Brassavola perrinii Hybrid orchid	Increased seedlings growth, favorable characteristics in the leaf anatomy of the orchid seedlings	Soares et al. (2012)
Cattleya loddigesii	Increased the number of shoots	Soares et al. (2011)
	Increased growth traits of seedlings	Soares et al. (2013)
Coleus hybridus	Maintained regenerative potential of vitroplantlets	Radovet et al. (2008)
Cotoneaster wilsonii	Reduced hyperhydricity	Sivanesan et al. (2011)
Dendrobium moniliforme	Increased resistance to low temperature	Duan et al. (2013)
Fragaria x *ananassa*	Increased biomass, thickness of leaf tissues and epicuticular wax deposition	Braga et al. (2009)
Musa sp. 'Grande Naine'	Well-developed stomata, Increased epicuticular wax layer in leaves	Asmar et al. (2013a)
Musa sp. 'Maca' banana	Increased chlorophyll content, biomass and seedlings growth	Asmar et al. (2011)
	Improved leaf anatomy	Luz et al. (2012)
	Improved morpho-physiological leaf characteristics of seedlings	Asmar et al. (2013b)
Ornithogalum dubium	Reduced hyperhydricity and improving leaf structure and *ex vitro* survival during acclimatization	Ziv (2010)
Oryza sativa	Improved callus induction, and plant regeneration	Islam et al. (2005)
	Improved structural stability of cell walls	He et al. (2013)
Perilla frutescens	Increased the growth rate and content of anthocyanins	Zhong et al. (1992)
Phalaenopsis hybrid	Increased shoot growth	Zhou (1995)
Phragmites australis	Stimulated callus growth, somatic embryogenesis and root formation	Mathe et al. (2012)
Picea abies	Ameliorated the effect of Al	Prabagar et al. (2011)
Psidium guajava	Inhibiting phenol based browning	Youssef et al. (2010)
Salvia splendens	Increased resistance to NaCl	Soundararajan et al. (2013)
Solanum tuberosum var. *Gersa*	Maintained regenerative potential of vitroplantlets	Radovet-Salinschi and Cachita-Cosma (2012)
S. tuberosum	Increased growth, biomass and tolerance to NaCl	Qing et al. (2005)
Viola x *wittrockiana*	Increased growth, biomass and chlorophyll content	Lim et al. (2012)
Vitis vinifera x *V. labrusca*	Increased the survival rate of callus under low temperature	Moriguchi et al. (1988)

Ziv (2010) investigated the effect of silicon on hyperhydricity in Ornithogalum dubium. Addition of Na_2SiO_3 to MS liquid medium containing BA, NAA and 6% sucrose in bioreactors, significantly reduced induction of hyperhydric shoots, and increased plant firmness and mechanical strength. Si treatment significantly reduced the content of hydrogen peroxide and activity of oxidative reductive enzymes such as APX, ascorbate oxidase and GPX in leaves of the regenerated shoots of O. dubiumwhen compared with the control (Table 1). Similarly, addition of Si as K_2SiO_3 to MS medium reduced the hyperhydricity in Cotoneaster wilsonii by decreasing the content of MDA in the regenerated shoots when compared with the control (Sivanesan et al., 2011). The authors observed the presence of Si in the in non-hyperhydric plants, but not in the hyperhydric leaf samples of C. wilsonii by energy dispersive X-ray analysis. Thus, the problem of hyperhydricity can be reduced by the inclusion of Si to both liquid and solid culture medium. Phenolic oxidative tissue browning is one of the bottlenecks in woody plant tissue culture. In guava, tissue browning was completely prevented by sealing the nodal explants cut ends with Si (Youssef et al., 2010) and there was no detrimental effect of Si on the subsequent steps of in vitro propagation. The authors suggested that Si could be used during explants preparation to control phenolic tissue browning in other plants. The morphological, anatomical and physiological characteristics of plantlets can improve in vitroby incorporating Si in the culture medium. However, further studies required to evaluate the effect of different source and concentration of silicon on the growth and development of various plants.

Abiotic Stress Tolerance

Duan et al. (2013) reported that Si enhance cold resistance of Dendrobium moniliforme by increasing the content of free proline, soluble sugar and soluble protein and decreasing MDA content. Si treatment improved the survival rate of grape 'Kyoho' and 'koshusanjaku' calli under low temperature by preventing browning (Moriguchi et al., 1988). In vitro storage of Coleus hybridus 'Jupiter' and Solanum tuberosum var. Gersa under silicone oil significantly reduced the growth and maintained their regenerative potential (Radovet et al., 2008; Radovet-Salinschi and Cachita-Cosma, 2012). These results reveal that Si can be used as cryoprotectant and included in the cryoprotective mixture for minimizing the toxicity of cryoprotectants. The ameliorating effect of Si on salt stress in vitro has been reported in A. multiflora (Sivanesan and Jeong, 2014), Salvia splendens 'Hot Jazz' (Soundararajan et al., 2013) and S. tuberosum (Qing et al., 2005). Si alleviates salt stress in plants by limiting NaCl uptake, maintenance of ultrastructure of stomata, improving photosynthetic activity, reducing free proline content and altering the production of antioxidant enzymes (Qing et al., 2005; Soundararajan et al., 2013; Sivanesan and Jeong, 2014). Prabagar et al. (2011) investigated the effect of Si on aluminium (Al) tolerance in Picea abies suspension cultures. Al toxicity was reduced when the liquid medium was supplemented with Si and the effect was increased at pH 5.0 than pH 4.2. Si supplementation protected P. abies cells and against Al toxicity by reducing the concentration of free Al in the cell wall. Si is also reported to enhance drought tolerance, alleviate lead toxicity and increase resistance to radiation and temperature stresses (Balakhnina and Borkowska, 2013; Zhu and Gong, 2014). The molecular mechanisms of Si on stress tolerance are poorly understood. Thus, more studies are needed to find out the role of Si in abiotic tolerance on various plants.

Future Prospects

Recent studies have shown the beneficial effects of Si in plant tissue culture (Table 1). However, further studies on a wide variety of plant species are needed to confirm the role of Si in plant tissue culture. In vitro culture is a useful system for studying physiological and biochemical functions of Si in plants at molecular level. Further, in vitro cell suspension culture systems provide an opportunity to study roles of Si at the single cell level. The inclusion of silicone A to Linsmaier and Skoog liquid medium also enhances cell growth and anthocyanins content in cell suspension culture of Perilla frutescens (Zhong et al., 1992). Thus, Si can also be used for the stimulation of secondary metabolites in the plant cell, tissue and organ cultures. We strongly recommend the inclusion of Si as a beneficial nutrient in the tissue culture medium to solve various micropropagation problems, and to increase tissue culture success.

Conflict of Interest Statement

The authors declare that the research was conducted in the absence of any commercial or financial relationships that could be construed as a potential conflict of interest.

ACKNOWLEDGMENT

This article was supported by the KU Research Professor Program of Konkuk University.

REFERENCES

1. Asmar, S. A., Pasqual, M., de Araujo, A. G., Silva, R. A. L., Rodrigues, F. A., and Pio, L. A. S. (2013a). Morphophysiological characteristics of acclimatized "Grande Naine" banana plants in response to in vitro use of silicon. Cienc. Agrar. 34, 73–82.
2. Asmar, S. A., Castro, E. M., Pasqual, M., Pereira, F. J., and Soares, J. D. R. (2013b). Changes in leaf anatomy and photosynthesis of micropropagated banana plantlets under different silicon sources. Sci. Hortic. 161, 328–332. doi: 10.1016/j.scienta.2013.07.021
3. Asmar, S. A., Pasqual, M., Rodrigues, F. A., de Araujo, A. G., Pio, L. A. S., and e Silva, S. O. (2011). Sources of silicon in the development of micropropagated seedlings of banana "Maçã ." Cienc. Rural. 41, 1127–1131. doi: 10.1590/S0103-84782011005000086
4. Balakhnina, T., and Borkowska, A. (2013). Effects of silicon on plant resistance to environmental stresses: review. Int. Agrophys.27, 225–232. doi: 10.2478/v10247-012-0089-4
5. Braga, F. T., Nunes, C. F., Favero, A. C., Pasqual, M., Carvalho, J. G., and Castro, E. M. (2009). Anatomical characteristics of the strawberry seedlings micropropagated using different sources of silicon. Pesqui. Agropecu. Bras. 44, 128–132. doi: 10.1590/S0100-204X2009000200003
6. Duan, X., Tang, M., and Wang, W. (2013). Effects of silicon on physiology and biochemistry of dendrobium moniliforme plantlets under cold stress. Agric. Biotechnol. 2, 18–21. doi: 10.1007/s13593-011-0039-8
7. Epstein, E. (1999). Silicon. Annu. Rev. Plant. Physiol. Plant Mol. Biol. 50, 641–664. doi: 10.1146/annurev.arplant.50.1.641
8. He, C., Wang, L., Liu, J., Liu, X., Li, X., Ma, J.,et al. (2013). Evidence for "silicon" within the cell walls of suspension-cultured rice cells. New Phytol. 200, 700–709. doi: 10.1111/nph.12401
9. Islam, M. M., Ahmed, M., and Mahaldar, D. (2005). In vitro callus induction and plant regeneration in seed explants of rice (Oryza sativa L.). Res. J. Agri. Biol. Sci. 1, 72–75.
10. Lim, M. Y., Lee, E. J., Jana, S., Sivanesan, I., and Jeong, B. R. (2012). Effect of potassium silicate on growth ad leaf epidermal characteristics of begonia and pansy grown in vitro. Kor. J. Hort. Sci. Technol. 30, 579–585.
11. Luz, J. M. Q., Asmar, S. A., Pasqual, M., de Araujo, A. G., Pio, L. A. S., and Resende, R. F. (2012). Modification in leaf anatomy of banana plants cultivar "Maca" subjected to different silicon sources in vitro. Acta Hort. 961, 239–243.
12. Ma, J. F. (2004). Role of silicon in enhancing the resistance of plants to biotic and abiotic stresses. Soil Sci. Plant Nutr. 50, 11–18. doi: 10.1080/00380768.2004.10408447
13. Mathe, C., Mosolygo, A., Suranyi, G., Beke, A., Demeter, Z., Toth, V. R.,et al. (2012). Genotype and explants-type dependent morphogenesis and silicon response of common reed (Phragmites australis) tissue cultures. Aquat. Bot. 97, 57–63. doi: 10.1016/j.aquabot.2011.11.005
14. Moriguchi, T., Kozaki, I., Matsuta, N., and Yamaki, S. (1988). Plant regeneration from grape callus stored under a combination of low temperature and silicone treatment. Plant Cell Tissue Organ Cult. 15, 67–71. doi: 10.1007/BF00039890
15. Prabagar, S., Hodson, M. J., and Evans, D. E. (2011). Silicon amelioration of aluminium toxicity and cell death in suspension cultures of Norway spruce (Picea abies L.). Environ. Exp. Bot. 70, 266–276. doi: 10.1016/j.envexpbot.2010.10.001

16. Qing, W., Huiying, H., and Jinwen, Z. (2005). Effect of exogenous silicon and proline on potato plantlet in vitro under salt stress.China Vegetables 9, 16–18.

17. Radovet, D., Cachita-Cosma, D., and Petrus, A. (2008). The micropropagation of Coleus hybridus Jupiter vitrocultures under different paraffin or silicon oil stratums. Studia Universitatis "Vasile Goldiş", Seria Ştiinte Vieţii (Life Sciences Series) 18, 61–70.

18. Radovet-Salinschi, D., and Cachita-Cosma, D. (2012). Testing the regenerative capacity of Solanum tuberosum var. Gersa explants after 24 weeks storage in living collection. Analele Universităţii din Oradea, Fascicula: Ecotoxicologie, Zootehnie şi Tehnologii de Industrie Alimentară? 11, 423–430.

19. Reed, B. M., Wada, S., DeNoma, J., and Niedz, R. P. (2013). Mineral nutrition influences physiological responses of pear in vitro. In Vitro Cell. Dev. Biol. Plant 49, 699–709. doi: 10.1007/s11627-013-9556-2

20. Sivanesan, I., and Jeong, B. R. (2014). Silicon promotes adventitious shoot regeneration and enhances salinity tolerance ofAjuga multiflora Bunge by altering activity of antioxidant enzyme. Sci. World J. 2014, 10 doi: 10.1155/2014/521703

21. Sivanesan, I., Song, J. Y., Hwang, S. J., and Jeong, B. R. (2011). Micropropagation of Cotoneaster wilsonii Nakai—a rare endemic ornamental plant. Plant Cell Tissue Organ Cult. 105, 55–63. doi: 10.1007/s11240-010-9841-2

22. Soares, J. D. R., Pasqual, M., de Araujo, A. G., De Castro, E. M., Pereira, F. J., and Braga, F. T. (2012). Leaf anatomy of orchids micropropagated with different silicon concentrations. Acta Sci. Agron. 34, 413–421. doi: 10.4025/actasciagron.v34i4.15062

23. Soares, J. D. R., Pasqual, M., Rodrigues, F. A., Villa, F., and de Araujo, A. G. (2011). Silicon sources in the micropropagation of the Cattleya group orchid. Acta Sci. Agron. 33, 503–507.

24. Soares, J. D. R., Villa, F., Rodrigues, F. A., and Pasqual, M. (2013). Concentrations of silicon and GA3 in in vitro propagation of orchids under natural light. Sci. Agrar. Parana. 12, 286–292.

25. Soundararajan, P., Sivanesan, I., Jo, E. H., and Jeong, B. R. (2013). Silicon promotes shoot proliferation and shoot growth of Salvia splendens under salt stress in vitro. Hort. Environ. Biotechnol. 54, 311–318. doi: 10.1007/s13580-013-0118-7

26. Van Bockhaven, J., De Vleesschauwer, D., and Höfte, M. (2013). Towards establishing broad-spectrum disease resistance in plants: silicon leads the way. J. Exp. Bot. 64, 1281–1293. doi: 10.1093/jxb/ers329

27. Voogt, W., and Sonneveld, C. (2001). "Silicon in horticultural crops grown in soilless culture," in Silicon in Agriculture, eds L. E. Datnoff, G. H. Snyder, and G. H. Korndorfer (Amsterdam: Elsevier Science), 115–131.

28. Youssef, M. A., El-Helw, M. R., Taghian, A. S., and El-Aref, H. M. (2010). Improvement of Psidium guajava L. using micropropagation. Acta Hort. 849, 223–230.

29. Zhong, J. J., Seki, T., Kinoshita, S. I., and Yoshida, T. (1992). Effects of surfactants on cell growth and pigment production in suspension cultures of Perilla frutescens. World J. Microbiol. Biotechnol. 8, 106–108. doi: 10.1007/BF01195826

30. Zhou, T. S. (1995). The detection of the accumulation of silicon in Phalaenopsis (Orchidaceae). Ann. Bot. 75, 605–607. doi: 10.1006/anbo.1995.1065

31. Zhu, Y., and Gong, H. (2014). Beneficial effects of silicon on salt and drought tolerance in plants. Agron. Sustain. Dev. 34, 455–472. doi: 10.1007/s13593-013-0194-1

32. Ziv, M. (2010). Silicon effects on growth acclimatization and stress tolerance of bioreactor cultured Ornithogalum dubiumplants. Acta Hort. 865, 29–36.

CHAPTER 3

PLANT REGENERATION FROM LEAF-DERIVED PROTOPLASTS WITHIN THE DAUCUS GENUS: EFFECT OF DIFFERENT CONDITIONS IN ALGINATE EMBEDDING AND PHYTOSULFOKINE APPLICATION

Katarzyna Mac´kowska, Agata Jarosz and Ewa Grzebelus

Institute of Plant Biology and Biotechnology, Faculty of Horticulture, University of Agriculture in Krakow, Al. 29 Listopada 54, 31-425 Krako´w, Poland

ABSTRACT

Protoplasts of four Daucus carota subspecies and three wild Daucus species were isolated from 2-week-old shoot cultures during overnight incubation in an enzyme mixture composed of 1 % (w/v) cellulase Onozuka R-10 and 0.1 % (w/v) pectolyase Y-23. Before the culture, they were embedded in autoclave- or filter-sterilized alginate solution. Modified thin alginate layer (TAL) and extra thin alginate film (ETAF) techniques were applied for protoplast immobilization. A rich mineral-organic medium based on the formulation of Kao and Michayluk supplemented with 0.1 mg l^{-1} 2,4-dichlorophenoxyacetic acid, 0.2 mg l^{-1} zeatin, and optionally 100 nM phytosulfokine (PSK), a peptidyl plant growth factor, was used for protoplast culture. Plating efficiency was genotype-dependent and in 40-day-old cultures, it varied from 10 % for Daucus pusillus to 73 % for D. carota subsp. sativus. A considerably higher ability in colony formation was observed in the modified TAL culture system using filter-sterilized alginate and in the presence of PSK in the protoplast culture medium. Plant regeneration through somatic embryogenesis stimulated by PSK application occurred for five out of the seven Daucus accessions used in the present study. We believe our data may facilitate the use of wild Daucus in somatic hybridization with cultivated carrot.

INTRODUCTION

The genus *Daucus* belongs to the Apiaceae family, one of the most important families of angiosperms from an economical point of view. Within the genus, 24 *Daucus* species and nine *Daucus carota* subspecies are recognized, according to the most recent estimates (Grzebelus et al. 2011). Among them, the best known and agronomically important is cultivated carrot (*D. carota* L. subsp. *sativus* Hoffm., 2n = 2x = 18). It is grown for its edible storage root, which cumulates α- and ß-carotene, the primary source of pro-vitamin A in the human diet. *Daucus* is widespread all over the world, being most frequently associated with the Mediterranean, Central Asia, and New World regions, and even Australia (Rubatzky et al. 1999). Wild *Daucus* germplasm, due to its worldwide origin and adaptation to a wide range of environmental conditions, could be a valuable natural source of genetic variability for developing new carrot cultivars with resistance/tolerance against adverse biotic and abiotic factors, as well as desirable culinary, nutritional, or processing traits (Camadro et al. 2008). Usually, to transfer germplasm from wild species to cultivated crops, intra- or inter-specific crosses are performed. Till now, two types of such generative hybridization have greatly contributed to the improvement of modern carrot cultivars, through the introduction of: (1) the petaloid type of cytoplasmic

Citation: Katarzyna Maćkowska, Agata Jarosz, Ewa Grzebelus, Plant regeneration from leaf-derived protoplasts within the Daucus genus: effect of different conditions in alginate embedding and phytosulfokine application, DOI: 10.1007/s11240-014-0436-1.

male sterility (CMS) from wild *D. carota* subspecies; and (2) resistance to carrot fly from *Daucus capillifolius*(Grzebelus et al. 2011). Despite these successful crosses, utilization of wild germplasm in carrot breeding programs is limited, probably due to the relatively poor characterization of*Daucus* genetic resources and different flowering biology, including usually shorter and only partially overlapping flowering periods and less inflorescences or very small flowers in comparison with carrot (Camadro et al. 2008; Grzebelus et al. 2011). It seems that somatic hybridization based on protoplast fusion could help to overcome these problems, but up to today, only protoplasts isolated from one wild species, i.e., *D. capillifolius*, have been fused in both a symmetric and an asymmetric way with carrot-derived protoplasts (Dudits et al. 1977; Ichikawa et al. 1987). The ability to regenerate plants from fused protoplasts is fundamental in somatic hybridization. Therefore, before applying this procedure to plant breeding programs, successful protocols for plant regeneration from protoplasts of both partners should be elaborated. For carrot, well known as a model species for plant tissue culture systems, efficient procedures for protoplast isolation and plant regeneration from different types of source tissue, such as suspension cultures, petioles, leaves, and hypocotyls have been presented (Grambow et al. 1972; Dirks et al.1996; Grzebelus et al. 2012a). However, to our knowledge, similar studies with regard to wild *Daucus* species do not exist.

To stimulate or improve protoplast development into plants, different culture systems have been employed over the years. Among them, incubation of protoplasts in liquid medium is the most simple to establish but results in unfavorable cell agglutination (Davey et al. 2005a). Therefore, for many species, techniques ensuring physical separation of protoplasts based on their embedding in semi-solid media, with agar being first used as the gelling agent (Nagata and Takebe 1971), have been preferred (Davey et al. 2005b). Immobilization of protoplasts in such media before culture decreases the production of polyphenols that may prolong cell viability, support cell wall regeneration, and promote mitotic divisions. Besides agar, agarose was proved to be superior as a solidifying agent, possibly because of its neutrality and lower gelling temperature (Davey et al. 2005a, c). Alginate, a natural binary copolymer of ß-d-mannuronic acid and α-l-guluronic acid extracted from cell walls of brown algae (Draget 2000), is another and useful alternative to agar and agarose, especially for heat-sensitive protoplasts, since gelling is induced by exposure to calcium ions (Davey et al. 2005a). Additionally, in systems using alginate, the beneficial effect of calcium ions on cell wall re-synthesis may be observed. Protoplasts embedded in alginate matrix can be dispensed as beads (Damm and Willmitzer 1988), blocks (Pan et al. 2003), or layers with different thickness such as thin alginate layers (TAL, Damm et al. 1989) and extra thin alginate films (ETAF, Pati et al. 2005), and then cultured in liquid media. Alginate has been successfully used as a gelling agent in protoplast cultures of many species; among others, in *Lotus covniculatus* and *Nicotiana tabacum* (Pati et al. 2005), *Citrus sinensis* (Niedz 2006), *Helianthus annuus* (Rákosy-Tican et al. 2007), *D. carota* (Grzebelus et al. 2012a), *Beta vulgaris* (Grzebelus et al.2012b), and *Brassica oleracea* (Kiełkowska and Adamus 2012).

Protoplasts from different species and even from different tissues of the same species generally vary in their nutritional requirements. Over the years, different medium formulations with defined and undefined supplements such as coconut milk have been tested. Among them, the most important are growth regulators, particularly auxins and cytokinins, normally essential for sustained divisions of protoplast-derived cells and their further differentiation into plants (Davey et al. 2005b). Recently, oligopeptides have been recognized as signaling molecules playing an important role in plant cell growth and development (Matsubayashi et al. 1997; Ryan and Pearce 2001). One of them is phytosulfokine (PSK), a plant-specific disulfated pentapeptide isolated for the first time from conditioned medium of an *Asparagus officinalis* mesophyll cell culture (Matsubayashi and Sakagami 1996). Since PSK is involved in nanomolar concentrations in the initial steps of cellular dedifferentiation, proliferation, and re-differentiation, its biological function is similar to that of plant hormones (Igasaki et al. 2003; Matsubayashi et al. 2004). PSK can not only induce cell division in suspension and protoplast cultures (Matsubayashi and Sakagami 1996; Matsubayashi et al. 1997), but appears to play a role in diverse aspects of plant development and reproduction, including the formation of tracheary elements and adventitious roots (Yamakawa et al. 1998; Matsubayashi et al.1999), somatic embryogenesis (Kobayashi et al. 1999; Igasaki et al. 2003), and pollen germination (Chen et al. 2000). It seems that besides the widely used auxins and cytokinins, PSK can be a promising supplement for in vitro plant cultures, especially in the case of recalcitrant species. To date, PSK has been applied to *Oryza sativa* and *Beta vulgaris* protoplast cultures and, as a result, increased plating efficiency was observed (Matsubayashi et al. 1997; Grzebelus et al. 2012b).

The main objective of the present work was to evaluate the effect of: (1) alginate sterilization method; (2) different systems of cell culture in alginate; and (3) PSK application on plating efficiency and plant regeneration from leaf-derived protoplasts, for the first time, for some *Daucus* species, to facilitate their subsequent use in somatic hybridization with cultivated carrot.

MATERIALS AND METHODS

Plant Material

Seven accessions of *Daucus* genus were used as a source of protoplasts, including three wild subspecies of *D. carota* (*D. carota* subsp. *azoricus,* subsp. *drepanensis,* subsp.*maritimus*), three wild species of *Daucus* (*D. aureus, D. pusillus, D. montevidensis*) and, as a control, cultivated carrot (*D. carota* L. subsp. *sativus* Hoffm., breeding line 9304B). Seeds were derived from research centers and gene banks (Table 1). Protoplasts were isolated from in vitro grown shoots produced from seeds. In order to obtain aseptic material, donor seeds were disinfected using a three-step procedure, as described by Grzebelus et al. (2012a). To make the germination of wild accessions possible, decontaminated seeds were left to swell in sterile water. After 1 day of incubation, seed coats were removed under a stereomicroscope and the embryos in endosperm tissue were placed in 9 × 1.5 cm Petri dishes on solid Murashige and Skoog (MS) medium with vitamins (Murashige and Skoog 1962), supplemented with 30 g l^{-1} sucrose and 6.5 g l^{-1}plant agar (Biocorp, Poland) at 18 ± 2 °C in the dark.

Seven or 30 days later, seedlings derived from cultivated and wild *Daucus* accessions, respectively, were transferred to glass jars with regeneration medium (R) composed of MS macro- and micro-elements, 0.1 mg l^{-1} thiamine HCl, 0.1 mg l^{-1} piridoxine HCl, 0.5 mg l^{-1} nicotinic acid, 3.0 mg l^{-1} glycine, 100 mg l^{-1} myo-inozytol, 20 g l^{-1}sucrose, and 2.5 g l^{-1} phytagel (Sigma) and kept in a climate room at 26 ± 2 °C under a 16-h photoperiod and light intensity of 55 μmol m^{-2} s^{-1}. Because of the limited number of seeds, for wild accessions, shoot cultures were established by transfer of leaf rosettes to fresh R medium every 2 weeks. In the case of carrot (line 9304B), new aseptic plantlets were regularly produced from seeds before each isolation experiment.

Table 1: Origin, seed source, and somatic chromosome number (2n) of *Daucus* accessions used for protoplast cultures

Accession	Origin, seed source	Reference no.	2n
D. carota subsp. *sativus*	USA, USDA carrot inbred line	9304B	18
D. carota subsp. *azoricus*	Germany, HRI 6667	WD102	18
D. carota subsp. *drepanensis*	Denmark, PI 279762	D19	18
D. carota subsp. *maritimus*	Germany, HRI 7996	WD114	18
D. aureus	Israel, PI 319403	D4	22
D. montevidensis	South America, JKI AS162	WD28	22
D. pusillus	Uruguay, PI 349267	D16	22

USDA = United States Department of Agriculture; PI = USDA Plant Introduction numbers; HRI = Horticulture Research Institute, Genetic Resources Unit (Wellesbourne, UK) numbers; JKI = Julius Kuhn Institute (Germany) numbers

Protoplast Isolation

Protoplasts of cultivated and wild *Daucus* accessions were isolated from expanded leaves with petioles of 2-week-old in vitro grown plantlets or shoot cultures (Fig. 1a), respectively. All steps of protoplast isolation were performed according to a previously described protocol (Grzebelus et al. 2012a). Briefly, about 1 g of source tissue was cut in 9 × 1.5 cm glass Petri dish into small pieces in the presence of a 0.5 M mannitol solution (pH = 5.6–5.8), and incubated for 1 h at 26 ± 2 °C in the dark to plasmolyze the cells. After that, the mannitol solution was replaced by an enzyme mixture consisting of 1 % (w/v) cellulase Onozuka R-10 (Duchefa Biochemie, Haarlem, The Netherlands), 0.1 % (w/v) pectolyase Y-23 (Duchefa Biochemie), 20 mM 2-(*N*-morpholino)ethanesulfonic acid (MES, Sigma-Aldrich, St. Louis, MO, USA), 5 mM $CaCl_2$, and 0.6 M mannitol (Sigma-Aldrich), pH = 5.6, filter-sterilized (0.22 μm, Millipore, Billerica, MA, USA). Leaf tissue was digested for 12–16 h at 26 ± 2 °C in the dark on a gyratory shaker (30 rpm). The released protoplasts were initially purified by filtration through a 100-μm nylon mesh (Millipore) and centrifuged at 100*g* for 5 min in order to remove the enzyme solution. The pellet was resuspended in 8 ml of 0.5 M sucrose with 1 mM MES, very slowly overlaid with 2 ml of W5 salt solution (Menczel et al. 1981), and centrifuged at 145*g* for 10 min. Viable protoplasts localized at the interface between two solutions of different concentrations as a green band were transferred into a fresh tube using a Pasteur pipette and washed twice by centrifugation at 100*g* for 5 min in carrot petiole protoplast medium (CPP), which consisted of macro- and micro-elements, and organic acids according to Kao and Michayluk (1975), vitamins according to Gamborg B5 medium (Gamborg et al. 1968), 74 g l^{-1} glucose, 250 mg l^{-1} casein enzymatic hydrolysate (Sigma-Aldrich), 0.1 mg l^{-1} 2,4-dichlorophenoxyacetic acid (2,4-d), and 0.2 mg ml^{-1} zeatin (pH = 5.6, filter-sterilized). After purification, protoplasts were resuspended in 1 ml of culture medium and counted using a Fuchs-Rosenthal hemocytometer. The working density before embedding in alginate matrix was adjusted to 8 × 10^5 cells per ml.

Figure. 1: Plant regeneration from leaf-derived protoplasts within the *Daucus* genus. a Source tissue of *D. carota* subsp. *maritimus*; b leaf protoplasts; c apple-green fluorescence of viable protoplasts after fluorescein diacetate (FDA) staining; d–f first mitotic divisions and multi-cell protoplast-derived aggregates in 10-, 20-, and 40-day-old cultures of *D. aureus*, respectively; g plasmolysis (*arrow*) of protoplast-derived cells in 40-day-old cultures of *D. pusillus*; h somatic embryos overgrowing alginate matrix; i, j protoplast-derived cells of *D. carota* subsp. *sativus* cultured in extra thin alginate film (ETAF) and thin alginate layer (TAL), respectively; k protoplast-derived proembryogenic masses and green somatic embryos in the cotyledonary stage, on filter paper placed on solid regeneration medium, 1 week after release from the alginate matrix; l protoplast-derived non-regenerating callus of *D. carota* subsp. *sativus*; m completely regenerated plant of *D. carota* subsp. *azoricus*(*bars* in b, c, g 50 µm, d–f 100 µm, i, j 1 cm)

Protoplast Embedding and Culture Conditions

For protoplast culture, two embedding systems in alginate matrix were applied, i.e., TAL and ETAF techniques, as described by Grzebelus et al. (2012a) and Pati et al. (2005), respectively, with some modifications. In both cases, a filter- or autoclave-sterilized (20 min at 121 °C, 0.1 MPa) alginate solution containing 2.8 % (w/v) Na-alginate (Sigma-Aldrich) and 0.4 M mannitol (pH = 5.8) was used. Before filtration through syringe disposable membrane filters (0.22 µm, Millipore), Na-alginate was dissolved in mannitol solution by overnight stirring on a gyratory shaker (100 rpm). Before forming TALs or ETAFs, equal volumes of sterile alginate solution and protoplast suspension at working density were gently mixed, reducing the density of embedded protoplasts to a final value of 4 × 10^5 per ml. In both techniques, polymerization of the protoplast-alginate mixture took place in 6 × 1.5 cm Petri dishes on Ca-agar plates [1 % (w/v) agar (Biocorp), 20 mM $CaCl_2$, 0.4 M mannitol]. TALs were obtained by spreading 300 µl of protoplast-alginate suspension on the surface of Ca-agar plates following gentle rotation of Petri dishes to minimize layer thickness, while to prepare ETAFs, 100 µl of protoplast-alginate suspension was placed on Ca-agar plates and immediately covered by a sterile coverglass (22 × 22 mm). After 1-h or 15-min incubation at room temperature, solidification of alginate matrix with embedded protoplasts occurred for TAL (Fig. 1j) and ETAF (Fig. 1i), respectively. Solid alginate sheets with immobilized protoplasts were gently transferred to 6 × 1.5 cm Petri dishes containing 4 ml of liquid culture medium. Two types of culture media were used: CPP medium and CPP supplemented with 100 nM PSK (phytosulfokine-α, PeptaNova GmbH, Germany). In order to maintain aseptic conditions of the cultures, 400 mg l^{-1} cefotaxime (Polfa Tarchomin SA, Poland) and 200 mg l^{-1} timentin (GlaxoSmithKline, Poland) were applied to each Petri dish. Cultures were incubated at 26 ± 2 °C in the dark. Protoplast medium with all supplements was renewed for the same one after 10 days of culture.

Plant Regeneration

After 2 months of culture in liquid medium, both protoplast-derived tissue [callus tissue and proembryonic masses (PEM)] and somatic embryos overgrown or emerged from alginate matrix, respectively, were used in the regeneration process. The regeneration procedure was slightly different for the two protoplast embedding techniques. In the case of TALs, protoplast-derived material was freed from alginate gel by its depolymerization through exposure to a 20 mM sodium citrate solution with 0.2 M mannitol (pH = 5.8) for 1 h at room temperature (Grzebelus et al. 2012a). Then, the released tissue was washed twice by centrifugation at 100*g* for 5 min. After a second wash in CPPD medium containing 0.1 mg l^{-1} α-naphthaleneacetic acid (NAA), 0.2 mg l^{-1} zeatin, pH = 5.6 (Dirks et al. 1996), 1 ml aliquots of protoplast-derived tissue and somatic embryos were spread on filter paper placed on solidified R medium in glass jars. The procedure for regeneration from ETAFs was performed without the alginate depolymerization step. ETAFs were washed in CPPD medium and transferred to glass jars on R medium. During the regeneration process, cultures were maintained in a climate room at 26 °C with a 16-h photoperiod under a light intensity of 55 μmol m^{-2} s^{-1}. Two weeks later, material from TAL-derived cultures was transferred to fresh R medium without filter paper and kept in the same environmental conditions. After 5 weeks of culture on solid R medium, regeneration capacity in TAL and ETAF protoplast embedding systems was evaluated.

Data Collection and Statistical Analysis

Protoplast yield was determined using a hemocytometer and presented as the protoplast number per gram fresh weigh (FW). The viability of protoplasts was evaluated just after immobilization of the cells in alginate matrix, by staining with fluorescein diacetate (FDA), as described by Grzebelus et al. (2012a), and expressed as a percentage of protoplasts with green fluorescence out of total observed cells. Plating efficiency, defined as the ability of single cells to form colonies through continuous mitotic divisions, was assessed on the 10th, 20th, and 40th days of culture and presented as the number of cell colonies per the total number of observed protoplasts (×100). All microscopic observations were performed under an inverted Axiovert S100 microscope (Carl Zeiss, Göttingen, Germany) equipped with the filter set appropriate for fluorescein visualization (λ_{Ex} = 485 nm, λ_{Em} = 515 nm). In order to describe regeneration efficiency after 5 weeks of culture on R medium, the number of normal and completely formed plants, plants with morphological abnormalities, and non-regenerating callus clumps were counted.

As repetitions, at least two to eight independent protoplast isolation experiments with a single treatment represented by three Petri dishes were carried out. Microscopic observations were performed in four or five microscopic fields on 100–200 cells per Petri dish. The overall effect of treatments was determined using multivariate analysis of variance (ANOVA) in Statistica ver. 9.0 (StatSoft. Inc. 2009). Duncan honestly significant difference test was used for mean separation.

RESULTS

Protoplast Yield and Viability

Highly chloroplast-rich protoplasts (Fig. 1b) were successfully isolated from the leaves of 2-week-old shoot cultures in all accessions, reaching on average 2.8 ± 0.3 × 10^6protoplasts per g FW (Table 2). The mean yield of *D. carota* subspecies-derived protoplasts was over two times higher (4.0 ± 0.4 × 10^6) than that from wild *Daucus*species (1.8 ± 0.3 × 10^6, $P < 0.001$). The applied enzyme mixture led to the release of large and workable cell populations (i.e., >10^6) for all tested accessions. The efficiency of isolation varied from 1.4 × 10^6 (*D. montevidensis*) to 4.5 × 10^6 (*D. carota* subsp. *sativus*and *D. carota* subsp. *drepanensis*) (Table 2). The differences in protoplast yield observed between tested accessions were significant ($P < 0.01$).

Table 2: Yield and viability of leaf-derived protoplasts within the *Daucus* genus

Accession	Protoplast yield (×10^6/g FW)		Protoplast viability (%)	
	Mean ± SE	n	Mean ± SE	n
D. carota subsp. *sativus*	4.5 ± 1.2 a	2	66.6 ± 2.6 bc	4
D. carota subsp. *azoricus*	2.9 ± 1.2 ab	2	92.6 ± 3.2 a	2
D. carota subsp. *drepanensis*	4.5 ± 0.6 a	3	73.9 ± 6.6 ab	6
D. carota subsp. *maritimus*	4.0 ± 0.4 a	4	75.8 ± 9.6 ab	5
D. aureus	2.6 ± 0.7 ab	3	63.0 ± 0.3 bc	3
D. montevidensis	1.4 ± 0.4 b	5	48.9 ± 1.6 c	4
D. pusillus	1.7 ± 0.3 b	5	72.2 ± 5.7 ab	6
Mean/total	2.8 ± 0.3	24	69.7 ± 2.9	30

FW fresh weight, *n* number of independent protoplast isolations. Means within columns with the same letters were not significantly different at $P \leq 0.05$

Protoplasts stained with FDA (Fig. 1c) just after embedding in alginate matrix exhibited an acceptable level of viability, yielding on average 70 % (Table 2). A slightly higher proportion of viable cells ($P < 0.01$) was observed for protoplasts isolated from *D. carota*subspecies (74.9 ± 3.9 %) in comparison with wild *Daucus* species (62.9 ± 3.8 %). Different numbers of living cells were also recorded for individual accessions ($P < 0.05$). The highest viability, ranging from 72 to 93 %, was observed for *D. pusillus*, *D. carota*subsp. *drepanensis*, *D. carota* subsp. *maritimus*, and *D. carota* subsp. *azoricus*, while the lowest was observed for D. *montevidensis* (49 %), *D. aureus* (63 %) and, unexpectedly, for*D. carota* subsp. *sativus* (67 %).

Plating Efficiency

Protoplasts of all accessions revealed to ability to undergo cell division. The first cell divisions (Fig. 1d) occurred between the 4th (*D. carota* subsp. *sativus* and *D. carota*subsp. *drepanensis*) and 8th days after protoplast plating (*D. carota* subsp. *azoricus*) (data not shown). In the subsequent days, continuous mitotic divisions took place and multi-cell aggregates were formed (Figs. 1e, f), reaching on average 32, 41, and 46 % in 10-, 20- and 40-day-old cultures, respectively. At subsequent points of evaluation, the proportion of colony formation highly depended on the accession, method of alginate sterilization, culture technique, and PSK application (Table 3, $P < 0.001$). In 10-, 20-, and 40-day-old cultures, mitotic division of protoplast-derived cells of *D. carota* subspecies was two- to three-fold more frequent in comparison with that of cells isolated from wild*Daucus* species ($P < 0.001$). In 40-day-old cultures, the highest numbers of cell aggregates were recorded for *D. carota* subsp. *sativus* and *D. carota* subsp. *drepanensis* (72 and 67 %, respectively), while the lowest numbers were recorded for *D. montevidensis* and *D. pusillus* (14 and 10 %, respectively). At all points of evaluation, on average, over 10 % more cell aggregates were developed after immobilization of protoplasts in a filter-sterilized than in an autoclave-sterilized alginate solution (Table 3). A positive influence of alginate filter-sterilization on plating efficiency was recorded for all tested accessions but significant differences between the two methods of sterilization were confirmed for *D. carota* subsp. *drepanensis*, *D. carota* subsp. *maritimus*, and *D. pusillus* (Fig. 2). The culture system with embedding of protoplasts in TALs better promoted division of protoplast-derided cells, resulting in a plating efficiency about 20 % higher in comparison with the ETAF system (Table 3). Again, such a tendency was observed for all accessions; however, for *D. aureus*, the differences were not statistically significant (Fig. 2). Supplementation of protoplast culture medium with PSK at the concentration of 100 nM showed a beneficial effect on plating efficiency in comparison with the control (Table 3). PSK application stimulated cell division in protoplast-derived cells in all analysed accessions and for most of them, excluding *D. carota* subsp. *azoricus* and *D. aureus*, the increase of plating efficiency was significant ($P \leq 0.05$, Fig. 2).

Table 3: Mean effect of accession, method of alginate sterilization, culture technique, and phytosulfokine (PSK) application on plating efficiency in protoplast cultures of different*Daucus* species

Treatment	Plating efficiency (% ± SE)		
	Age of culture (days)		
	10	20	40
Accession			
D. carota subsp. *sativus*	50.4 ± 3.6 a	61.5 ± 3.2 a	72.4 ± 2.5 a
D. carota subsp. *azoricus*	30.0 ± 7.0 c	52.6 ± 9.3 ab	54.0 ± 9.5 c
D. carota subsp. *drepanensis*	44.5 ± 4.8 ab	58.6 ± 5.5 a	67.3 ± 5.6 ab
D. carota subsp. *maritimus*	33.4 ± 4.0 c	46.0 ± 5.0 bc	55.1 ± 5.2 bc
D. aureus	35.5 ± 5.0 bc	40.5 ± 5.2 c	46.2 ± 7.2 c
D. montevidensis	16.4 ± 2.1 d	14.8 ± 2.5 d	13.7 ± 2.9 d
D. pusillus	11.2 ± 2.0 d	11.6 ± 1.9 d	9.7 ± 2.1 d
Alginate sterilization			
Autoclave	23.8 ± 2.3 b	30.0 ± 2.9 b	35.1 ± 3.4 b
Filter	34.9 ± 2.5 a	43.5 ± 3.0 a	46.9 ± 3.5 a
Culture technique			
ETAF	18.1 ± 2.1 b	24.6 ± 2.7 b	27.7 ± 3.3 b
TAL	38.2 ± 2.3 a	46.5 ± 2.9 a	51.6 ± 3.3 a
PSK concentration (nM)			
0	22.5 ± 2.2 b	29.3 ± 2.8 b	34.9 ± 3.4 b

Treatment	Plating efficiency (% ± SE)		
	Age of culture (days)		
	10	20	40
100	36.5 ± 2.5 a	44.8 ± 3.0 a	47.6 ± 3.5 a

In each section of the table, means with the same letters did not differ significantly ($P \leq 0.001$) within the columns

TAL thin alginate layer; *ETAF* extra thin alginate film

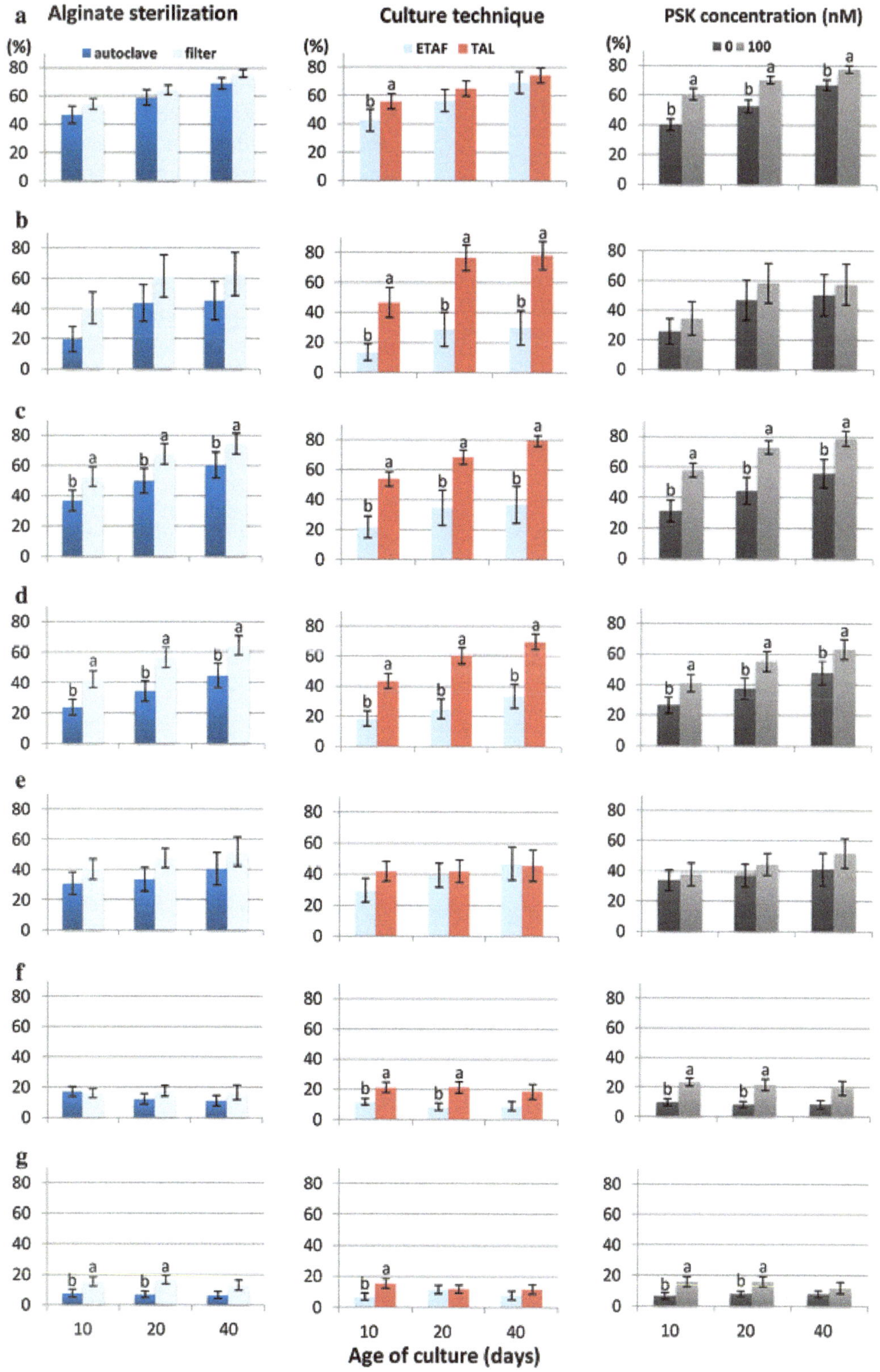

Figure. 2: Effect of alginate sterilization method (*left column*), culture technique (*middle column*), and phytosulfokine (PSK) application (*right column*) on plating efficiency in protoplast cultures of *Daucus carota* subsp. *sativus* (**a**), *D. carota* subsp. *azoricus* (**b**), *D. carota*subsp. *drepanensis* (**c**), *D. carota* subsp. *maritimus* (**d**), *Daucus aureus* (**e**), *Daucus montevidensis* (**f**), and *Daucus pusillus* (**g**). *Bars* represent standard error. For subsequent points of evaluation, means denoted with different letters are significantly different ($P \leq 0.05$)

Plant Regeneration

During 2 months of culture in liquid medium, some of the protoplast-derived cells showed plasmolysis symptoms such as contraction of the cytoplasm before first division or after several mitotic cycles (Fig. 1g), while others underwent sustained divisions, leading to micro- and macro-colony development following the formation of micro- and macro-calli of up to 1 mm in size. At that time, somatic embryos were also observed occasionally (Fig. 1h). After transfer of protoplast-derived tissues to solid regeneration medium, yellow-green and friable PEMs or green and gray compact calli were produced (Figs. 1k, l). From PEM somatic embryos developed, which usually regenerated into morphologically normal and rooted plants (Fig. 1m). However, some disorders in development pattern were recorded. Compact callus tissue failed to differentiate into plants.

For five out of seven accessions, normal plants, abnormal plants, and non-regenerating calli were all obtained (Table 4). The number of normal plants depended highly on the accession ($P = 0.001$) and varied on average from 8 (*D. carota* subsp. *maritimus*) to 73 (*D. aureus*, Table 4). Accessions differed also in the production of abnormal plants ($P = 0.01$)—the highest number of such plants (31) developed from protoplast-derived tissue of *D. carota* subsp. *drepanensis*. Compact and non-regenerating callus tissue most frequently formed in cultures of *D. carota* subsp. *sativus* (45), while for the other four accessions, the number of such callus clumps was lower, reaching from 7 (*D. carota* subsp. *maritimus*) to 18 (*D. aureus*). In *D. montevidensis* and *D. pusillus*, protoplast cultures led to the formation of micro-calli of up to 140 μm in size after several cycles of cell division; than mitoses were arrested and further development was not observed. The method of alginate sterilization had no influence on the regeneration capacity of protoplast-derived tissue. The two applied systems of protoplast embedding in alginate matrix (i.e., TAL and ETAF) affected the growth of abnormal plants and compact calli ($P < 0.01$, Table 4). Less deformed plants and callus clumps were obtained from tissues produced in ETAFs and in TALs, respectively. On the other hand, supplementation of protoplast cultures with PSK increased the yield of regenerated plants four-fold in comparison with the control experiments ($P = 0.01$).

Table 4: Effect of accession and different conditions during protoplast culture on regeneration capacity from 2-month-old protoplast-derived tissue within the *Daucus* genus

Treatment	No. (mean ± SE)		
	Normal plants	**Abnormal plants**	**Clumps of non-regenerating callus**
Accession			
D. carota subsp.*sativus*	20.5 ± 4.4 b	12.2 ± 2.5 b	45.4 ± 16.3 a
D. carota subsp.*azoricus*	9.2 ± 3.7 b	1.3 ± 0.6 b	9.7 ± 5.8 b
D. carota subsp.*drepanensis*	18.3 ± 4.2 b	30.7 ± 9 a	10.1 ± 3.1 b
D. carota subsp.*maritimus*	8.3 ± 2.4 b	0.5 ± 0.4 b	7.3 ± 2.8 b
D. aureus	73.2 ± 33.7 a	6.5 ± 1.9 b	17.6 ± 7.7b
D. montevidensis	0.0 ± 0.0 b	0.0 ± 0.0 b	0.0 ± 0.0 b
D. pusillus	0.0 ± 0.0 b	0.0 ± 0.0 b	0.0 ± 0.0 b
Alginate sterilization			
Autoclave	13.7 ± 4.3	4.5 ± 1.0	8.6 ± 3.8
Filter	17.1 ± 6.1	10.2 ± 3.3	13.9 ± 3.1
Culture technique			
ETAF	20.2 ± 8.3	2.0 ± 0.5 b	21.3 ± 5.3 a
TAL	11.7 ± 1.7	11.7 ± 3.0 a	3.5 ± 0.9 b
PSK concentrations (nM)			
0	6.2 ± 2.9 b	8.0 ± 3.0	6.6 ± 1.8
100	24.9 ± 6.8 a	6.8 ± 1.9	16.1 ± 4.6

In each section of the table, means with the same letters did not differ significantly ($P \leq 0.05$) within the columns

TAL thin alginate layer; *ETAF* extra thin alginate film

DISCUSSION

Since genetically modified organisms are not fully accepted by the public, there is an interest in exploiting protoplast-to-plant technologies in the production of novel germplasm for plant breeding (Eeckhaut et al. 2013). One of the most common protoplast-based approaches uses protoplast fusion to generate unique plants in commercial crops with desirable agronomically relevant traits such as resistance to pathogens, stress tolerance, or CMS (Matsumoto 1991; Akagi et al. 1995; Kisaka and Kameya 1998). Somatic hybridization through protoplast fusion overcomes pre- and post-zygotic barriers among incompatible species, allowing plants with novel genetic combinations within the species, between different species of the same genus, or even between different genera to be obtained (Davey et al. 2005a). However, the application of somatic hybridization in practical breeding processes requires the establishment of a protoplast-to-plant system for the fusion parents. Such studies are reasonable, since it is known that the regenerative capacity of different genotypes varies significantly even for closely related species or individual lines within the same species (e.g., Davey et al. 2005c). To our knowledge, the present study is the first attempt to assess the regeneration ability of leaf protoplasts isolated from wild *Daucus* species and *D. carota* subspecies other than cultivated carrot (*D. carota* subsp. *sativus*).

The availability of a large number of protoplasts with high viability is the first step for successful protoplast culture. The efficiency of protoplast isolation and their quality, expressed by cell viability, are governed by several factors including: genotype; type of source tissue; mechanical procedures on source tissue (slicing or removal of the epidermis); conditioning of source tissue before enzyme maceration (dark or osmotic treatment); conditions of tissue digestion (i.e., composition of enzyme mixture, temperature and duration of enzyme incubation, pH of the enzyme solution, gentle agitation), and the method of protoplast isolation (Frearson et al. 1973; Rao and Prakash1995; Ortin-Parraga and Burgos 2003; Sinha et al. 2003; Davey et al. 2005c; Kiełkowska and Adamus 2012). When large populations of protoplasts are required, which is the norm for fusions, from 1 g of FW, between 10^5 and 10^7 viable cells should be released (Davey et al. 2010). Though the protoplast yield obtained in the present study, varying from 1.4 to 4.5 × 10^6, was genotype-dependent, for all accessions it was even high enough to perform protoplast fusion. Additionally, these results showed that the protocol of protoplast isolation based on slicing and plasmolysis of source tissue, overnight digestion with gentle shaking, and gradient centrifugation for protoplast purification, elaborated for protoplast isolation from leaves and hypocotyls of cultivated carrot (Grzebelus et al. 2012a), may be also applied to their wild relatives. The induction and maintenance of shoot cultures for wild *Daucus* are a long and laborious process; nevertheless, the isolation of protoplasts from short-term hypocotyl cultures seems to be impossible, due to the limited number of available seeds. The viability of protoplasts released from wild accessions was satisfactory and not significantly lower than that of cultivated carrot, but varied greatly, ranging from 50 to 93 %. Slightly differences in physiological status of source tissue or seasonal variation may explain the different numbers of viable protoplasts observed in analyzed accessions, since shoot cultures were maintained over the year. Keskitalo (2001) reported that protoplast isolation from cultured shoots of *Tanacetum* was most successful during winter and spring, suggesting the persistence of the seasonal clock in vitro. We observed a different dependence, with higher values of isolation efficiency during spring and summer and lower values at autumn and winter time (data not shown). Moreover, protoplast isolation from shoot cultures even 1 day older than 2 weeks resulted in reductions in both their yield and viability.

Synthesis of the cell wall, cell cycle re-entry, and sustained mitotic divisions are key steps in the production of daughter cells and further plant regeneration from protoplast-derived tissues (Eeckhaut et al. 2013). Improvements in protoplast-to-plant systems have been achieved by testing different culture techniques and medium formulations with various supplements (Anthony et al. 1997; Matsubayashi et al. 1997; Oh and Clouse 1998; Davey et al. 2005a). Since over 40 years ago, when fertile plants from protoplast-derived cells of tobacco were regenerated for the first time (Nagata and Takebe 1971), many systems of protoplast culture have been verified to improve regeneration ability from a wide range of plant species (reviewed by Davey et al. 2005a, b, c; Eeckhaut et al. 2013). Generally, a higher rate of cell colony formation was noted in the case of protoplasts embedded before culture, especially in alginate matrix, as compared to non-embedded protoplasts. Such a positive effect of embedding on cells may be explained by membrane stabilization through lipid peroxidase inhibition, the prevention of leakage of cell wall precursors or other metabolites, lower ethylene level (Bajaj 1989), or improved signaling cascade before the first cytokinesis (Sheahan et al. 2007; Zaban et al. 2013). Moreover, it was documented that protoplast separation ensured a better oxygen supply, resulting in avoidance of browning (Majewska-Sawka et al. 1994; Pati et al. 2005; Eeckhaut et al. 2013). The beneficial effect of protoplast immobilization in alginate matrix on plating efficiency was recorded for, among others, soybean (Tricoli et al. 1986), sweet orange (Niedz 1993), *N. tabacum* (Dovzhenko et al. 1998), and *Arabidopsis thaliana* (Dovzhenko et al. 2003). According to very early studies on some properties of alginate solutions and gels, the method of sterilization, a necessary step when working with in vitro cultures, may influence polymerization of the alginate molecules (Leo et al. 1990). The commonly used method of sterilization of heat and high pressure treatment (autoclaving) can modify such properties of the gel as mechanical strength, abrasion resistance, diffusion within the gel, and development of entrapped biomass. It seems that only few authors working with alginate-embedded protoplasts took this fact into consideration, applying filter-sterilization (Draget et al. 1988; Hall et al. 1997; Dovzhenko et al. 1998; Dovzhenko et al.2003). Very often, the method of alginate solution sterilization is not given in the papers. Moreover, it is difficult to find numerical data confirming the beneficial effect of filter-sterilized versus autoclave-sterilized alginate on protoplast development. Nevertheless, several authors suggested the positive role of filter-sterilized alginate on plating efficiency, e.g., for *Brassica napus* (Draget et al. 1988), *B. vulgaris* (Hall et al. 1997), or *N. tabacum*(Dovzhenko et al. 1998). Although filter sterilization of alginate solution is quite a difficult and long process (preceded by overnight stirring on a gyratory shaker to dissolve), the values noted for plating efficiency in the present study confirmed its stimulatory effect on mitotic activity of protoplast-derived cells. Additionally, due to better clarity of filter-sterilized alginate, as compared to autoclave-sterilized, microscopic observations were made easily and precisely.

Immobilization of protoplasts in gel matrices not only enhances plating efficiency but is sometimes crucial for mitotic activity by the protoplast-derived cells (Ebrahimzadeh et al.2001; Grzebelus et al. 2012b). A general conclusion was proposed based on data from many reports, suggesting that the thinner the matrix, the higher the frequency of colony formation, probably as a consequence of increased compound diffusion from the medium to the protoplasts (Dovzhenko et al. 1998, 2003; Pati et al. 2005; Eeckhaut et al. 2013). Reduction of the alginate gel thickness from droplet to layer was achieved originally by inserting, before polymerization, a polypropylene grid or coverglass in the TAL or ETAF system, respectively (Golds et al. 1992; Pati et al. 2005). In the present study, we used a modified TAL system, described in previous reports (Grzebelus et al. 2012a, b), based on: (1) application of a reduced volume of protoplast-alginate mixture to form the alginate layer (300 µl instead the 625 µl proposed in earlier studies); (2) excluding polypropylene mesh grid from the system, as it is difficult to obtain; and (3) minimizing layer thickness by rotation of protoplast-alginate mixture during application. In the ETAF technique, in comparison with the original protocol (Pati et al. 2005), we modified the polymerization step—this was performed on Ca-agar plates, as in the TAL technique, instead of on microscope slides. The proposed modifications for both TAL and ETAF simplify the systems, reducing handling and costs. The values of plating efficiency observed in *N. tabacum* and *Lotus corniculatus* protoplast cultures indicated that ETAF was on a par with the TAL technique (Pati et al. 2005). In all *Daucus* accessions evaluated in the present study, mitotic activity of protoplast-derived cells was observed both in TAL and ETAF cultures; however, contrary to expectation, using ETAF led to an approximately two-fold decrease of plating efficiency at all points of evaluation. It could be due to fact that the ETAFs, in comparison with TALs, were produced from a three-fold smaller volume of protoplast-alginate mixture (100 vs. 300 µl, respectively) and were represented by one-cell layers. Although in both systems the density of embedded protoplasts was the same, the cross-stimulating effect of dividing adjacent cells was higher in the multi-cell layers obtained in the TAL technique.

Lately, PSK has been recognized as a chemical nursing agent that drastically promotes cell divisions in rice and sugar beet protoplast cultures (Matsubayashi et al. 1997; Grzebelus et al. 2012b). Similarly, protoplast cultures of all seven *Daucus* accessions analyzed in this study responded positively to PSK supplementation at the concentration of 100 nM, leading to, on average, an above 10 % higher level of plating efficiency in comparison with PSK-free control. However, for two wild *Daucus* species, stimulation of cell activity was observed only in early cultures; it gradually stopped and, as a result, calli were not formed. This may suggest that the applied concentration of PSK was too low for these two accessions to promote sustained division of the protoplast-derived cells, while for the remaining *Daucus* species, it was sufficient to stimulate a higher level of somatic embryogenesis, since four-fold fewer plants were regenerated from PSK-free cultures. Similar promotion of somatic embryogenesis as a result of PSK supplementation was observed in suspension cultures of plants as diverse as carrot (dicotyledons) and*Cryptomeria japonica* (conifers) (Hanai et al. 2000; Igasaki et al. 2003).

The differential response of wild relatives in comparison with cultivated crop to the same culture conditions is a known phenomenon. In protoplast cultures, both plating efficiency and plant regeneration may be genotype-dependent. In an early study on mesophyll protoplasts of *A. thaliana*, Damm and Willmitzer (1988) noticed between 10- and 20-fold differences in the level of plating efficiency among the analyzed races. Similarly, differences in the ability to form colony and/or plant regeneration between wild and cultivated representatives of the same genus were observed in protoplast cultures of tomato (Mühlbach 1980), *Helianthus* (Bohorova et al. 1986) and, lately, *Cichorium*(Deryckere et al. 2012). In the present study, all wild *D. carota* subspecies showed high potential both for colony formation and plant regeneration, while two out of three wild*Daucus* species exhibited recalcitrant status, since mitotic divisions of protoplast-derived cells were arrested in early cultures. The factors preventing protoplasts from expressing totipotency may: (1) be related to the oxidative stress generated during protoplast isolation and subsequent culture; and/or (2) result from a lack of signaling molecules synthesized in plants and involved in the developmental process. The oxidative stress induced during the isolation procedure may be long-term and individual species may respond to it differently, among others, producing and cumulating peroxides and other active oxygen species (Watanabe et al. 2002; Davey et al. 2005c). It was postulated that suppressed expression of totipotency in protoplasts was correlated with reduced activity of the cellular antioxidant machinery (Papadakis et al. 2001). Protection from oxidative damage and, in consequence, overcoming protoplast recalcitrance may be possibly achieved by application of an exogenous antioxidant such as ascorbic acid, glutathione, tocopherol, polyamines, or n-propyl gallate.

In conclusion, to our knowledge, this is the first report to show successful plant regeneration from leaf protoplast cultures of wild representatives within the *Daucus*genus. We demonstrated that high efficiency of cell colony formation and plant regeneration could be induced after protoplast embedding in a modified thin layer of filter-sterilized alginate and supplementation of culture medium with PSK. The protocol can be easily applied in rarely exploited protoplast fusion between cultivated carrot and wild *Daucus* species.

ACKNOWLEDGMENTS

The authors thank Prof. P. W. Simon, Prof. D. Grzebelus, Prof. R. Baranski, Dr. C. Allender, and Dr. T. Nothnagel for providing seeds. This work was supported by the Polish National Science Centre, Grant No. NN310 440 238, awarded to E. G.

REFERENCES

1. Akagi H, Taguchi T, Fujimura T (1995) Stable inheritance and expression of the CMS traits introduced by asymmetric protoplast fusion. Theor Appl Genet 91:563–567
2. Anthony P, Davey MR, Power JB, Lowe KC (1997) Enhanced mitotic division of cultured*Passiflora* and *Petunia* protoplasts by oxygenated perfluorocarbon and haemoglobin. Biotechnol Tech 11:581–584

3. Bajaj Y (1989) Plant protoplasts and genetic engineering. Springer, Heidelberg

4. Bohorova NE, Cocking EC, Power JB (1986) Isolation, culture and callus regeneration of protoplasts of wild and cultivated *Helianthus* species. Plant Cell Rep 5:256–258

5. Camadro EL, Cauhépé MA, Simon PW (2008) Compatibility relations between the edible carrot *Daucus carota* and *D. pusillus*, a related wild species from the Argentinian Pampas. Euphytica 159:103–109

6. Chen YF, Mastubayashi Y, Sakagami Y (2000) Peptide growth factor phytosulphokine-α contributes to the pollen population effect. Planta 211:752–755

7. Damm B, Willmitzer L (1988) Regeneration of fertile plants from protoplasts of different*Arabidopsis thaliana* genotypes. Mol Gen Genet 213:15–20

8. Damm B, Schmidt R, Willmitzer L (1989) Efficient transformation of *Arabidopsis thaliana*using direct gene transfer to protoplasts. Mol Gen Genet 217:6–12

9. Davey MR, Anthony P, Power JB, Lowe KC (2005a) Plant protoplast technology: current status. Acta Physiol Plantarum 27:117–129

10. Davey MR, Anthony P, Power JB, Lowe KC (2005b) Plant protoplasts: status and biotechnological perspectives. Biotechnol Adv 23:131–171

11. Davey MR, Anthony P, Power JB, Lowe KC (2005c) 2004 SIVB congress symposium proceedings "Thinking outside the cell": plant protoplast technology: status and applications. In Vitro Cell Dev Biol Plant 41:202–212

12. Davey MR, Anthony P, Patel D, Power JB (2010) Plant protoplasts: isolation, culture and plant regeneration. In: Davey MR, Anthony P (eds) Plant cell culture essential methods. Wiley-Blackwell, New York, pp 153–173

13. Deryckere D, Eeckhaut T, Van Huylenbroeck J, Van Bockstaele E (2012) Low melting point agarose beads as a standard method for plantlet regeneration from protoplasts within the *Cichorium* genus. Plant Cell Rep 31:2261–2269

14. Dirks R, Sidorov V, Tulmans C (1996) A new protoplast culture system in *Daucus carota* L. and its applications for mutant selection and transformation. Theor Appl Genet 93:809–815

15. Dovzhenko A, Bergen U, Koop HU (1998) Thin-alginate-layer technique for protoplast culture of tobacco leaf protoplasts: shoot formation in less than two weeks. Protoplasma 204:114–118

16. Dovzhenko A, Dal Bosco C, Meurer J, Koop HU (2003) Efficient regeneration from cotyledon protoplasts in *Arabidopsis thaliana*. Protoplasma 222:107–111

17. Draget KI (2000) Alginates. In: Phillips GO, Williams PA (eds) Handbook of hydrocolloids. Woodhead Publishing Limited, Cambidge, pp 380–395

18. Draget KI, Myhre S, Østgaard K (1988) Regeneration, cultivation and differentiation of plant protoplasts immobilized in Ca-alginate beads. J Plant Physiol 132:552–556

19. Dudits D, Hadlaczky G, Lévi E, Fejér O, Haydu Z, Lázár G (1977) Somatic hybridisation of*Daucus carota* and *D. capillifolius* by protoplast fusion. Theor Appl Genet 51:127–132

20. Ebrahimzadeh H, Noori-Daloii MR, Karamian R (2001) Shoot regeneration from saffron protoplasts immobilized in Ca-alginate beads. J Sci R Iran 11:69–73

21. Eeckhaut T, Lakshmanan PS, Deryckere D, Van Bockstaele E, Van Huylenbroeck J (2013) Progress in plant protoplast research. Planta. doi:10.1007/s00425-013-1936-7

22. Frearson EM, Power JB, Cocking EC (1973) The isolation, culture and regeneration of*Petunia* leaf protoplasts. Dev Biol 33:130–137

23. Gamborg OL, Miller RA, Ojima K (1968) Nutrient requirements of suspension cultures of soybean root cells. Exp Cell Res 50:151–158

24. Golds TJ, Babczinsky J, Rauscher G, Koop HU (1992) Computer controlled tracking of single cell development in *Nicotiana tabacum* L. and *Hordeum vulgare* L. protoplasts embedded in agarose/alginate films. J Plant Physiol 140:582–587

25. Grambow HJ, Kao KN, Miller RA, Gamborg OL (1972) Cell division and plant development from protoplasts of carrot cell suspension cultures. Planta 103:348–355

26. Grzebelus D, Baranski R, Spalik K, Allender C, Simon PW (2011) Daucus. In: Kole Ch (ed) Wild crops relatives: genomic and breeding resources, vegetables. Berlin, Springer, pp 91–113

27. Grzebelus E, Szklarczyk M, Baranski R (2012a) An improved protocol for plant regeneration from leaf and hypocotyl-derived protoplasts of carrot. Plant Cell Tiss Organ Cult 109:101–109

28. Grzebelus E, Szklarczyk M, Greń J, Śniegowska K, Jopek M, Kacińska I, Mrożek K (2012b) Phytosulfokine stimulates cell divisions in sugar beet (*Beta vulgaris* L.) mesophyll protoplast cultures. Plant Growth Regul 67:93–100

29. Hall RD, Riksen-Bruinsma T, Weyens G, Lefèbvre M, Dunwell JM, van Tunen A, Krens FA (1997) Sugar beet guard cell protoplasts demonstrate a remarkable capacity for cell division enabling applications in stomatal physiology and molecular breeding. J Exp Bot 28:255–263

30. Hanai H, Matsumo T, Yamamoto M, Matsubayashi Y, Kobayashi T, Kamada H, Sakagami Y (2000) A secreted peptide growth factor, phytosulfokine, acting as a stimulatory factor of carrot somatic embryo formation. Plant Cell Physiol 41:27–32

31. Ichikawa H, Tanno-Suenaga L, Imamura J (1987) Selection of *Daucus* cybrids based on metabolic complementation between X-irradiated *D. capillifolius* and iodoacetamide-treated *D. carota* by somatic cell fusion. Theor Appl Genet 74:746–752

32. Igasaki T, Akashi N, Ujino-Ihara T, Matsubayashi Y, Sakagami Y, Shinohara K (2003) Phytosulfokine stimulates somatic embryogenesis in *Cryptomeria japonica*. Plant Cell Physiol 44:1412–1416

33. Kao KN, Michayluk MR (1975) Nutritional requirements for growth of Vicia hajastana cells and protoplasts at

a very low population density in liquid media. Planta 126:105–110

34. Keskitalo M (2001) Can protoplast production from in vitro cultured shoots of *Tanacetum*vary during season? Agric Food Sci Finl 10:145–151
35. Kiełkowska A, Adamus A (2012) An alginate-layer technique for culture of *Brassica oleracea* L. protoplasts. In Vitro Cell Dev Biol 48:265–273
36. Kisaka H, Kameya T (1998) Cold and salt tolerance of somatic hybrid calli between barley (*Hordeum vulgare* L.) and carrot (*Daucus carota* L.). Breeding Sci 48:11–15
37. Kobayashi T, Eun CH, Hanai H, Matsubayashi Y, Sakagami Y, Kamada H (1999) Phytosulphokine-α, a peptidyl plant growth factor, stimulates somatic embryogenesis in carrot. J Exp Bot 50:1123–1128
38. Leo WJ, McLoughlin AJ, Malne DM (1990) Effects of sterilization treatments on some properties of alginate solutions and gels. Biotechnol Prog 6:51–53
39. Majewska-Sawka A, Nakashima H, Mori K (1994) Isolation and culture of suspension-derived protoplasts of *Beta vulgaris* L. Biol Plant 36:9–13
40. Matsubayashi Y, Sakagami Y (1996) Phytosulfokine, sulfated peptides that induce the proliferation of single mesophyll cells of *Asparagus officinalis* L. Proc Natl Acad Sci USA 93:7623–7627
41. Matsubayashi Y, Takagi L, Sakagami Y (1997) Phytosulfokine-α, a sulfated pentapeptide, stimulates the proliferation of rice cells by means of specific high- and low-affinity binding sites. Proc Natl Acad Sci USA 94:13357–13362
42. Matsubayashi Y, Takagi L, Omura N, Morita A, Sakagami Y (1999) The endogenous sulfated pentapeptide phytosulfokine-α stimulates tracheary element differentiation of isolated mesophyll cells of *Zinnia*. Plant Physiol 120:1043–1048
43. Matsubayashi Y, Goto T, Sakagami Y (2004) Chemical nursing: phytosulfokine improves genetic transformation efficiency by promoting the proliferation of surviving cells on selective media. Plant Cell Rep 23:155–158
44. Matsumoto E (1991) Interspecific somatic hybridization between lettuce (*Lactuca sativa*) and wild species *L. virosa*. Plant Cell Rep 9:531–534
45. Menczel L, Nagy F, Kiss Z, Maliga P (1981) Streptomycin resistant and sensitive somatic hybrids of *Nicotiana tabacum* + *Nicotiana knightiana*: correlation of resistance to *N. tabacum* plastids. Theor Appl Genet 70:590–594
46. Mühlbach HP (1980) Different regeneration potentials of mesophyll protoplasts from cultivated and a wild species of tomato. Planta 148:89–96
47. Murashige T, Skoog F (1962) A revised medium for rapid growth and bio assays with tobacco tissue culture. Physiol Plant 18:100–127
48. Nagata T, Takebe I (1971) Plating of isolated tobacco mesophyll protoplasts on agar medium. Planta 99:12–20
49. Niedz RP (1993) Culturing embryogenic protoplasts of 'Hamlin' sweet orange in calcium alginate beads. Plant Cell Tiss Organ Cult 34:19–25
50. Niedz RP (2006) Regeneration of somatic embryos from sweet orange (*C. sinensis*) protoplasts using semi-permeable membranes. Plant Cell Tiss Organ Cult 84:353–357
51. Oh MH, Clouse SD (1998) Brassinolide affects the rate of cell division in isolated leaf protoplasts of *Petunia hybrida*. Plant Cell Rep 17:921–924
52. Ortin-Parraga F, Burgos L (2003) Isolation and culture of mesophyll protoplasts from apricot. J Hort Sci Biotechnol 78:624–628
53. Pan ZG, Liu CZ, Murch SJ, El-Demerdash M, Saxena PK (2003) Plant regeneration from mesophyll protoplasts of the Egyptian medicinal plants *Artemisia judaica* L. and *Echinops spinosissimus* Turra. Plant Sci 165:681–687
54. Papadakis AK, Siminis CI, Roubelakis-Angelakis KA (2001) Reduced activity of antioxidant machinery is correlated with suppression of totipotency in plant protoplasts. Am Soc Plant Biol 126:434–444
55. Pati PK, Sharma M, Ahuja PS (2005) Extra thin alginate film: an efficient technique for protoplast culture. Protoplasma 226:217–221
56. Rákosy-Tican E, Aurori A, Vesa S, Kovacs K (2007) In vitro morphogenesis of sunflower (*Helianthus annuus*) hypocotyl protoplasts: the effects of protoplast density, haemoglobin and spermidine. Plant Cell Tiss Organ Cult 90:55–62
57. Rao KS, Prakash AH (1995) A simple method for the isolation of plant protoplasts. J Biosci 20:645–655
58. Rubatzky VE, Quiros CF, Simon PW (1999) Carrots and related vegetable *Umbelliferae*. CABI Publishing, New York
59. Ryan CA, Pearce G (2001) Polypeptide hormones. Plant Physiol 125:65–68
60. Sheahan M, Rose R, McCurdy D (2007) Actin-filament-dependent remodelling of the vacuole in cultured mesophyll protoplasts. Protoplasma 230:141–152
61. Sinha A, Wetten AC, Caligari PDS (2003) Effect of biotic factors on the isolation of*Lupinus albus* protoplasts. Austr J Bot 51:103–109
62. StatSoft, Inc (2009) STATISTICA (data analysis software system), version 9
63. Tricoli DM, Hein MB, Carnes MG (1986) Culture of soybean mesophyll protoplasts in alginate beads. Plant Cell Rep 5:334–337
64. Watanabe M, Suzuki K, Kawasaki H, Watanabe Y (2002) Differential responses of*Brassica napus* and *Petunia hybrida* to leaf protoplast isolation stress. Physiol Plant 114:645–651
65. Yamakawa S, Sakurai C, Matsubayashi Y, Sakagami Y, Kamada H, Satoh S (1998) The promotive effects of a peptidyl plant growth factor, phytosulfokine, on the formation of adventitious roots and expression of a gene for a root-specific cystatin in cucumber hypocotyls. J Plant Res 111:453–458
66. Zaban B, Maisch J, Nick P (2013) Dynamic actin controls polarity induction de novo in protoplasts. J Integr Plant Biol 55:142–159

CHAPTER 4

RESPONSE OF CARROT PROTOPLASTS AND PROTOPLAST-DERIVED AGGREGATES TO SELECTION USING A FUNGAL CULTURE FILTRATE OF ALTERNARIA RADICINA

Ewa Grzebelus Maria Kruk Alicja Macko-Podgo´rni Dariusz Grzebelus

Department of Genetics, Plant Breeding and Seed Science, Faculty of Horticulture, University of Agriculture in Krakow, Al. 29 Listopada 54, 31-425 Krako´w, Poland

ABSTRACT

Protoplasts isolated from three accessions of cultivated carrot and 5-day-old protoplast-derived aggregates were subjected to selection to identify somaclonal variants with enhanced tolerance to the fungal disease black rot incited by Alternaria radicina. Different concentrations [1, 2, 3.5, 5, 10, 20, 35 and 50 % (v/v)] of a fungal culture filtrate (FCF) from 2-week-old liquid cultures of A. radicina were used. Protoplasts and aggregates were subjected to short-term selection for a period of 10 days. All FCF concentrations added to the cultures on the day of isolation decreased protoplast survival frequency and plating efficiency, while FCF applied 5 days later inhibited cell divisions in 5–50 % concentrations. The responses of protoplasts to the treatment were genotype dependent. Most R0 plants were regenerated in all accessions from cell lines grown with 1 % FCF, while only a few plants were produced from 2 to 3.5 % FCF-treated cultures of 'Dolanka' and the breeding line '9304B', respectively. Nineteen-percent of putative stress-tolerant regenerants were tetraploids, while only 5 % tetraploids were observed in the control. The incidence of unique random amplified polymorphic DNA fragments indicating possible chromosomal rearrangements was low and did not differ among regenerants after selection and those derived from the control. Mobilization of miniature inverted repeat transposable elements was not observed. Some R0 individuals regenerated both from FCF-treated and untreated cultures showed lower susceptibility to A. radicina in a laboratory assay in comparison to control plants grown from seed. Regenerants from FCF-treated cultures showed lower frequency of flowering plants and a higher rate of male sterility. Pollen viability of the putative stress-tolerant regenerants varied over a wide range (6–98 %), independently of in vitro selection conditions. Our data suggest that A. radicina FCF may be feasible for the in vitro selection to generate plants with superior phenotypic performance against A. radicina.

INTRODUCTION

Cultivated carrots (a rich source of a- and b-carotene, precursors of vitamin A). Carrot roots also contain other compounds such as: disacharides, phenolics, terpenes, saponins, fatty acids, and proteins (Grzebelus et al. 2011). Black rot, caused by Alternaria radicina Meier, Drechsler and E.D. Eddy, a seed-borne fungal pathogen, is one of the most serious diseases found wherever carrots are commercially produced (Pryor et al. 1998). The pathogen attacks carrots in the field, at any stage of growth and development, and also during storage. In the course of vegetation, dry, black, necrotic lesions on leaf petioles, laminas, and root crown can be observed. Large rotten spots might be formed along the roots under post-harvest conditions, eventually leading to total yield loss. A. radicina is frequently

Citation: Ewa Grzebelus Maria Kruk Alicja Macko-Podgo´rni Dariusz Grzebelus, Response of carrot protoplasts and protoplast-derived aggregates to selection using a fungal culture filtrate of Alternaria radicina, DOI 10.1007/s11240-013-0353-8.

present on carrot seeds, reducing its quality through decreasing seed germination and causing seedling death. Seeds are the main A. radicina source, but the pathogen can attack plants if present in the soil (Pryor et al. 1998). Once introduced, the fungus can survive in soil for years, even in the absence of carrot host plants. The use of fungicides is very often ineffective, and in addition some chemically synthesized fungicides can cause environmental pollution. Therefore, successful control of black rot has been difficult, costly, labor- and resource-intensive, depending on integration of the most appropriate methods for distinct carrot production regions (Tylkowska 1992; Pryor and Gilbertson 2001; Farrar et al. 2004). Cultivation of varieties resistant to A. radicina could overcome this problem. So far, resistant cultivars have not been developed using conventional breeding methods due to the absence of useful sources of resistance genes in cultivated and wild Daucus germplasm. Hence, application of biotechnological tools in breeding could be an alternative to develop black rot-tolerant carrot lines. In vitro selection might offer a feasible and cost-effective tool for developing such plants (Rai et al. 2011). That technique can be successfully applied to carrot, which is a model species for plant tissue culture systems and plant regeneration studies (Baranski 2008).

Tissue culture conditions impose severe physiological stress that may be further increased by addition of a selective agent to the medium. It may result in a range of changes in the genome, including polyploidization and chromosomal rearrangements. It was also shown that transposable elements (TEs) can be mobilized in the course of the culture (Kikuchi et al. 2003). In vitro selection exploits that genetic variation, the so-called somaclonal variation (Larkin and Scowcroft 1981), which can result in a range of genetically stable changes, some of them being potentially useful for crop improvement. At the DNA level, the occurrence of these genomic changes can be detected by distinct molecular markers (Gao et al. 2009). To perform in vitro selection, different explants, such as plant cells (protoplasts and cell suspensions), tissues (callus), or organs (shoots and somatic embryos) are cultured on media supplemented with the selective agents. Exposure of plant explants to selective agents generates selection pressure allowing the survival and growth only of cells with the desired stress-tolerant phenotype. For fungal diseases, a broad range of agents can be used to select resistant lines via in vitro cultures including fungal culture filtrates (FCF), pure phytotoxins, or intact pathogens (Sˇva´bova´ and Lebeda 2005; Rai et al. 2011). These selective agents elicit typical reactions such as the production and accumulation of antimicrobial phytoalexins in the cells (Sˇva´bova´ and Lebeda 2005). A high correlation between in vitro and in vivo resistance to pathogenic fungi was reported (Kumar et al. 2008; Savita et al. 2011), however, resistance to pathotoxins at the cellular level may not always lead to whole plant resistance to the pathogen (Sharma et al. 2010). Despite the fact that carrot is a model plant for in vitro culture, very few reports describing the in vitro selection for improving its resistance to pathogens can be found (Baranski et al. 1997). To our knowledge, there is no information on applying that technique to select carrot lines with improved levels of resistance to black rot. In the present study, the phytotoxic effect of the fungal culture filtrate from A. radicina on protoplasts and protoplastderived cell aggregates was analyzed. The main objective of our research was to regenerate carrot plants from protoplast cultures under A. radicina culture filtrate pressure, and subsequently to obtain and evaluate these regenerants.

MATERIALS AND METHODS

Plant Material

Three male fertile accessions of carrot were used as donors for protoplast isolation: two Polish open-pollinated varieties 'Dolanka' and 'Amsterdamska' and the breeding line '9304B' (kindly provided by P.W. Simon, USDA, USA). All of them were susceptible to black rot. Protoplasts were isolated from in vitro grown plantlets derived from seeds. The procedure and conditions of seed disinfection, germination, and seedling maintenance were as described by Grzebelus et al. (2012). Briefly, seeds were incubated in a water bath at 40 C, then transferred to a 0.2 % (v/v) solution of fungicide 'Bravo' (Syngenta, USA) and finally immersed in a 20 % (w/v) solution of chloramin T (sodium N-chlorotoluene-4-sulphonamide) (30 min each). After washing in sterile water and air-drying, the seeds were germinated in 9 9 1.5 cm Petri dishes on solid Murashige and Skoog (MS) medium with vitamins (Murashige and

Skoog 1962) supplemented with 30 g l^{-1} sucrose and 6.5 g l^{-1} plant agar (Biocorp, Poland) at 26 ± 2 C in the dark. After about 7 days of culture, the seedlings were transferred to glass jars containing regeneration medium (R) composed of MS macro- and micro-elements, 0.1 mg l^{-1} thiamine HCl, 0.1 mg l^{-1} piridoxine HCl, 0.5 mg l-1 nicotinic acid, 3.0 mg l^{-1} glycine, 100 mg l^{-1} myo-inozytol, 20 g l^{-1} sucrose, and 2.5 g l^{-1} phytagel (Sigma-Aldrich, St. Louis, MO, USA). Cultures were kept in a climate room at 26 ± 2 C under a 16-h photoperiod and light intensity of 55 lmol m-2 s^{-1} .

Isolation and Maintenance of A. Radicina

Alternaria radicina was isolated from naturally infected roots of carrot. After surface disinfection in 70 % ethanol, root discs with infected tissue were excised and cultured for 2 weeks on potato-dextrose agar (PDA, Duchefa Biochemie, Haarlem, The Netherlands) at 26 ± 2 C in the dark. After microscopic verification, single spore isolates were multiplied on PDA in 9 9 1.5 cm Petri dishes at 26 ± 2 C in the dark. Stock isolates of A. radicina were maintained at 8 ± 2 C and renewed periodically. P

Preparation of the Fungal Culture Filtrates from A. Radicina

FCF was produced from A. radicina isolate KGHN3. One mycelial disc (approx. 0.5 cm2) of a 2-week-old fungal culture was transferred to 100-ml Erlenmeyer flasks with 40 ml of potato-dextrose broth (PDB, Duchefa Biochemie). The flasks were placed on a gyratory shaker (150 rpm) and incubated at 26 ± 2 C in the dark for 2 weeks. The mycelium and spores were then filtered from the liquid cultures and the filtrate was passed through a glass fiber prefilter (1.0 lm, Millipore, Billerica, MA, USA). The purified filtrate was adjusted to pH 5.6 and filter-sterilized through syringe disposable membrane filters (0.22 lm, Millipore) and added to the protoplast culture medium as a selective agent.

Protoplast Isolation and Culture

Protoplasts were isolated from leaves of 4-week-old in vitro grown plantlets. All steps of protoplast isolation and culture were performed according to the established protocol of Grzebelus et al. (2012). Briefly, leaves were cut into fine pieces in a glass Petri dish in the presence of preplasmolysis solution (0.5 M mannitol), incubated for 1 h at 26 ± 2 C in the dark, and then digested overnight in the enzyme solution [1 % (w/v) cellulase Onozuka R-10 (Duchefa Biochemie), 0.1 % (w/v) pectolyase Y-23 (Duchefa Biochemie), 20 mM 2-(N-Morpholino) ethanesulfonic acid (MES, Sigma-Aldrich), 5 mM CaCl2, and 0.6 M mannitol (Sigma-Aldrich), pH 5.6, filter-sterilized (0.22 lm, Millipore)] on a gyratory shaker (30 rpm). The released protoplasts were purified from the undigested tissue by filtration through a 80-lm nylon sieve (Millipore), and then from the enzyme solution by centrifugation at 1009 g for 5 min. The protoplast pellet was resuspended in the solution of 0.5 M sucrose with 1 mM MES, and overlaid with W5 medium containing 154 mM NaCl, 125 mM CaCl2, 5 mM KCl, and 5 mM glucose, pH 5.6 (Menczel et al. 1981). After centrifugation at 1459 g for 10 min, the viable protoplasts were collected in the interphase band, transferred to a fresh tube and washed twice by centrifugation at 1009 g for 5 min in W5 solution and carrot petiole protoplast medium (CPP), respectively. CPP culture medium consisted of macro-, micro-elements and organic acids according to Kao and Michayluk (1975), vitamins according to Gamborg B5 medium (Gamborg et al. 1968), 74 g l^{-1} glucose, 250 mg l^{-1} casein enzymatic hydrolysate (SigmaAldrich), 0.5 lM 2,4-dichlorophenoxyacetic acid (2,4-D), and 0.9 lM zeatin (pH 5.6, filter-sterilized).

The working density was adjusted to 8 9 10^5 protoplasts ml^{-1} via a Fuchs Rosenthal hemocytometer chamber. Before the culture, protoplasts were embedded in the alginate matrix. A modified procedure of the thin calcium alginate layer technique (Grzebelus et al. 2012) was used. Equal volumes of protoplasts resuspended in the CPP medium and 2.8 % (w/v) alginic acid sodium salt (SigmaAldrich) in 0.4 M mannitol solution were mixed carefully to obtain a final density of 4 9 10^5 ml^{-1} of protoplasts in the culture. Polymerization of the alginate matrix took place after spreading of the protoplasts-alginate mixture (in 300-ll aliquots) onto 1 % (w/v) agar (Biocorp) containing 20 mM CaCl2 and 0.4 M mannitol in 6 9 1.5 cm Petri dishes. Solid sheets of alginate with embedded protoplasts formed after a 1 h incubation at 26 ± 2 C. The layers were then carefully transferred from the agar plates to 4 ml of liquid CPP culture medium in 6 9 1.5 cm Petri dishes. Protoplast cultures were incubated in the dark at 26 ± 2 C for about 2 months.

Protoplast Selection

The selection medium was prepared by adding filter-sterilized volumes of FCF and autoclaved double-distilled water to the double-strength CPP medium to provide concentrations of the selective agent in the culture at 1, 2, 3.5, 5, 10, 20, 35, or 50 % (v/v). Protoplasts were treated with the filtrate in the first (type A experiments) or in the fifth (type B experiments) day of culture. In the control combination, FCF was not applied. After 10 days of selection (i.e., in 10 and 15-day-old cultures for type A and B experiments, respectively), the culture medium with FCF was removed and replaced by the fresh medium without the FCF. Each selection experimental set was established from protoplasts isolated from a single plant. To assess the effect of FCF on protoplast growth, cell viability and the ability to form aggregates were defined. Protoplast viability was analyzed after fluorescein diacetate (FDA) staining as described by Anthony et al. (1999) with modification of Grzebelus et al. (2012). In viable cells, the apple-green fluorescence was observed as the result of elimination of acetate groups from FDA by functional esterases present in the intact membrane of active cells. Fluorescein accumulation in the cells was visualized under an Axiovert S100 microscope (Carl Zeiss, Go¨ttingen, Germany) equipped with an excitation-emission filter set at k = 485–515 nm. The viability of protoplasts was estimated immediately after isolation (1 h) and in the 24th and 48th hour of culture for type A experiments, and only in the first hour of culture for type B experiments, and was expressed as the number of cells with apple-green fluorescence per the total number of observed cells (9100). Plating efficiency was estimated 10 days after FCF application, i.e., on the 10th day of culture for type A experiments and on the 5th and 15th day of culture for type B experiments, and was expressed as the number of cell aggregates per the total number of plated protoplasts (9100).

Plant Regeneration and Acclimatization

After 2 months of culture in the dark at 26 ± 2 C (including the selection period), both proembryonic masses (PEM) and somatic embryos emerged from the alginate matrix in FCFtreated and control combinations, were regenerated into plants (R0) and were acclimatized to ex vitro conditions according to Grzebelus et al. (2012). Briefly, after depolymerization of Ca-alginate layers in 20 mM sodium citrate solution containing 0.2 M mannitol, released PEM and embryos were washed by centrifugation (5 min at 1009 g) in the CPPD medium containing 0.5 lM NAA and 0.9 lM zeatin, pH 5.6 (Dirks et al. 1996), plated in glass jars on the solidified R medium, and finally placed in a climate room at 26 ± 2 C under a 16-h photoperiod and light intensity of 55 lmol m^{-2} s^{-1}. Rooted plants were transplanted to the moss-coconut fiber substrate (Ceres International, Poland) and acclimatized to ex vitro conditions by regularly reducing the humidity in a climate chamber at 20 ± 1 C under a 16-h photoperiod and light intensity of 30 lmol m^{-2} s^{-1}. M

MOLECULAR AND CYTOGENETIC ANALYSIS OF R0 PLANTS

Cytogenetic Analyses

Cytometric analysis of acclimatized plants was applied to determine the ploidy level. Measurement of the relative nuclear DNA content was performed on Partec CA II (Partec, Mu¨nster, Germany) following a protocol described by Kielkowska and Adamus (2010). Additionally, mitotic chromosome spreads from root tips were prepared to verify any abnormal numbers of chromosomes (Nowicka et al. 2012). Briefly, root tips were collected from ex vitro grown plants, treated with 2 mM 8-hydroxyquinoline for 3–4 h at 26 ± 2 C and then fixed in freshly prepared methanol – glacial acetic acid (3:1) solution. Isolated meristems were macerated in an enzyme mixture consisting of 4 % (w/v) cellulase Onozuka R10 (Serva Electrophoresis, Heidelberg, Germany) and 2

% (w/v) pectolyase Y23 (Duchefa Biochemie), in distilled water, pH 4.8 for 45 min at 37 C, and then carefully washed in distilled water, fixed again, spread on microscopic slides, and chopped with fine forceps. Chromosome number was scored under a phasecontrast microscope (Axiovert S100, Carl Zeiss, Go¨ttingen, Germany).

Random Amplified Polymorphic DNA

DNA amplifications were carried out using 12 arbitrary 10-mer primers (Eurofins MWG Operon), viz. OPA01, OPA02, OPA03, OPA09, OPA10, OPA13, OPA15, OPA16, OPB04, OPB05, OPF06, and OPN09. Polymerase chain reaction was performed in 10 ll containing 10 ng of DNA, 1 lM of primers, 0.25 mM of dNTPs (Thermo Scientific, Massachusetts, USA), 0.5 U of Dream Taq polymerase (Thermo Scientific) and 1x Dream Taq buffer. The cycle parameters were: 94 C for 1 min, followed by 40 cycles of 30 s at 93 C, 1 min at 42 C, 90 s at 68 C, and the final extension time of 5 min at 72 C. Products were analyzed by electrophoresis in 1 % agarose gels (Prona, Hispanagar, Burgos, Spain) supplemented with 10 ll of 5 % ethidium bromide solution per 100 ml of the gel. Min

Miniature Inverted repeat Transposable Elements Transposon Display

DNA digestion, ligation to adaptors, pre-selective, and selective amplification were performed according to Grzebelus et al. (2007) with modifications. Approximately 200 ng of DNA was digested in 10 ll with 5 U NdeI (50 - CA^TATG-30) (Thermo Scientific) for 3 h at 37 C. After fragmentation, thermal inactivation of the enzyme was conducted at 65 C for 20 min. A volume of 2.5 ll of the completely digested DNA was ligated to splinkerette-like adaptors modified from Devon et al. (1995). The ligation mixture (10 ll) was incubated overnight at 4 C with 1 lM of adaptors, 0.625 U of T4 DNA ligase (Thermo Scientific) and 1x T4 DNA ligase buffer. Adaptors used for ligation were prepared as follows: 40 lM SplAl (50 -CGA ATC GTA ACC GTT CGT ACG AGA ATG TCC TCT CCA 2 ACG AGC CAA GG-30), 40 lM SplA2 (50 -TAC CTT GGC TCG TTT TTT TTT GCA AAA A-3') and 1x adaptor annealing buffer (0.1 M Tris pH 7.5, 1 M NaCl, 0.01 M EDTA). The adaptor mixture was incubated for 10 min at 65 C and slowly cooled down to 24 C. Preselective amplification was set up using the adaptor-specific primer tdP1 (50 -CGA ATC GTA ACC GTT CGT ACG AGA A-30) and one of the MITE-specific primers (KRAKtd1: 50 -GCT GCT TTT HGA TTT ACC AAA CAR TTT-30 , DcStolD1M–50 -CCA CTT CAC CCA CTT TTC CT-30 , or DchATtd1: 50 -TTT CAG AAG TTC GGT CTT CGG T-30). Ligated samples were five-fold diluted and 1 ll was used for pre-selective amplification, prepared in 10 ll with 1 lM of MITE-specific primer, 1 lM of tdP1, 0.25 mM of dNTPs (Thermo Scientific), 0.5 U of Dream Taq polymerase (Thermo Scientific) and 1x Dream Taq buffer. The amplification profile was as follows: 94 C for 1 min, 20 cycles of 94 C for 30 s, 55 C for 30 s, 68 C for 2 min, and 68 C for 5 min. For selective amplification, nested primers were used [KRAKtd2: 50 -GAT TTA CCA AAC ART TTT TCA CCT GHT-30 , DcStolTD2M: 50 - TAA ACA ATT GGG TGG GAC GG-30 , or DchATTD2M: 50 -ACC GAA CCG AAT GCT CA-30 and TdMseI (NNN): 50 -TCC AAC GAG CCA AGG TAANNN-30 . For Krak, NNN stands for ACT; AGT;CAC; CTG; GTA; TGA; TGT, for DcSto, NN stands for AC; AG; GA, TdMseI (NN) for DchAT, NN stands for AC; AG; CA; CT; GA; GT; TC; TG]. Diluted pre-amplification (1:50) was used as a template for amplification set up in 10 ll containing 2 ll of diluted pre-amplified DNA, 1 lM of nested MITE-specific primer, 0.5 lM of TdMseI, 0.25 mM of dNTPs (Thermo Scientific), 0.5 U of Dream Taq polymerase (Thermo Scientific), and 1x Dream Taq buffer. Amplification was carried out starting using touch-down conditions in the first five cycles: 94 C (1 min), 5 cycles of 94 C (30 s), 60 C (30 s) decreased by 1 C per cycle, 68 C (2 min) followed by 35 cycles of 94 C (30 s), 55 C (30 s), 68 C (2 min), and 68 C (5 min). Products were separated for 4 h on agarose gels supplemented with 10 ll of 5 % ethidium bromide solution per 100 ml of the gel. Assessment of

Assessment of Response of R0 Plants to A. Radicina

Duetoanextendedplantregenerationperiodfor'Amsterdamska' and the line 9304B, only 'Dolanka'-derived regenerants were available for evaluation for resistance to black rot. For this purpose, greenhouse-grown R0 plants regenerated from FCF-treated and untreated cultures as well as field-grown control plants were applied to a laboratory petiole assay according to Grzebelus et al. (2003). The isolate KGHN3 of A. radicina was used in the experiments. Agar pieces overgrown by A. radicina mycelium were placed on a moistened filter paper in 9 9 1.5 cm Petri dishes. Then, the basal part of the petiole detached from the youngest fully developed carrot leaf per plant was attached to the mycelium. Following suggestions of Baranski et al. (2007), after 8 days of incubation in the dark at 26 ± 2 C the resistance was assessed by measuring length of the lesion along the petiole.

Analysis of Pollen Viability and seed Production from R0 Plants

After 4 months of growth in the greenhouse, the potted plants of all accessions were vernalized for 3 months at 10 ± 2 C to induce flowering. After that period, the plants were transferred back to the greenhouse. Alexander's procedure (1969) based on differential staining of aborted (colored as light-green) and non-aborted grains (colored as magenta-red) was used to evaluate pollen viability. Subsequently, the flowers with receptive stigmas were selfpollinated by hand to produce R1 progeny. Seeds were harvested and counted about 2 months later.

Statistical Analysis

The experiments were organized in a completely randomized design, with three replications, each treatment being represented by three Petri dishes. For protoplast viability and plating efficiency, counting was carried out in four or five microscopic fields on 100–200 cells per Petri dish. The mean values and standard errors were calculated. The overall effect of treatments was assessed using the analysis of variance (ANOVA) and Tukey's honestly significant difference (HSD) test in Statistica 9.0 (StatSoft. Inc. 2009). Test for equivalence of two-proportions was used to compute the significance level for the differences between proportions.

Results

Quality and development of carrot protoplasts in CCP medium not containing A. radicina FCF In type A experiments, the level of protoplast viability just after isolation was satisfactory and very similar for all tested accessions (P = 0.966), ranging from 72 ('Amsterdamska') to 75 % ('Dolanka', Table 1). In the next 2 days, no decrease in protoplast viability was observed (P = 0.932). After cell wall reconstruction, cell division occurred regularly and on the 10th day of culture two-, four-, and multi-cell aggregates were visible under the microscope. At that time the examined accessions responded similarly to the culture conditions (P = 0.083) reaching a plating efficiency of 50 % for 'Amsterdamska' to 80 % for the line 9304B (Table 1). In type B experiments, protoplast viability was as high as in the type A experiments yielding from 72 ('Dolanka') to 76 % ('Amsterdamska' and the line '9304B', P = 0.827). Similar to the 10th day, on the 15th day of culture, accessions did not differ in plating efficiency (P = 0.581), but the values varied from 68 (line '9304B') to 76 % ('Dolanka', Table 1).

EFFECT OF FCF ON PROTOPLAST VIABILITY AND DEVELOPMENT

Exposure of Protoplasts to FCF at the day of Isolation

Supplementation of CPP medium with the FCF on the first day of culture (type A experiments) reduced protoplast viability on average from 73 to 36 and 27 % after 24 and 48 h of culture, respectively (P\0.001). Cell viability strongly depended on the concentration of FCF in the culture medium (P\0.001). All FCF concentrations lowered the protoplast survival frequency on average from 55 % at 1 % FCF to 39 % at 50 % FCF (Fig. 1a). Such a significant impact of the FCF concentration on protoplast viability was confirmed both after 24 and 48 h of culture (P\0.001, Table 2). After 2 days of culture, in the presence of higher concentrations of FCF in the medium (from 20 to 50 %), extremely low levels of viable cells were observed (from 8 to 1 %, respectively). The carrot accessions exhibited slightly different levels of protoplast viability (P = 0.003) reaching from 45.4 ± 4.5 % ('Amsterdamska') to 52.9 ± 4.1 % (line '9304B').

FCF very strongly affected cell divisions and aggregate formation (P\0.001, Fig. 1b) reducing the plating effi- ciency in comparison to the control. The degree of that reduction was directly related to the concentration of FCF in the culture medium. After 10 days of culture, the plating efficiency decreased from 67 ± 6.1 % in the control to 24.5 ± 11.5 % at 1 % FCF, and to 0.2 ± 0.1 % at 5 % FCF. At higher FCF concentrations in the medium, cell divisions were completely inhibited. Protoplasts isolated from different accessions responded differently to the applied treatments (P\0.001). In 1 % FCF-treated cultures, a decrease of plating efficiency from 79 to 59 % and from 71 to 15 % was observed for the line '9304B' and 'Dolanka', respectively (Table 2). 'Amsterdamska' was the most sensitive to FCF and formed aggregates only in the control treatment.

Exposure of Protoplasts to FCF on the 5th day of Culture

In type B experiments, on the 5th day of culture mainly twoand four-cell aggregates were observed. Multi-cell aggregates were visible only occasionally. At that stage, plating efficiency reached on average 33 % and varied from 25.1 ± 1.5 % for 'Amsterdamska' to 39.4 ± 2.7 % for the line '9304B' (P\0.001). At that point, FCF was applied to the culture. Data analysis showed that further cell divisions highly depended on the FCF concentration in the culture medium (P\0.001). Ten days after FCF application in 15-day-old cultures treated only with lower concentrations of FCF (1–3.5 %) plating efficiency was at the same level as in the control (72 %) and varied from 71 to 58 % (Fig. 2). Higher concentrations of FCF inhibited cell divisions, decreasing plating efficiency up to five-fold in 50 % FCFtreated cultures. In 10–50 % FCF-treated cultures, plasmolysis of the cells was observed, explaining the lower number of cell aggregates in comparison to that observed in 5-day-old cultures. Similar to type A experiments, protoplasts isolated from different accessions responded differently to FCF treatments (P\0.001). A significant decrease of aggregate formation was observed in 3.5 and 10 % FCFtreated cultures between the line '9304B' and cultivar 'Dolanka' (Table 3), while for 'Amsterdamska' in 10–50 % FCF-treated cultures, cells in all aggregates were plasmolysed and new aggregates were not formed.

Plant Regeneration from FCF-Treated Protoplasts

In the course of 2 months of culture, steady growth of cell aggregates into microcalli, and the subsequent formation of proembryonic masses and regeneration into somatic embryos were observed. This growth pattern was characteristic for all accessions on FCF-free medium, as well as in 1 % FCF-treated cultures, and both time points of the selective agent application (Table 4). Somatic embryogenesis occurred in type A and B experiments for 2 % FCF-treated cultures of line '9304B' and in type B experiments for 2 and 3.5 % FCF-treated cultures of 'Dolanka'. In all other FCF treatments, the development of aggregates was not observed as a result of complete arrest of mitotic divisions following contraction and clustering of the cytoplasm and organelles.

Table 1: Protoplast viability and plating efficiency of donor accessions in the CPP medium without fungal culture filtrate

Experiment type	Accession	Protoplast viability (% ± SE)			Plating efficiency[a] (% ± SE)
		Age of culture (h)			
		1	24	48	
A	Dolanka	74.6 ± 3.7	77.8 ± 4.4	71.9 ± 8.1	70.9 ± 5.1
	Amsterdamska	71.7 ± 11.3	68.7 ± 9.1	64.5 ± 6.3	49.9 ± 8.9
	9304B	73.6 ± 6.8	76.8 ± 1.8	80.8 ± 2.4	79.2 ± 1.2
B	Dolanka	71.4 ± 1.9	n.a.	n.a.	76.4 ± 4.4
	Amsterdamska	75.5 ± 3.4	n.a.	n.a.	71.6 ± 5.6
	9304B	75.7 ± 8.5	n.a.	n.a.	67.8 ± 6.1

CPP carrot petiole protoplast culture medium, n.a. not applicable

[a] For type A and B experiments plating efficiency scored after 10 and 15 days of culture, respectively

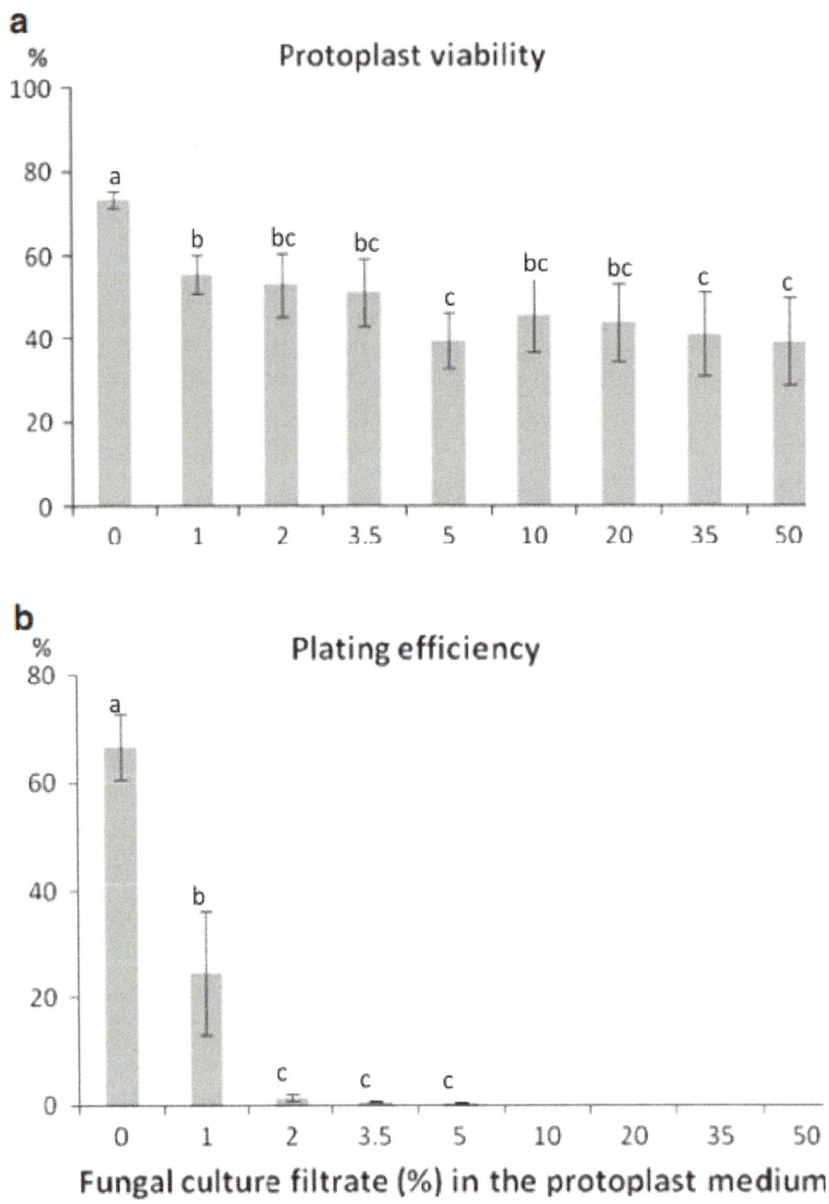

Figure. 1: Effect of the fungal culture filtrate applied on the isolation day (type A exp.) on protoplast viability a and plating efficiency b, averaged over all accessions. Bars represent standard error. Means denoted with different letters were significantly different (P B 0.05)

In control cultures, germination of protoplast-derived somatic embryos was very high, and a precise estimation of the numbers was not possible. Thus, the efficiency of somatic embryogenesis was not recorded for FCF-treated nor for untreated cultures. Plants were regenerated from all cultures where PEM and somatic embryos had formed (Table 4). The derived population represented all accessions, but the number of plants from control cultures was intentionally limited. In total, 820 R0 plants were obtained including 265 control plants and 555 plants from FCFtreated cultures. Within the latter group, regenerants from 1 % FCF-treated cultures were the most frequent, while only a few plants were regenerated from 2 to 3.5 % FCFtreated cultures. In most cases, derived plants appeared normal in growth and shape, however, some individuals with fennel-like leaves were observed.

Cytogenetic and Molecular Characterization of R0 Plants

Cytometric analysis was performed on 457 plants regenerated after FCF treatment and on 139 control plants (Table 4). Nineteen percent of regenerants obtained under the stress of exposure to FCF and 5 % of control plants were tetraploids, the remaining regenerants being diploids. More tetraploid plants were produced both from cultures exposed to FCF immediately after protoplast isolation and from those exposed to FCF treatment after 5 days of culture (P\0.05). Chromosomes were counted for 109 plants, of which 83 represented plants regenerated in the presence of A. radicina FCF. The results of chromosome counting were consistent with the cytometric data, indicating that no aneuploids were produced in the course of protoplast cultures (data not shown). RAPD-PCR profiles of the regenerated plants were almost uniform and identical to those obtained for the donor plants. Three to nine additional unique bands per one round of selection were identified in the profiles of some individuals, constituting 0.06–0.19 % of all observed bands, respectively. However, they occurred at a similar rate for plants regenerated from protoplasts treated with FCF and those of the control, suggesting that addition of the selective agent to the culture did not increase the frequency of genomic rearrangements detectable with the RAPD-PCR system. The low incidence of putative rearrangements identified in the regenerated plants could, therefore, be attributed solely to the protoplast culture. None of the investigated MITE families, i.e., Krak, DcSto, and Dc-hAT, were mobilized in the course of protoplast cultures. All transposon display assays produced completely uniform banding patterns, identical to those of the donor plants, regenerants from the control, and regenerants from the cultures treated with the selective agent.

Response of R0 Plants to Alter aria Radicina

Eight days after inoculation of detached leaf petioles with A. radicina, black lesions typical of black rot were visible. When selective pressure was applied on the first day of culture, both regenerants from FCF-treated and untreated protoplasts displayed lesion progress significantly slower in comparison to the field control. In the case of a later application of selective pressure to the culture, only 3.5 % FCF-derived plants exhibited smaller lesions on the petioles than plants from the field control (Table 5). Generally, regenerants obtained from FCF-treated protoplast cultures did not differ significantly from those from untreated cultures with respect to their petiole assay score. However, within the FCF-derived plants 49–65 % individuals developed lesions smaller than the mean values of the in vitro control.

Table 2: Mean effect of the fungal culture filtrate (FCF) applied on the isolation day (type A exp.) on protoplast viability and plating efficiency

Concentration of FCF (%)	Protoplast viability (averaged over all accessions) (% ± SE)		
	Age of culture (h)		
	1	24	48
0	73.3 ± 3.6 a	74.4 ± 3.2 a	72.4 ± 4.0 a
1	73.3 ± 3.6 a	52.4 ± 5.2 ab	37.2 ± 9.0 b
2	73.3 ± 3.6 a	33.5 ± 6.2 bc	19.9 ± 3.1 bc
3.5	73.3 ± 3.6 a	28.4 ± 4.8 bc	17.8 ± 6.1 bcd
5	73.3 ± 3.6 a	25.5 ± 3.0 bc	15.7 ± 3.1 cd
10	73.3 ± 3.6 a	21.8 ± 2.7 bc	13.2 ± 3.5 cd
20	73.3 ± 3.6 a	20.0 ± 2.2 bc	8.3 ± 3.1 cd
35	73.3 ± 3.6 a	14.2 ± 0.3 c	2.7 ± 2.0 d
50	73.3 ± 3.6 a	8.8 ± 1.6 c	0.9 ± 0.5 d
	Plating efficiency (after 10 days of culture) (% ± SE)		
	Accession		
	Dolanka	Amsterdamska	9304B
0	70.9 ± 5.1 a	49.9 ± 8.9 a	79.2 ± 1.1 a
1	15.1 ± 11.2 b	0.0 ± 0.0 b	58.6 ± 2.4 b
2	1.0 ± 0.5 b	0.0 ± 0.0 b	3.2 ± 1.2 c
3.5	1.0 ± 0.4 b	0.0 ± 0.0 b	0.9 ± 0.2 c
5	0.7 ± 0.2 b	0.0 ± 0.0 b	0.0 ± 0.0 c
10	0.0 ± 0.0 b	0.0 ± 0.0 b	0.0 ± 0.0 c
20	0.0 ± 0.0 b	0.0 ± 0.0 b	0.0 ± 0.0 c
35	0.0 ± 0.0 b	0.0 ± 0.0 b	0.0 ± 0.0 c
50	0.0 ± 0.0 b	0.0 ± 0.0 b	0.0 ± 0.0 c

Means within columns followed by different letters were significantly different at P B 0.05

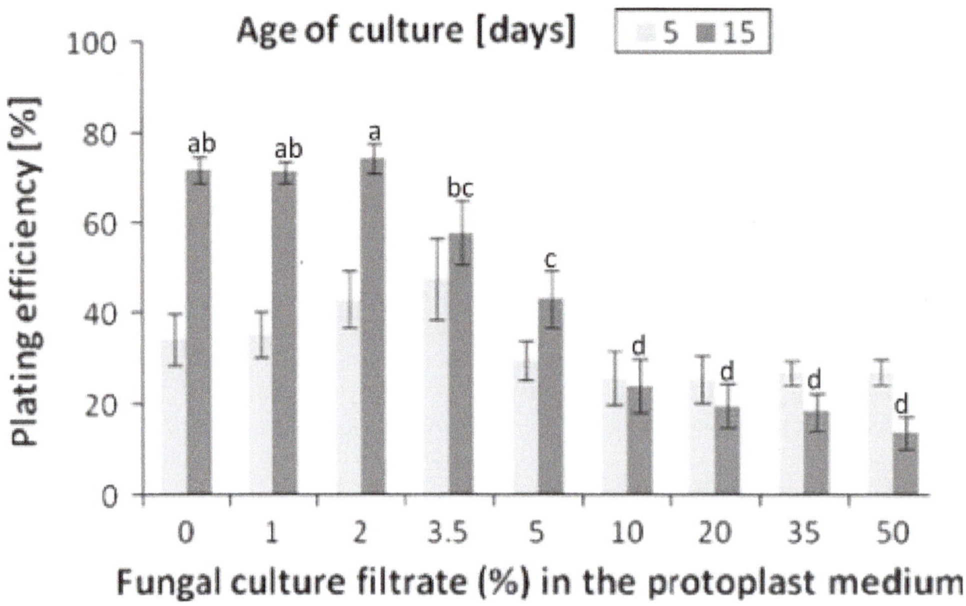

Figure. 2: Effect of the fungal culture filtrate applied on the fifth day of protoplast culture (type B exp.) on plating efficiency, averaged over all accessions. Bars represent standard error. Means for plating efficiency evaluated after 15 days of culture denoted with different letters were significantly different (P B 0.05)

Pollen Viability and Seed Production from R0 Plants

Populations of 558 plants regenerated from FCF-treated cultures and 200 control plants were vernalized. After 1–2 months of growth in the greenhouse, 75 and 91 %, respectively, started to flower. For the plants obtained from protoplasts treated with FCF immediately after isolation, significantly lower levels of flowering were observed in comparison to the control population (P\0.001). Pollen viability was the most variable trait, both for the regenerants from the FCF-treated cultures and those from the control (Table 6). In both populations, plants producing only sterile pollen grains were indentified. In all cases, male sterility was accompanied by the presence of brown anthers. A higher frequency of male sterile plants (37 vs. 20 % of control plants) was recorded only for regenerants derived from protoplasts treated with FCF on the 5th day of culture (P = 0.01). For individuals forming flowers with typical yellow anthers, the level of viable pollen grains reached on average 59.9 ± 1.1 %. The differences in pollen viability among regenerants derived from different accessions were noted (P\0.001). The highest proportion of viable pollen grains, reaching 73.1 ± 1.2 %, was noted for 'Amsterdamska' regenerants while the lowest (45.5 ± 1.7 %) was observed for 'Dolanka' regenerants. That trait did not depend either on the application time of FCF to the culture medium or the concentration of FCF in the culture medium. However, pollen viability varied over a broad range from very low (6–7 %) to very high (91–98 %) both for regenerants derived from FCF-treated and untreated cultures (Table 6).

Table 3: Effect of the fungal culture filtrate (FCF) applied on the fifth day of carrot protoplast culture (type B exp.) on plating efficiency scored at the fifteenth day of culture

Concentration of FCF (%)	Plating efficiency (% ± SE)		
	Accession		
	Dolanka	Amsterdamska	9304B
0	76.4 ± 4.4 a	71.6 ± 5.5 a	67.8 ± 6.1 a
1	70.9 ± 4.4 a	72.7 ± 5.7 a	70.5 ± 5.7 a
2	79.2 ± 4.3 a	77.8 ± 1.7 a	65.0 ± 7.7 a
3.5	64.6 ± 6.2 a	76.2 ± 1.6a	27.1 ± 2.5 b
5	44.5 ± 17.7 ab	47.2 ± 20.8 a	39.2 ± 4.7 ab
10	28.7 ± 4.7 b	0.0 ± 0.0 b	35.2 ± 5.0 b
20	23.0 ± 6.1 b	0.0 ± 0.0 b	29.2 ± 1.6 b
35	25.4 ± 3.3 b	0.0 ± 0.0 b	23.6 ± 0.3 b
50	19.4 ± 5.9 b	0.0 ± 0.0 b	17.3 ± 1.9 b

Means within columns followed by different letters were significantly different at P B 0.05

Table 4: Plant regeneration from fungal culture filtrate-treated cultures and ploidy analysis of derived regenerates

Accession	Concentration of FCF (%)	FCF applied on the 1st day of culture					FCF applied on the 5th day of culture				
		Regeneration ability[a]	No. of plants		Ploidy		Regeneration ability	No. of plants		Ploidy	
			Regenerated	Analysed	2x	4x		Regenerated	Analysed	2x	4x
Dolanka	0	pem + se	32	31	30	1	pem + se	37	18	18	0
	1	pem + se	112	107	73	34	pem + se	19	19	17	2
	2	pl	–	–	–	–	pem + se	74	67	44	23
	3.5	pl	–	–	–	–	pem + se	10	10	10	0
Amsterdamska	0	pem + se	81	21	20	1	pem + se	14	10	10	0
	1	pem + se	241	166	151	15	pem + se	15	12	11	1
	2	pl	–	–	–	–	pl	–	–	–	–
	3.5	pl	–	–	–	–	pl	–	–	–	–
9304B	0	pem + se	68	41	39	2	pem + se	33	18	15	3
	1	pem + se	49	44	36	8	pem + se	27	25	23	2
	2	pem + se	3	3	1	2	pem + se	5	4	3	1
	3.5	pl	–	–	–	–	pl	–	–	–	–
Total	0	pem + se	181	93	89	4	pem + se	84	46	43	3
	1–3.5	pem + se, pl	405	320	261	59	pem + se, pl	150	137	108	29

FCF fungal culture filtrate

[a] pem ? se: proembryonic mass and somatic embryos formated as a result of development of cell aggregates derived in the first 10–15 days of culture, pl: plasmolysis of cells in aggregates, no further development observed

Seeds were produced from 351 individuals including 251 regenerants from FCF-treated cultures and 100 control plants (Table 6). The yield of seed varied from very low (0.1 g) to high (5.3 g per plant) reaching on average 0.8 ± 0.1 g and 0.9 ± 0.3 g for regenerants from the FCFtreated cultures and the control plants, respectively.

DISCUSSION

The present study was the first attempt to assess the potential of using A. radicina FCF for in vitro selection against black rot in carrot protoplast cultures. Methods that utilize in vitro selection to induce tolerance to biotic stresses were very rarely used for carrot, which has been recognized as a model species for in vitro culture systems.

Table 5: Lesion length on detached petioles of 'Dolanka'-derived R0 plants after inoculation with Alternaria radicina

Origin of R0 plants		No. of tested plants	Lesion length		% plants below mean of	
Application of FCF to culture	Concentration of FCF (%)		Range	Average	Field control	In vitro control
On 1st day	Field control	6	22.4–72.0	44.4 ± 8.9 a	–	–
	0	18	16.2–31.5	25.7 ± 0.9 b	100	38.9
	1	26	13.2–33.9	22.6 ± 1.1 b	100	65.4
	Field control	6	22.4–72.0	44.4 ± 8.9 a	–	–
On 5th day	0	9	22.3–44.8	29.8 ± 2.2 ab	88.9	55.6
	1	11	19.7–35.8	31.4 ± 4.3 ab	90.9	54.5
	2	65	19.3–45.5	31.1 ± 0.9 ab	96.9	49.2
	3.5	7	20.9–35.3	27.9 ± 2.2 b	100	57.1

A single petiole was collected from each plant. Plants were treated as replicates

Means within columns followed by *different letters* were significantly different at $P \leq 0.05$

FCF fungal culture filtrate

One of the few reports of the development of carrot calli in the presence of a crude extract was from Erwinia carotovora cultures (Baranski et al. 1997). Many such attempts have been reported for major crops including maize, sugarcane, cotton, potato, and rice (reviewed in Rai et al. 2011). To enhance disease tolerance in the presence of selective agents, callus tissue was often used as an explant (as reviewed in S˘va´bova´ and Lebeda 2005). Protoplasts may be an attractive alternative allowing the use of much larger cell populations at a physiologically uniform stage, as compared to other types of explants, thus, offering the opportunity to develop protocols for efficient selection both in controlled conditions and using less space. Protoplasts or small cell aggregates are usually much more sensitive to pathotoxins than other types of plant material, being very suitable for use in bioassays (Gentile et al. 1992). Another advantage of using protoplasts for in vitro selection, especially after embedding in a gel matrix to prevent cell agglutination, is the adequate exposure of all cells to the selective agent. In such conditions, diffusion of compounds from the medium to the protoplasts is increased (Dovzhenko et al. 1998), and cells escaping from the selection agent are minimized. In contrast, clumps of calli are represented by heterogenic and multi-cell layers where susceptible cells may survive selective pressure by escaping full exposure to the selective agent or by proximal protection of resistant cell lines (Remotti et al. 1997; Mahlanza et al. 2013). However, the number of species with protoplasts capable of reconstituting cell walls and undergoing mitotic divisions resulting in callus formation and normal plant regeneration is still limited. Often, only carefully selected genotypes within the species are capable of producing protoplast-derived plants (Davey et al. 2005). Such recalcitrance of many species, sometimes combined with the complexity of protoplast isolation and culture, is probably the main reason that protoplast cultures are rarely used for in vitro selection. From that point of view, the carrot should be considered as especially suitable, as fast and highly efficient regeneration protocols for protoplasts isolated from different tissue sources have been developed (Dudits et al. 1977; Dirks et al. 1996; Grzebelus et al. 2012). Exposure of protoplasts to biotic stresses was reported for other species including alfalfa, grape vine, hop, lemon, orange tree, rape, tobacco, and tomato (reviewed in Sˇva´bova´ and Lebeda 2005).

Pathogen culture filtrates and phytotoxins are most commonly used for in vitro selection and regeneration of disease-tolerant plants in many crops (reviewed by Sˇva´- bova´ and Lebeda 2005; Rai et al. 2011). Culture filtrates are known to contain a number of phytotoxic compounds, which may play an important role in pathogenicity. Explants from susceptible and tolerant materials cultured in vitro vary in response to these toxic metabolites, thus, making such phytotoxins effective selective agents for in vitro selection strategies towards disease tolerance (Binarova et al. 1990; Remotti et al. 1997; Kumar et al. 2008; Flores et al. 2012). Radicinin, radicinol, and epi-radicinol are the main phytotoxins involved in the pathogenicity of A. radicina against carrots (Solfrizzo et al. 2004). These were identified in naturally infected carrot roots, but also produced in vitro on carrot discs or in liquid potato-dextrose medium after inoculation with A. radicina mycelium (Shakir 1999; Solfrizzo et al. 2004, 2005). Higher levels of these phytotoxins were observed in carrot cultivars more susceptible to A. radicina (Solfrizzo et al. 2004). It was reported that the culture filtrate of A. radicina caused carrot wilting, leaf burning, necrosis on seedlings, and inhibited root elongation of germinated seeds, thus, confirming its toxicity on the intact plant material (Shakir 1999; Tylkowska et al. 2003). However, to our knowledge it was never previously evaluated at the cellular level. FCF toxicity can be verified with a range of staining assays that detect living cells, such as those based on the use of FDA (Widholm 1972). Survival of cells in the presence of the selective agent may be the major resistance criterion. Exposure of cells to different concentrations of toxins present in the FCF results in death, reduced growth or viability, so that surviving cells can be identified (Griffith and Anderson 1987). In our experiments, the percentage of surviving protoplasts on media containing various concentrations of FCF decreased with an increase of the selective agent concentration, reaching 1 at 50 % FCF after 2 days of culture. A similar tendency in response to different doses of the fungal filtrate after viability staining was documented in protoplast cultures of different Brassica accessions (Sjo¨din and Glimelius 1989), lemon and orange (Gentile et al. 1992), and Coffea arabica (Nyange et al. 1997).

Table 6: Pollen viability and seed production of R0 plants derived from fungal culture filtrate-treated cultures

Accession	Application of FCF to culture	Concentration of FCF (%)	No. of plants			Pollen viability		Seed production after self-pollination (g)		
			Evaluated	Sterile	Fertile	Range	Mean ± se	No. of plants	Range	Mean ± se
Dolanka	On 1st day	0	48	19	29	6.7–93.0	42.0 ± 4.3	26	0.1–2.3	0.8 ± 0.1
		1	92	40	52	5.7–82.1	38.6 ± 3.1	39	0.1–1.9	0.6 ± 0.1
	On 5th day	0	36	10	26	25.3–74.8	50.9 ± 3.6	13	0.4–2.1	0.9 ± 0.2
		1	12	2	10	6.6–75.8	41.8 ± 6.4	12	0.3–1.8	0.7 ± 0.1
		2	63	31	32	29.1–80.7	57.5 ± 2.1	24	0.3–2.0	0.8 ± 0.1
		3.5	9	7	2	51.3–63.5	57.4 ± 6.1	2	0.5–0.7	0.6 ± 0.1
Amsterdamska	On 1st day	0	30	4	26	29.4–91.4	68.3 ± 3.1	23	0.4–5.3	2.1 ± 0.3
		1	128	6	122	33.8–97.5	73.3 ± 1.4	120	0.3–4.4	1.5 ± 0.1
	On 5th day	0	12	3	9	53.3–92.6	74.4 ± 7.1	7	0.1–0.6	0.2 ± 0.1
		1	12	3	9	68.6–92.7	83.3 ± 2.9	9	0.1–0.5	0.1 ± 0.05
9304B	On 1st day	0	31	8	23	17.4–82.7	59.1 ± 3.3	15	0.1–1.0	0.4 ± 0.1
		1	30	8	22	7.1–87.5	54.3 ± 4.3	20	0.1–1.3	0.5 ± 0.1
		2	3	1	2	55.0–72.3	63.6 ± 8.7	1	0.5	0.5
	On 5th day	0	24	2	22	23.8–84.0	63.8 ± 4.0	16	0.1–3.1	1.1 ± 0.2
		1	26	3	23	25.8–92.6	56.1 ± 4.0	23	0.1–4.2	1.2 ± 0.3
		2	2	0	2	70.6–79.6	75.1 ± 4.5	1	1.5	1.5
Total	On 1st day	0	109	31	78	6.7–91.4	56.5 ± 7.7	64	0.1–5.3	1.1 ± 0.5
		1–2	253	55	198	5.7–97.5	57.5 ± 7.4	180	0.1–4.4	0.8 ± 0.2
	On 5th day	0	72	15	57	23.8–92.6	63.0 ± 6.8	36	0.1–3.1	0.7 ± 0.3
		1–3.5	124	46	78	6.6–92.7	61.9 ± 6.1	71	0.5–4.2	0.8 ± 0.2
Total	Total	0	181	46	135	6.7–93.0	59.8 ± 4.8	100	0.1–5.3	0.9 ± 0.3
		1–3.5	377	101	276	5.7–97.5	60.1 ± 4.5	251	0.1–4.4	0.8 ± 0.1

FCF fungal culture filtrate

Another method used for the quantitative determination of pathogen filtrate toxicity is the evaluation of cell growth expressed by plating efficiency (Breiman and Galun 1981). In the present study, we presented a significant inhibition of mitotic divisions in carrot protoplast cultures following treatment with varying doses of FCF, which may be correlated with susceptibility to the pathogen. The decrease in plating efficiency was observed for both times of selective agent application. However, a much stronger negative response of cells was observed when the FCF was added on the first day of culture. This can be explained by the fact that cells with reconstructed cell walls forming small protoplast-derived aggregates present in 5-day-old cultures exhibited slightly higher tolerance to the fungal toxin in comparison to the cell wall-free 1-day-old protoplasts. Similar observations were reported in Brassica protoplast cultures (Sjo¨din and Glimelius 1989). This may suggest that the presence of the cell wall constitutes active or passive barriers to the phytotoxic agents (Breiman and Galun 1981). However, higher concentrations of FCF completely inhibited cell growth in both types of experiments. Similar to the findings of Nyange et al. (1997) we observed protoplast death, i.e., contraction and clustering of the cytoplasm and organelles, suggesting that this damage was caused by toxins present in the FCF. Analysis of plating efficiency and ability to regenerate plants from FCF-treated cultures showed the differential response of carrot accessions to the selective agent, likely correlated with different susceptibility levels to A. radicina. The maximum selection pressure that allowed us to recover tolerant cell lines was low and stayed in the range of 1.0–3.5 % FCF depending on the protoplast donor accession and the time of FCF application. Similar low concentrations of A. alternata filtrate were appropriate for selection of tobacco protoplasts (Ishida and Kumashiro 1988). Two in vitro selection variants were proposed: (1) stepwise long-term treatment where cultures were exposed to stress with a gradual increase in concentrations of the selective agent, and (2) shock treatment where cultures were directly subjected to a high concentration of the selective agent (Rai et al. 2011). The first variant helped to avoid escapes and decreased the level of mosaicism through regular elimination of sensitive cells. This was preferred for explants consisting of many cell layers such as plant organs, callus tissue, or suspension cultures. Successful selection with stepwise treatment was reported in the callus tissue of carnation (Thakur et al. 2002), chrysanthemum (Kumar et al. 2008) and citrus (Savita et al. 2011). However, prolonged exposure to toxins can result in reduced regeneration ability and occurrence of undesired mutations. On the contrary, shorter selection cycles may yield more regenerants and healthier plants (Remotti et al. 1997; Sharma et al. 2010). Therefore, in the present study we applied a one-step procedure with a short period of selection pressure (10 days), and single cells or small aggregates were subjected to selection in the liquid medium after embedding protoplasts in the alginate matrix. These conditions ensured adequate infiltration of the FCF to each cell present in the culture. Shorter selection cycles lasting from 2 to 4 days were employed with tobacco protoplasts (Ishida and Kumashiro 1988), while longer (21 days) cycles were used with cell aggregate cultures of Brassica accessions (Sjo¨din and Glimelius 1989).

The in vitro selection techniques are based on the induction of genetic variability in the explants cultured under stress pressure. Somaclonal variation generated in such conditions may be identified on morphological, physiological, molecular, or cytological levels (reviewed by Bairu et al. 2011). In many cases, plants regenerated under biotic stress did not show phenotypic variation and remained morphologically identical to the donor plants (Thakur et al. 2002; Ganesan and Jayabalan 2006; Kumar et al. 2008; Savita et al. 2011), or the frequency of changed plants was low, i.e., 2 % in FCF-challenged embryogenic callus cultures of sugarcane (Mahlanza et al. 2013). Our results are consistent with these observations. Occasional variation in leaf morphology and abnormal development of storage roots in R0 plants (data not shown) are typical negative effects caused by in vitro culture conditions and occur commonly in carrots (Baranski 2008). Here, we also noted a high incidence of male-sterile plants among 'Dolanka' regenerants. Although it has never been reported, we frequently observed the phenomenon in regenerants obtained from 'Dolanka' donors, and the cultivar itself is characterized by a high share of partially or fully malesterile plants. Any stress in the course of in vitro cultures can induce somaclonal variation. Ploidy level or chromosome number variation is a direct and strong evidence of changes in the genetic composition of an organism (Bairu et al. 2011). It has been hypothesized that toxic filtrates used for selection may affect the ploidy of regenerated plants, and that the level of resistance can be simply amplified in polyploid somaclones (Latunde Dada and Lucas 1983; Hartman et al. 1984). Our results, in part, agree with these speculations as more tetraploid plants were obtained from the FCF-treated cultures. Most tetraploids included in the petiole assay exhibited smaller lesions after inoculation with A. radicina in comparison to the mean for the control plants (data not shown). This may indicate a lower susceptibility of tetraploids to black rot. Other types of chromosomal mutations, e.g. chromosome breakage and chromosome loss have been documented in plants derived from in vitro cultures (Larkin and Scowcroft 1981; Van den Bulk 1991; Bairu et al. 2011). We determined the chromosome number to identify aneuploids within R0 plants. Our results were consistent with previous observations (Dudits et al. 1976; Palmer and Widholm 1975; Grzebelus et al. 2012), suggesting that the chromosome number was stable in carrot suspensions or protoplast cultures, except for a possible full genome duplication. In addition, the level of rearrangements revealed by RAPDPCR was very low and FCF treatment did not increase its frequency. In vitro cultures and application of the selective agent did not drive any mobilization events of the three MITE families under investigation, likely because of the short time of the culture. It was reported that TE activation in carrot callus cultures resulting in somaclonal variation required a long-term exposure to culture conditions (Ozeki et al. 1997).

An important aspect of crop improvement via in vitro selection strategy is that traits selected at the cellular level must be expressed in the regenerated plants. Since tolerance showed by single cells may vary from that of the whole plant (Van den Bulk 1991), tests to screen the putative tolerant lines and regenerants to the pathogen in planta are necessary. In the carrot, the best reliable evaluation concerning the level of tolerance to black rot of R0 regenerants should be performed after storage. However, this was not possible, as plants regenerated from in vitro cultures produce abnormal, deformed and forked roots, not suitable for storage. Adaptation of in vitro cultured cells to the selective agent can be epigenetic (Rai et al. 2011), and analysis of R0 plants is not sufficient. Thus, to assess the level of tolerance to A. radicina, a preliminary laboratory assay on petioles was carried out including only a population of regenerants from one donor accession. Some individuals with higher tolerance to black rot were observed, both in the population derived from FCF-treated and untreated cultures. This observation may suggest that a desirable characteristic was an effect of in vitro conditions in general. It is known that results of laboratory tests may not correlate with the plant response in the field (Szczeponek et al. 2006), thus, a detailed analysis of resistance to A. radicina should be performed on R1 progenies both in the course of vegetation and in storage. In conclusion, we applied a large-scale selection based on in vitro cultures to develop carrot materials with improved tolerance to black rot. To our knowledge, this is the first report describing the development of plants from D. carota protoplasts maintained under the selective pressure of toxins from A. radicina FCF. We proved its negative effect on protoplast viability and plating efficiency suggesting its usefulness as an effective selective agent. The analysis of protoplast viability presented here may also be used as a quick assay to determine the phytotoxicity of FCF prepared from different isolates of A. radicina. The basic advantage of the established protocol is its simplicity and low cost compared with disease-resistant transgenic plants, which still are not accepted by the consumers (Costa-Font et al. 2008). Application of a long-term selection to the culture might possibly increase the level of tolerance to black rot of treated cells and protoplast-derived plants.

ACKNOWLEDGMENTS

The authors thank Mirosława Gładysz for her excellent technical assistance and Krystyna Strycharczuk for cytometric analysis. This work was supported by the Polish Ministry of Science and Higher Education (grant no. N310 047 32/2394).

REFERENCES

1. Alexander MP (1969) Differential staining of aborted and nonaborted pollen. Stain Technol 44:117–122
2. Anthony P, Otoni W, Power JB, Lowe TC, Davey MR (1999) Protoplast isolation culture and plant regeneration from Passi- flora. In: Hall RD (ed) Methods in molecular biology—plant cell culture protocols. Humana Press, Totowa, pp 169–179
3. Bairu MW, Aremu AO, Van Staden J (2011) Somaclonal variation in plants: causes and detection methods. Plant Growth Regul 63:147–173
4. Baranski R (2008) Genetic transformation of carrot (Daucus carota) and other Apiaceae species. Transgenic Plant J 2:18–38
5. Baranski R, Nowak E, Grzebelus D, Szklarczyk M, Michalik B (1997) Development of carrot callus in the presence of the crude extract from Erwinia carotovora culture. J. Appl. Genet. 38A:91–99
6. Baranski R, Klocke E, Nothnagel T (2007) Enhancing resistance of transgenic carrot to fungal pathogens by the expression of Pseudomonas fluorescence microbial factor 3 (MF3) gene. Physiol Mol Plant Pathol 71:88–95
7. Binarova P, Nede˘lnik J, Fellner M, Nedba´lkova´ B (1990) Selection for resistance to filtrates of Fusarium spp. in embryogenic cell suspension culture of Medicago sativa L. Plant Cell Tiss Organ 22:191–196
8. Breiman A, Galun E (1981) Plant protoplasts as tools in quantitative assays of phytotoxic compounds from culture filtrates of Phytophthora citrophthora. Physiol Plant Path 19:181–191
9. Costa-Font M, Gil JM, Traill WB (2008) Consumer acceptance,valuation of and attitudes towards genetically modified food: review and implications for food policy. Food Policy 33:99–111
10. Davey MR, Anthony P, Power JB, Lowe KC (2005) Plant protoplast: status and biotechnological perspectives. Biotechnol Adv 23:131–171
11. Devon RS, Porteous DJ, Brookes AJ (1995) Splinkerettes—improved vectorettes for greater efficiency in PCR walking. Nucleic Acids Res 23:1644–1645
12. Dirks R, Sidorov V, Tulmans C (1996) A new protoplast culture system in Daucus carota L. and its applications for mutant selection and transformation. Theor Appl Genet 93:809–815
13. Dovzhenko A, Bergen U, Koop HU (1998) Thin-alginate-layer technique for protoplast culture of tobacco leaf protoplasts: shoot formation in less than two weeks. Protoplasma 204:114–118
14. Dudits D, Kao KN, Constabel F, Gamborg OL (1976) Embryogenesis and formation of tetraploid and hexaploid plants from carrot protoplasts. Can J Bot 54:1063–1067
15. Dudits D, Hadlaczky G, Le´vi E, Feje´r O, Haydu Z, La´za´r G (1977)
16. Somatic hybridisation of Daucus carota and D. capillifolius by protoplast fusion. Theor Appl Genet 51:127–132
17. Farrar JJ, Pryor BM, Davis RM (2004) Alternaria diseases of carrot. Plant Dis 88:776–784
18. Flores PS, Otoni WC, Dhingra OD, de Souza Diniz SPS, dos Santos TM, Bruckner CH (2012) In vitro selection of yellow passion fruit genotypes for resistance to Fusarium vascular wilt. Plant Cell Tiss Organ Cult 108:37–45
19. Gamborg OL, Miller RA, Ojima K (1968) Nutrient requirements of suspension cultures of soybean root cells. Exp Cell Res 50:151–158
20. Ganesan M, Jayabalan N (2006) Isolation of disease-tolerant cotton (Gossypium hirsutum L. cv. SVPR 2) plants by screening somatic embryos with fungal culture filtrate. Plant Cell Tiss Organ Cult 87:273–284
21. Gao D-Y, Vallejo VA, He B, Gai Y-C, Sun L-H (2009) Detection of DNA changes in somaclonal mutants of rice using SSR markers and transposon display. Plant Cell Tiss Organ Cult 98:187–196
22. Gentile A, Tribulato E, Continella G, Vardi A (1992) Differential responses of citrus calli and protoplasts to culture filtrate and toxin of Phoma tracheiphila. Theor Appl Genet 83:759–764

23. Griffith HM, Anderson AJ (1987) Response of Phaseolus vulgaris protoplasts to chemical components of fungal origin. Can J Bot 65:63–68

24. Grzebelus D, Baranski R, Reby E (2003) A laboratory petiole assay of carrot susceptibility to Alternaria radicina. Folia Hort 15(2):41–48

25. Grzebelus D, Jagosz B, Simon PW (2007) The DcMaster Transposon Display maps polymorphic insertion sites in the carrot (Daucus carota L.) genome. Gene 390:67–74

26. Grzebelus D, Baranski R, Spalik K, Allender Ch, Simon PW (2011) Daucus. In: Kole Ch (ed) Wild crops relatives: genomic and breeding resources. Vegetables, Springer, pp 91–113

27. Grzebelus E, Szklarczyk M, Baranski R (2012) An improved protocol for plant regeneration from leaf- and hypocotyl-derived protoplasts of carrot. Plant Cell Tiss Organ Cult 109:101 109

28. Hartman CL, McCoy TJ, Knows TR (1984) Selection of alfalfa (Medicago sativa) cell lines and regeneration of plants resistant to the toxins produced by Fusarium oxysporum f. sp. medicaginis. Plant Sci Lett 34:183–194

29. Ishida Y, Kumashiro T (1988) Expression of tolerance to the hostspecific toxin of Alternaria alternata (AT toxin) in cultured cells and isolated protoplasts of tobacco. Plant Dis 72:892–895

30. Kao KN, Michayluk MR (1975) Nutritional requirements for growth of Vicia hajastana cells and protoplasts at a very low population density in liquid media. Planta 126:105–110

31. Kiełkowska A, Adamus A (2010) In vitro culture of unfertilized ovules in carrot (Daucus carota L.). Plant Cell Tiss Organ Cult 102:309–319

32. Kikuchi K, Terauchi K, Wada M, Hirano HY (2003) The plant MITE mPing is mobilized in anther culture. Nature 421:167–170

33. Kumar S, Kumar S, Negi SP, Kanwar JK (2008) In vitro selection and regeneration of chrysanthemum (Dendranthema grandiflorum Tzelev) plants resistant to culture filtrate of Septoria obesa Syd. In Vitro Cell Dev Biol Plant 44:474–479

34. Larkin PJ, Scowcroft SC (1981) Somaclonal variation—a novel source of variability from cell culture for plant improvement. Theor Appl Genet 60:197–214

35. Latunde Dada AO, Lucas JA (1983) Somaclonal variation and reaction to Verticillium wilt in Medicago sativa L. plants regenerated from protoplasts. Plant Sci Lett 32:205–211

36. Mahlanza T, Rutherford RS, Snyman SJ, Watt MP (2013) In vitro generation of somalconal variant plants of sugarcane for tolerance to Fusarium sacchari. Plant Cell Rep 32:249–262

37. Menczel L, Nagy F, Kiss Z, Maliga P (1981) Streptomycin resistant and sensitive somatic hybrids of Nicotiana tabacum ? Nicotiana knightiana: correlation of resistance to N. tabacum plastids. Theor Appl Genet 70:590–594

38. Murashige T, Skoog F (1962) A revised medium for rapid growth and bioassayas with tobacco tissue culture. Physiol Plant 18:100–127

39. Nowicka A, Grzebelus E, Grzebelus D (2012) Fluorescent in situ hybridization with arbitrarily amplified DNA fragments differentiates carrot (Daucus carota L.) chromosomes. Genome 55:205–213

40. Nyange NE, Williamson B, Lyon GD, McNicol RJ, Connolly T (1997) Responses of cells and protoplasts of Coffea Arabica genotypes to partially purified culture filtrates produced by Colletotrichum kahawae. Plant Cell Rep 16:763–769

41. Ozeki Y, Davies E, Takeda J (1997) Somatic variation during long term subculturing of plant cells caused by insertion of a transposable element in a phenylalanine ammonia-lyase (PAL) gene. Mol Gen Genet 254:407–416

42. Palmer JE, Widholm J (1975) Characterization of carrot and tobacco cell cultures resistant to p-fluorophenylalanine. Plant Physiol 56:233–238

43. Pryor BM, Gilbertson RL (2001) A PCR-based assay for detection of Alternaria radicina on carrot seed. Plant Dis 85:18–23

44. Pryor BM, Davis RM, Gilbertson RL (1998) Detection of soilborne Alternaria radicina and its occurrence in California carrot fields. Plant Dis 82:891–895

45. Rai MK, Kalia RK, Singh R, Gangola MP, Dhawan AK (2011) Developing stress tolerant plants through in vitro selection—an overview of the recent progress. Environ Exp Botany 71:89–98

46. Remotti PC, Lo¨ffler HJM, van Vloten-Doting L (1997) Selection of cell-lines and regeneration of plants resistant to fusaric acid from Gladiolus x grandiflorus cv. 'Peter Pears'. Euphytica 96: 237–245

47. Savita, Virk GS, Nagpal A (2011) In vitro selection of calli of Citrus jambhiri Lush. for tolerance to culture filtrate of Phytophthora parasitica and their regeneration. Physiol Mol Biol Plants 17: 41–47

48. Sjo¨din C, Glimelius K (1989) Differences in response to the toxinirodesmin PL produced by Phoma lingam (Tode ex Fr.) Desm.

49. on protoplasts, cell aggregates and intact plants of resistant and susceptible Brassica accessions. Theor Appl Genet 77:76–80

50. Shakir AS (1999) Studies on seed-borne fungi of some vegetables grown in Punjab with special reference to Alternaria radicina on carrot. Ph.D. Thesis, University of Agriculture, Faisalabad, Pakistan

51. Sharma A, Rathour R, Plaha P, Katoch V, Khalsa GS, Patial V, Singh Y, Pathania NK (2010) Induction of Fusarium wilt (Fusarium oxysporum f. sp. pisi) resistance in garden pea using induced mutagenesis and in vitro selection techniques. Euphytica 173: 345–356

52. Solfrizzo M, Vitti C, De Girolamo A, Visconti A, Logrieco A, Fanizzi FP (2004) Radicinols and radicinin phytotoxins produced by Alternaria radicina on carrots. J Agric Food Chem 52: 3655–3660

53. Solfrizzo M, De Girolamo A, Vitti C, Tylkowska K, GrabarkiewiczSzczesna J, Szopinska D, Dorna H (2005) Toxigenic profile of Alternaria alternata and Alternaria radicina occurring on umbelliferous plants. Food Addit Contam 22:302–308

54. StatSoft, Inc (2009) STATISTICA (data analysis software system), version 9. www.statsoft.com

55. Sˇva´bova´ L, Lebeda A (2005) In vitro selection for improved plant resistance to toxin-producing pathogens. J Phytopathol 153: 52–64

56. Szczeponek A, Laszczak P, Wesołowska M, Grzebelus D, Michalik B (2006) Carrot infection by Alternaria radicina in field conditions and results of laboratory tests. Comm Agr Appl Biol Sci 71/3b: 1125–1132

57. Thakur M, Sharma DR, Sharma SK (2002) In vitro selection and regeneration of carnation (Dianthus caryophyllus L.) plants resistant to culture filtrate of Fusarium oxysporum f. sp. dianthi. Plant Cell Rep 20:825–828

58. Tylkowska K (1992) Carrot seed-borne diseases caused by Alternaria species. In: Chelkowsky J, Visconti A (eds) Alternaria biology, plant diseases and metabolites. Elsevier Science Publisher, Amsterdam, pp 337–352

59. Tylkowska K, Grabarkiewicz-Szczesna J, Iwanowska H (2003) Production of toxins by Alternaria alternata and A. radicina and their effects on germination of carrot seeds. Seed Sci Technol 31:309–316

60. Van den Bulk RW (1991) Application of cell and tissue culture and in vitro selection for disease resistance breeding—a review. Euphytica 56:269–285

61. Widholm JM (1972) The use of fluorescein diacetate and phenosafranine for determining viability of cultured plant cells. Stain Technol 47:189–194

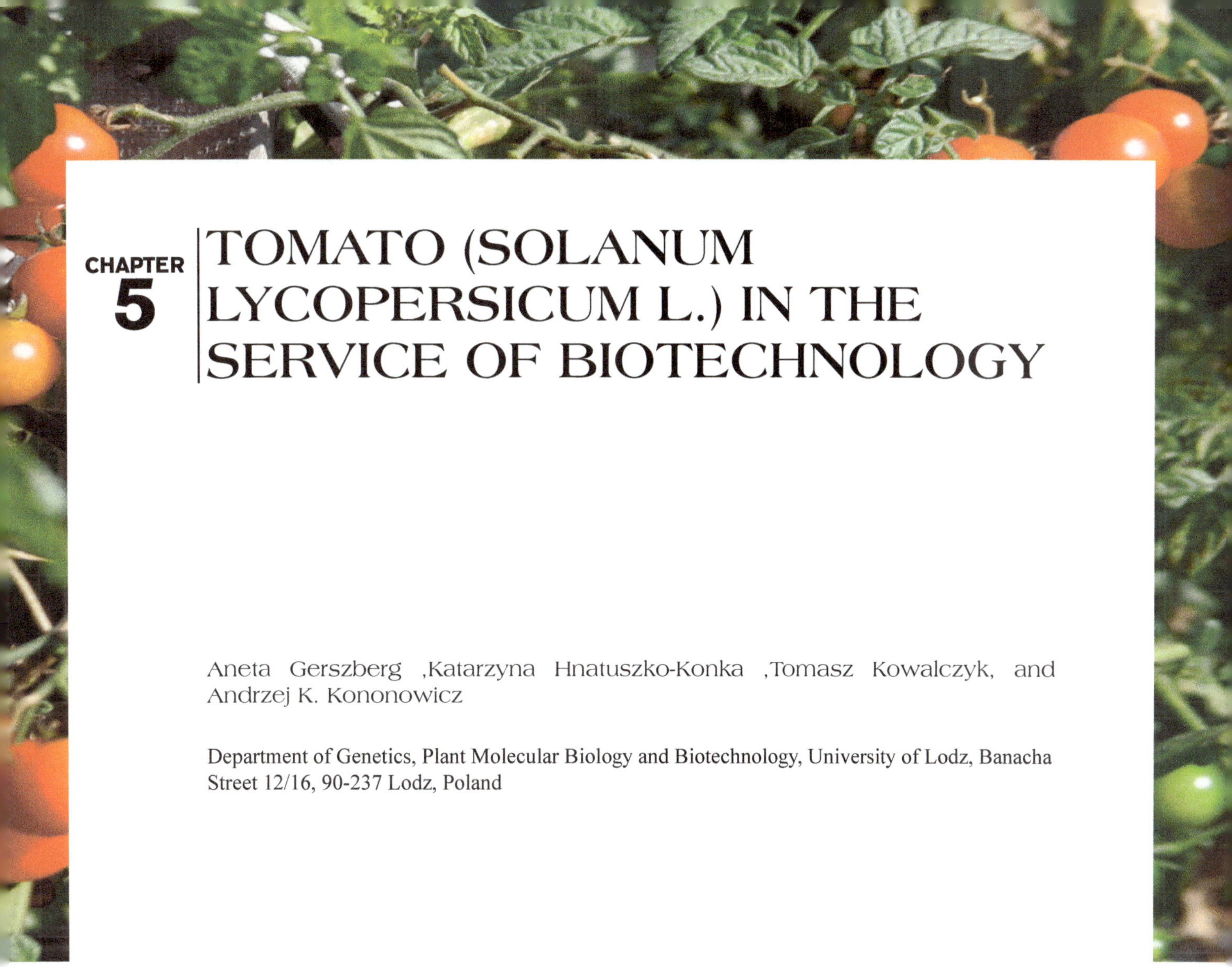

CHAPTER 5

TOMATO (SOLANUM LYCOPERSICUM L.) IN THE SERVICE OF BIOTECHNOLOGY

Aneta Gerszberg ,Katarzyna Hnatuszko-Konka ,Tomasz Kowalczyk, and Andrzej K. Kononowicz

Department of Genetics, Plant Molecular Biology and Biotechnology, University of Lodz, Banacha Street 12/16, 90-237 Lodz, Poland

ABSTRACT

Originating in the Andes, the tomato (Solanum lycopersicum L.) was imported to Europe in the 16th century. At present, it is an important crop plant cultivated all over the world, and its production and consumption continue to increase. This popular vegetable is known as a major source of important nutrients including lycopene, β-carotene, flavonoids and vitamin C as well as hydroxycinnamic acid derivatives. Since the discovery that lycopene has anti-oxidative, anti-cancer properties, interest in tomatoes has grown rapidly. The development of genetic engineering tools and plant biotechnology has opened great opportunities for engineering tomato plants. This review presents examples of successful tissue culture and genetically modified tomatoes which resistance to a range of environmental stresses improved, along with fruit quality. Additionally, a successful molecular farming model was established.

INTRODUCTION

Originating from the Andes, tomatoes (*Solanum lycopersicum* L.) were imported to Europe in the 16th century. At present, this plant is common around the world, and has become an economically important crop. Furthermore, this plant is a model species for introducing agronomically important genes into dicotyledonous crop plants (Paduchuri et al. 2010). The tomato is considered a protective food because of its particular nutritive value, as it provides important nutrients such as lycopene, beta-carotene, flavonoids, vitamin C and hydroxycinnamic acid derivatives. Furthermore, this crop has achieved tremendous popularity especially in recent years with the discovery of lycopene's anti-oxidative activities and anti-cancer functions (Wu et al. 2011; Raiola et al. 2014). Thus, tomato production and consumption are constantly increasing. It is noteworthy that tomatoes are not only sold fresh, but also processed as soups, sauces, juices or powder concentrates. The tomato ranks 7th in worldwide production after maize, rice, wheat, potatoes, soybeans and cassava, reaching a worldwide production of around 160 million tons on a cultivated area of almost 4.8 million hectares in 2011 (FAOSTAT 2011).

From botanical point of view, the tomato is a fruit. Nevertheless, it contains a much lower sugar content compared to other fruits. It is a diploid plant with $2n = 24$ chromosomes. The tomato belongs to the *Solanaceae* family, which contains more than 3,000 species, including plants of economic importance such as potatoes, eggplants, tobacco, petunias and peppers (Bai and Lindhout 2007). In 1753, Linnaeus placed the tomato in the*Solanum* genus (alongside with potato) under the specific name *S. lycopersicum*. In 1754, Philip Miller moved it to its own genus, naming it *Lycopersicum esculentum* (Foolad2007;

Citation: Aneta Gerszberg Katarzyna Hnatuszko-Konka Tomasz Kowalczyk Andrzej K. Kononowicz, Tomato (Solanum lycopersicum L.) in the service of biotechnology, DOI 10.1007/s11240-014-0664-4.

Perlata and Spooner 2007). Nevertheless, the designation of the tomato was for a long time a subject of consideration and discussion by many scientists. The use of molecular data (genome mapping) and morphological information allowed for the verification of of the Solanaceae classification when the genus *Lycopersicon* was re-introduced in the *Solanum* genus in the *Lycopersicon* section (Foolad 2007). Thus, after almost two centuries, the description of Linnaeus was confirmed. Due to the use numerous citations from recent references and in order to be consistent with much of the literature, the Linnaeus classification is followed in this review. Several reports indicate Peru as the centre of diversity for wild relatives of tomato. It seems justified to consider *S. lycopersicum cerasiforme* as an ancestor of the cultivated tomato because of its abundant existence in Central America (Bai and Lindhout 2007). Nevertheless, recent extensive genetic studies have revealed that the closest relative of the tomato is *Solanum pimpinellifolium* (The Tomato Genome Consortium 2012). It turned out that the genome sequences of both of the above-mentioned tomatoes as well as domesticated cultivars and*S. pimpinellifolium* showed only a 0.6 % nucleotide divergence. Domestication has triggered a range of traits (morphological and physiological) that distinguish domesticated crops from their wild ancestor. Studies on the domestication process, not only of tomatoes but also in the cases of maize and rice revealed that rapid phenotypic divergence is often controlled genetically by a relatively small number of loci (Koenig et al. 2013).

Among the members of family *Solanaceae*, many species of economic importance such as tomatoes, potatoes, tobacco, peppers and eggplants can be distinguished. In recent years, interest of scientists in the tomato as a model plant has significantly increased, also due to the fact that its genome has been sequenced (The Tomato Genome Consortium 2012). The tomato is an excellent model for both basic and applied research programs. This is due to it possessing a number of useful features, such as the possibility of growing under different cultivation conditions, its relatively short life cycle, seed production ability, relatively small genome (950 Mb), lack of gene duplication, high self-fertility and homozygosity, easy way of controlling pollination and hybridization, ability of asexual propagation by grafting and possibility to regenerate whole plants from different explants (Bai and Lindhout 2007; The Tomato Genome Consortium 2012). Among the existing tomato genotypes, cv. Micro Tom is considered to be a model system due to the aforementioned unique characteristics (Kobayashi et al. 2013). This dwarf tomato cultivar was created for ornamental purposes and originated by crossing two cultivars (Florida Basket and Ohio 4013-3), and shows small and ripened fruits as well as dark-green and rugose leaves. The phenotype of this cultivar is due to mutations in the *SELF-PRUNING* (SP), *DWARF* (D) and *MINATURE*(*mnt*) where the latter is likely associated with GA signaling (Marti et al. 2006). Additionally, in contrast to other model organisms such as *Arabidopsis* or rice, the tomato has many interesting features. For example, tomato plants produce fleshy fruits that are important for the human diet. The tomato has sympodial shoots, and it is the only model plant with compound leaves. Furthermore, there exists a large pool of tomato mutants, which were either spontaneous or induced by chemicals or irradiation, that are available at the Tomato Genetic Resource Center (Lozano et al. 2009), LycoTill platforms and TOMATOMA base (Minoia et al. 2010; Saito et al. 2011). Over the years, the Botanical and Experimental Garden in the Netherlands has collected the germplasm of the *Solanaceae*species and preserved extensive ex situ plant collections of tomatoes (Bai and Lindhout2007). This data is a crucial resource for breeders and scientists for improving the quality and yield of tomato cultivars. However, extensive knowledge concerning the molecular bases underlying these complex traits is necessary. This challenge requires comprehensive and genetically high-quality populations of mutants as well as the availability of these resources to the research community to promote functional analyses of tomatoes (isolation of key genes involved in development and growth regulation).**INTRODUCTION**

Tissue Cultures of Tomato

Traditional improvement methods are time-consuming and troublesome due to the time of breeding, and there is a problem with the choice of criteria appropriate for breeding purposes. Thus, the establishment of simple and efficient regeneration systems is a fundamental prerequisite of taking advantage of cell and tissue culture for genetic improvement (genetically transformed plants for commercial applications). The in vitro culture of the tomato has been successfully used in different biotechnological application including the clonal propagation of high-value commercial cultivars, virus-free plants, and genetic transformation (Namitha and Negi 2013; Hanus-Fajerska 2006; Li et al. 2011; Yarra et al. 2012).

The Flavr Savr tomato (also known as CGN-89564) was the first commercially grown, genetically engineered food to be granted a license for human consumption. The Food and Drug Administration approved the Flavr Savr tomato in 1994. Unfortunately, the tomatoes had a bland taste and they also were very delicate, proving difficult to transport. They were off the market by 1997. In China, the GM tomato Huafan No 1 (from Huzahong Agricultural University), which had long shelf life characteristics, was the first GM plant to be approved for commercialization in 1996. Other tomato varieties that have been authorized in some countries (USA, Japan, Mexico, Canada) include: 351N from Agritope Inc (Portland, USA), 8,338 and 5,345 from Monsanto (St. Louis, USA), 1345-4 from DNA Plant Technology Corp (Oakland, USA), B, Da, F from Zeneca Seeds. A detailed description is presented in Table 1.

Table 1: Transgenic tomato varieties approved for commercialization. Based on Yang et al. (2005) and Fukkuda-Parr (2012)

Company	Event	Trait	Year approved	Approved for	Country
Calgene	Flavr Savr CGN-89564	Delayed softening (developed by additional*PG* gene expressed)	1994	All uses in USA; Japan and Mexico for feed and for environment	USA
Calgene	Flavr Savr N 73 1436-11	Delayed ripening (developed by additional*PG* gene expressed)	1996	All uses in USA	USA
CAAS	About 10 events	Data not available	1998	Data not available	China
DNA plant technology	1345-4	Delayed ripening (developed by a truncated aminocyclopropane cyclase synthase (*ACC*) gene)	1994	All uses in USA; food in Canada and Mexico	USA
Zeneca and Petoseed	B, Da, F	Delayed ripening (developed by additional*PG* gene expressed)	1994	All uses in USA; food in Canada and Mexico	USA
Monsanto	8338	Delayed ripening (developed by introduction of 1-aminocyclopropane-1-carboxylic acid deaminase (*accd*) gene)	1995	All uses in USA	USA
Agritope	351N	Delayed ripening (developed by introduction the *S*-adenosylmethionine hydrolase (*SAM-K*) gene)	1995	All uses in USA	USA
Monsanto	5345	Insect resistant (developed by introduction of one*cry1Ac* gene)	1997	All uses in USA; food in Canada	USA
Huazhong Agriculture University (HZAU)	Hufan no 1	Delayed ripening (developed by introduction anti-sense*EFE* gene)	1996	Data not available	China
Beijing University	PK-TM8805R (8805R)	Delayed ripening	1999	Food, feed, cultivation in China	China

It should be noted that elaborating a cost-effective and productive protocol for the mass propagation of high-quality tomato plants (via tissue culture) could significantly help reduce the market value of seedlings. An efficient regeneration system is also crucial for the success of such techniques as haploid regeneration, micropropagation, somatic hybridization, mutation selection and germplasm storage. As many independent studies on the tomato show, plant regeneration achieved through organogenesis is affected by several factors such as genotype, explant source, age of explants, media composition and environmental conditions (Mamidala and Nanna 2011; Namitha and Negi 2013; Sherkar and Chavan 2014; Wayase and Shitole 2014). There are many reports regarding tomato transformation and in vitro plant regeneration from different explants (including seed-cut cotyledons, hypocotyls, leaves, stem sections, pedicels, petioles and inflorescences) via organogenesis (Khoudi et al. 2009; Yasmeen 2009; Goel et al. 2011; Koleva Gudeva and Dedejski 2012; Rai et al. 2013; Namitha and Negi 2013; Sherkar and Chavan 2014; Wayase and Shitole 2014). These reports also describe the recalcitrance of 'non-competent' tomato explants (partial or total inability to respond to in vitro culture) (Fuentes et al. 2008; Mamidala and Nanna 2011). Thus, the improvement of the adventitious shoot regeneration system using tissue culture methods for tomato plants is still important due to the diverse morphogenic potential of the different genotypes. As mentioned above, scientists have used different types of explant, but it should be emphasized that the type of explants determines not only the frequency of the explants' organogenesis but also determines the number of shoots produced per explant (Bahurpe et al. 2013; Jehan and Hassanein 2013). Namitha and Negi (2013) demonstrated that the efficiency of shoot regeneration ability followed the order hypocotyls > cotyledon > leaf. In earlier studies, Mamidala and Nanna (2011) reported that cotyledons explants showed organogenesis superiority over hypocotyls and leaf explants. It turned out that leaf explants showed effective regeneration only on one of the media tested (MS + 2 mg/L BAP, 6-Benzyloaminopurine + 0.1 mg/L IAA, Indole-3-acetic acid), which suggests that using this type of explants can minimize genotype-dependent variations. In contrast to this report, Chaudry et al. (2010) observed the higher regeneration potential of hypocotyls than of leaf explants. On the other hand, Harish et al. (2010) reported that shoot formation efficiency was greatest in the order: hypocotyls > leaf > stem. Moreover, they noticed significant differences in regeneration capacity between the six cultivars tested in terms of regeneration process duration as well as the number and size of the regenerated shoots. Ashakiran et al. (2011), who examined the effect of TDZ (Thidiazuron) on organogenesis induction from cotyledonary and leaf nodes, obtained similar results. On the other hand, Zhang et al. (2012) indicated that the location of the cutting wound in explants significantly affected callus induction and adventitious bud formation. They demonstrated that the highest frequency of bud induction occurred at the middle part of the cotyledon segment. They also proved that the way an explant was placed on a medium affected the differentiation rate of cotyledon buds, the back-up placing of the cotyledon onto the medium proved best.

It is well known that regeneration process depends on the age of explants. Moreover, it is reported that young explants have shown to give better morphogenic response than older ones (Harish et al. 2010). Dai et al. (1988) revealed that the regeneration capacity of tomato explants increased with their age. Depending on the type of explant, seedlings of different ages (7, 8, 10, or 14 days old) were used (Ishag et al. 2009; Kantor et al. 2010; Ali et al. 2012; Ajenifujah-Solebo et al. 2012; Bahurpe et al. 2013). Furthermore, much data suggests that the size of explants is essential for efficient plant regeneration in tomatoes. The optimal sizes for tomatoes are 0.7–2 cm long segments for hypocotyls and 5 mm × 5 mm for cotyledons (Ishag et al. 2009; Chaudry et al. 2010; Ajenifujah-Solebo et al. 2012).

There are two methods used to regenerate plantlets in vitro: somatic embryogenesis (direct or indirect) and organogenesis. From the point of view of conducting research on heredity or genetic engineering, the second pathway is more desirable as it allows the avoidance of genetic variation. Hence, most of the published procedures were based on direct organogenesis from intact explants (e.g. cotyledons, hypocotyl, leaf) or protoplast cultures or shoot development from meristematic cells (Ajenifujah-Solebo et al. 2012; Namitha and Negi 2013), while attempts to regenerate tomato via somatic embryogenesis are rather rare. However, Godishala et al. (2011) reported a simple and reproducible protocol for tomato cv S-22 regeneration via somatic embryogenesis. Additionally, Guan et al. (2012) demonstrated that shoot organogenesis and somatic embryogenesis occurred simultaneously during the in vitro regen-eration of transgenic cherry tomato (*Solanum esculentum* var. *cerasi-forme*) mutant leaf explants treated by 6-BA combined with IAA. Interestingly, only the somatic embryogenesis pathway was observed during the regeneration of non-transformed cherry tomato plants under the same cul-ture condition. Khuong et al. (2013) observed a similar effect on cv. Micro Tom using trans-zeatin (TZ) (1 mg/L) combined with IAA (0.1 ml/L). They noticed very little callus formation, indicating direct shoot differentiation as described earlier for the Rio Grande cultivar and, on the other hand, they also noticed indirect embryogenesis via callus formation with the same kind hormonal regimen but that they had different concentrations (respectively 1 and 2 mg/L).

Exogenous fitohormones in a medium play an important role in regulating callus induction and organ differentiation or rooting. According to numerous reports IAA, NAA (α-Naphthaleneacetic acid), 2,4-D (2,4-Dichlorophenoxyacetic acid), ZT and 6-BAP are the hormones commonly used in in vitro cultures of tomato to ameliorate callus induction and plant regeneration (Kantor et al. 2010; Mamidala and Nanna 2011; Ashakiran et al.2011; Zhang et al. 2012; Namitha and Negi 2013). KIN (Kinetin), 2iP (6-(γ,γ-dimethylallylamino) purine), TDZ, and IBA (Indole-3-butyric acid) are other plant growth regulators (PGR) that were tested (Ishag et al. 2009; Chaudry et al. 2010; Wu et al. 2011; Ashakiran et al. 2011). Tomato shoot induction from different types of explants was achieved in different cultivars through the modification of the media conditions. Moreover, it was shown that the type of basal medium used (e.g. MS or B5) (Murashige and Skoog 1962; Gamborg et al. 1968) may significantly affect the regeneration process rate (Ashakiran et al. 2011; Wu et al. 2011). Rashid and Bal (2010) demonstrated that MS fortified with kinetin 0.5 mg/L and BAP 0.5 mg/L was the optimal medium for inducing direct shoot regeneration. In contrast to these findings, Wu et al.

(2011) emphasized the superiority of B5 basal medium to the MS medium. The regeneration rate for shoots from MicroTom explants on B5 was considerably higher than on MS (+12 %), and the best variant for regeneration (from cotyledons and hypocotyls) was MS supplemented with 1.5 mg/L 6-BA and 0.05 mg/L IBA, reaching 95.8 and 60 % respectively. Interestingly, the regeneration frequency of the MicroTom explants decreased with increasing IBA concentration. Kantor et al. (2010) discovered that MS supplemented with 1 mg/L zeatin and 0.05 mg/L IAA stimulated the highest number of regenerants. Zhang et al. (2012) obtained similar results for cotyledon explants: that MS supplemented with 2 mg/L zeatin and 0.01 mg/L IAA turned out to be most effective. On the other hand, MS with the combination of BAP (2 mg/L) and IAA (0.1 mg/L) was found to be the best for inducing shoot regeneration (74 %) and multiple shoot formation per explants from hypocotyls (Namitha and Negi 2013). In contrast to the media indicated by many authors, Ali et al. (2012) suggested that a MS medium supplemented with combination of 1.0 mg/L kinetin and 1.0 mg/L BA to be optimal for producing the highest number of shoots per explant from hypocotyls and cotyledons in tomatoes.

In plant tissue cultures, the growth and regeneration of plants can be improved by a small quantity of organic nutrients. In general, these adjuvants can be a potential source of vitamins, amino acids, fatty acids, peptides, carbohydrates or natural PGR at different concentrations (e.g. zeatin). One of them is coconut milk (CM), which contains a complex combination of several compounds. CM is predominantly used in orchid tissue culture. Afroz et al. (2010) used it to enhance the in vitro regeneration efficiency of five varieties of tomato. CM is known to induce plant cell proliferation and their rapid growth. Afroz et al. (2010) successfully replaced zeatin with CM and kinetin. They showed that coconut water alone was insufficient to promote satisfactory multiplication, but the combination of CW with IAA and kinetin allowed the achieving of faster regeneration (12–15 days from leaf explants and 20–25 days from hypocotyls) with a maximum number of shoot primordia. Bhatia and Ashwath (2008) used other adjuvants such as activated charcoal, ascorbic acid and casein hydrolysate to improve shoot regeneration response from cotyledon explants. The results showed that activated charcoal as well as ascorbic acid could improve the quality of the regenerated tomato shoots while casein hydrolysate can be effectively utilized to reduce callus response underneath the shoots, consequently decreasing the chance of somaclonal variation. While most scientists use the MS medium supplemented with a combination of auxins and cytokinins in different concentrations, Plana et al. (2006) reported an alternative procedure to regenerate tomato plants where there is a deficiency of exogenous fitohormones. The medium they proposed contained MS salts, 4 mg/L thiamine, 100 mg/L myo-inositol and 3 % sucrose. In practice, this procedure combines the pre-culture and seed cuttings to promote organogenesis without callus development. The main advantages of this method are simplicity, time efficiency and, most importantly, the proposed procedure allowed the obtaining of a shoot formation without developmental/morphological abnormalities (e.g. leaves and shoots without apical meristem and vitrified structures).

Environmental conditions such as light or temperature were found to be crucial for tomato regeneration. As it is well known, light (by the length of exposure or its quality) influences explant growth and differentiation processes. The response of tomato explants to tissue culture depends on the quality and quantity of light used during growth of a mother plant. Glowacka (2004) investigated the influence of red, yellow, green, blue and natural light on the micropropagation of tomatoes. The study showed a distinct influence of red and yellow light on shoot and internode elongation, and the plantlets were easy-to-cut. On the other hand, blue and natural light inhibited shoot and internode elongation. Extending the regeneration period had no influence on the growth of the plantlets under red and yellow light. It turned out that red and yellow light had favorable influence on root formation. Since light is indispensable for the regeneration of tomato shoots, studies on tomato regeneration have exploited the 16 h photoperiod (Ali et al. 2012; Zhang et al. 2012; Namitha and Negi 2013). For example, Bhatia and Ashwath (2005) revealed that maximum shoot regeneration response (60 %) occurred in the explants exposed to 16 h of light and 8 h of darkness. The response decreased at 2 h dark (47 %) or 24 h (40 %) light. These results are in contrast to studies conducted by Tyburski and Tretyn (1999), who reported that tomatoes could be regenerated in the absence of light. The shoots that were regenerated under dark conditions were chlorotic, but they developed chlorophyll after exposure to a 16 h photoperiod. It was shown that the texture of the medium affected tomato regeneration. Velcheva et al. (2005) developed two distinct systems—solidified medium or liquid medium—for the regeneration of commercial tomato cultivars (Daniela 144, Brillante 179, Annan 3,017, Galina 3,019, and Bernadine 5,656) after a*Agrobacterium*-mediated transformation. In terms of the regeneration ability of different types of explants, their results are in agreement with the statement that hypocotyl explants are worse compared to the cotyledons when cultured on a solid medium and the liquid medium allowed the obtaining of similar regeneration efficiencies for both hypocotyls and cotyledons. Obviously, it cannot be ruled out that the physical parameters of a liquid culture (e.g. gas exchange, frequent passages or constant agitation of explants) played an essential role in efficient hypocotyl regeneration. Undoubtedly, the study by Velcheva et al. (2005) demonstrated some advantages of this procedure: regeneration is initiated from epidermal to subepidermal cells and the selection process in liquid media seems to be much more effective compared to similar selection performed on solid media.

Rooting is the final step of the regeneration protocol in plant tissue cultures. There are many factors affecting the rooting process (e.g. the physiological status of plantlets, medium composition, growth regulators). Mostly, MS or 1/2MS are used as a basal medium for rooting. Mensuali-Sodi et al. (1995), Rashid and Bal (2010) and Bahurpe et al. (2013) suggested that for root induction, the tomato does not require any exogenous plant growth regulators. However, in most cases, root formation would be achieved with auxins (IAA, NAA or IBA) alone with concentrations ranging from 0.1 to 1 mg/L (Chaudry et al.2010; Ashakiran et al. 2011; Mamidala and Nanna 2011; Zhang et al. 2012; Namitha and Negi 2013; Sherkar and Chavan 2014; Wayase and Shitole 2014). Abundant rooting is usually observed after 2 weeks.

GENETIC ENGINEERING OF TOMATOES

Methods of Tomato Transformation

The first *Agrobacterium*-mediated transformation of tomatoes was reported in 1986 (McCormick et al. 1986). Since then, several transformation protocols for different tomato cultivars have been developed using various explants (e.g. cotyledons, hypocotyls, leaves, fruits) (El-Siddig et al. 2011; Wu et al. 2011; Yarra et al. 2012; Garcia-Hurtado et al. 2012; Hasan et al. 2008; Orzaez et al. 2006, 2009, Orzaez and Granell 2009; Yasmeen et al.2009). The process of plant genetic transformation is very complex, with many factors playing an important role including the application of nurse cells, the addition of acetosyringon to the culture or preculture media, bacterial factors (*Agrobacterium* strain, culture density) and tissue-specific factors (the genotype and the type of the explants) as well as the plasmid vector, the composition of the culture medium (concentration of fitohormones), the type and concentration of antibiotics, the cocultivation time, etc. (Fuentes et al. 2008; Jabeen et al. 2009; Sharma et al. 2009; El-Siddig et al. 2011; Wu et al.2011; Guo et al. 2012; Chetty et al. 2013; Koul et al. 2014). Unfortunately *Agrobacterium*-mediated transformation is still not suitable for tomato varieties with low regeneration capacity. However, some attempts to establish an efficient transformation procedure for such cultivars (e.g. Cambell-28) have been made (Fuentes et al. 2008). The development of a system for stable genetic transformation of tomato plastids was a milestone in the transformation of the tomato (Ruf et al. 2001). This relatively new transformation technology allowed to investigate the possibility to elevate the pro-vitamin A content in tomatoes (Apel and Bock 2009). Lycopene β-cyclase genes from an eubacterium *Erwinia herbicola* and from a higher plant, a daffodil (*Narcissus pseudonarcissus*), were introduced into the tomato plastid genome in order to enhance carotenoid biosynthesis and induce the conversion of lycopene to provitamin A. This research gave unexpected results, namely that the transplastomic tomatoes also showed a 50 % increase in total carotenoid accumulation in plants expressing the lycopene β-cyclase from daffodils. Another example of tomato chloroplast transformation was given by Zhou et al. (2008), who demonstrated that the HIV antigens p24 and Nef could be expressed in a plastid of tomato plants.

Recently, several procedures for the stable transformation of tomato plants have been reported (Hasan et al. 2008; Sharma et al. 2009; El-Siddig et al. 2011). Notwithstanding, there is still lack of an efficient, simple and reliable protocol, which significantly hinders the functional analysis of transgenes. To overcome this problem, scientists use the transient transformation methodology. This alternative technology could provide a rapid tool for the functional analysis of the genes of interest (transgenes) (Wróblewski et al.2005). An important breakthrough in the fast reverse genetics was achieved by using a powerful tool—virus induced gene silencing (VIGS) technology (Orzaez and Granell 2009; Fernandez-Moreno et al. 2013). Jaberolansar et al. (2010) and Romero et al. (2011) successfully demonstrated that the Tobacco Rattle Virus (TRV)-based VIGS vector could be used in tomato to silence genes. On the other hand, Zhou et al. (2012) applied Potato Virus X as a tool for virus-induced gene complementation for revealing a transcription factor network in the modulation of tomato fruit ripening. Furthermore, Orzaez et al. (2006), in order to shorten the time of the functional analysis of genes in fruit development, used an *Agrobacterium*-mediated transformation by infiltrating tomato fruit tissue. This new procedure, called "fruit agroinjection", involves injecting*Agrobacterium* suspension into green fruits, resulting in complete fruit infiltration. The aforementioned method was found to be an invaluable tool for transient expression in fruits and also significantly facilitates functional research concerning that organ. Some of the data in the literature suggests fleshy fruits as an ideal target for genetic engineering (Spolaroe et al. 2001; Orzaez et al. 2006). Nevertheless, the identification and quantification of nonvisual phenotypes could be hindered by the irregular distribution of VIGS effects in fruit. For that reason, Orzaez et al. (2009) elaborated a simple visually traceable VIGS system for fruit. This methodology consists of two elements: (1) a tomato line expressing *Rosea1* and *Delia* transcription factors under the control E8 promoter that show a purple-fruited (anthocyianin-rich) phenotype, and (2) the agroinjection of a modified TRV VIGS vector incorporating partial *Rosea1* and *Delia* sequence into *Del/Ros1*plants, which was shown to be able to restore red fruit phenotype. Hasan et al. (2008), using the agroinjection procedure, obtained a stable transformed tomato fruit. The transformation efficiency ranged from 54 to 68 % in seedlings raised from seeds collected from the infiltrated fruits. Yasmeen et al. (2009) noticed that mature red fruit resulted in a higher frequency of transformation than immature green fruit, and the transformation efficiency was 40–42 %. Among the particularly popular methods there are those that avoid regeneration of the tissue culture as they allow the exclusion of complex, time consuming procedures, thus shortening the time of the entire process. The *in planta*transformation method is one of these. This method has been successfully used so far for various plant species (both monocotyledonous and dicotyledonous plants, including the tomato) to obtain transgenic plants. One of the versions of the *in planta* transformation method is the floral dip procedure. The floral dip method was used by Yasmeen et al. (2009) to obtain transgenic tomatoes. For this purpose, they tested two approaches: the transformation of unopened flowers before pollination and the transformation open flowers after pollination. Yasmeen et al. (2009) showed that the floral stage as well as the gene construct had a significant impact on transformation efficiency. The results revealed that flowers treated before pollination gave higher percentages of transformation (12 % for*LFY* gene construct and 23 % for *GUS* gene construct) compared to those treated after pollination. Although the transformation efficiency was promising, some undesirable changes were observed. Compared to the control plants, the transgenic plants carrying transgenes (*AP1) Apethala* gene from *A. thaliana* or *LFY (LEAFY* gene from *A. thaliana*) were phenotypically different. They were shorter, their stems were not as erect and their leaves were curled. Additionally, these plants produced normal flowers earlier than the control plants, but they were infertile and failed to bear fruit. To our best knowledge, this kind of protocol is not widespread with regards to the tomato.

Applicable Tomato Transformation

At present, GM technology is widely acclaimed as being able to produce "upgraded crops", including tomatoes, more rapidly and efficiently than selection breeding and therefore has the potential to reduce food shortages. Many plants important from economic point of view are genetically modified to resist a wider range of environmental conditions such as poor soil conditions (e.g. salinity and metal contamination) or drought, extreme temperatures (i.e. heat or cold). The genetic engineering allowed for increased productivity by enhancing efficiencies of metabolic or photosynthetic pathways.

Resistance to Abiotic Stresses

Transgenic approaches have been attempted to improve tolerance to abiotic stresses. A large number of genes either involved in signaling and regulatory pathways or encoding enzymes known to alleviate stress have been introduced to produce plants with increased stress resistance against salinity, high and low temperatures, oxidative stress, heavy metals or drought. Heavy metals that accumulate in soil are extremely harmful contaminants. Their toxic effects cause disturbances in cell membrane functioning, photosynthetic and mitochondrial electron transport, enzyme inactivation and basic cellular metabolism, thus leading to disturbances in the energy balance of a cell as well as hindering mineral management and growth suppression. For most, even a slight increase in the concentrations of metal ions in a cell is harmful, however in the course of evolution some species have developed mechanisms to protect themselves from the harmful effects of high concentrations of heavy metals present in the environment. Barabasz et al. (2012) demonstrated that the expression of HMA4 (P_{1B}-ATPase) from *Arabidopsis halleri* in plants could be a useful approach to engineer altered metal distribution in tissues, which could be useful for biofortification or phytoremediation. It turned out that the expression of the *AhHMA4* gene facilitated Zn translocation from root to shoot and also induced Zn uptake in a Zn supply-dependent manner.

Drought is defined as water deficit in the environment and it is closely correlated with soil salinity. The presence of salt in the soil causes ionic and osmotic stresses, which lead to metabolic imbalances and nutritional deficiencies and may also cause oxidative stress. High soil salinity may damage plants during the vegetation period. It is believed that these two kinds of stresses are among the most devastating abiotic stresses that limit crop productivity worldwide. For example, droughts in Poland can occur in different seasons of the year, but they are most common in spring, occurring every few years. It is a very serious economic problem for any country because of large yield losses, and thus farmer income decreases and food prices increase. More recently, global warming may have been worsening this situation in most agricultural regions around the world. The ultimate aim is to develop crop plants with improved water use efficiency that can minimize drought-induced yield losses. Furthermore, drought stress tolerance may not only ameliorate productivity the land already in use but may also allow for the exploitation of cultivable land with limited water supplies. Over the last two decades, the number of publications concerning genetically modified plants for drought resistance has increased, indicating their scientific and applied importance. In this literature, many metabolic systems and candidate genes were targeted to achieve drought resistance. As is commonly known, resistance to abiotic or biotic stresses is a multifactor trait involving several genes. Therefore, genetic engineering for developing stress-tolerant crops based on the introgression of genes known to be involved in stress response and putative tolerance is being developed. Hence, numerous researchers focused on one or a few genetic changes to modify key metabolites (e.g. glycine betaine and proline) (Goel et al. 2011; Álvarez-Viveros et al. 2013) or proteins *Late Embryogenesis Abundance* (LEA) (Muñoz-Mayor et al. 2012) for drought resistance. Another strategy to increase the level of drought and salinity tolerance in plants consists of a transfer of genes encoding different types of proteins involved in molecular responses to abiotic stress such as osmoprotectants, chaperones, detoxifying enzymes, transcription factors, signal transduction proteins (kinases and phosphatases) and heat-shock proteins (HSPs) (Wang et al. 2011; Mishra et al. 2012; Li et al. 2013). It is known that the mitogen-activated protein kinases are involved in tolerance-related signaling networks associated with various stressors, including drought stress. The results obtained by Li et al. (2013) using VIGS methodology confirmed this observation. It was found that *SpMPK1* (the mitogen-activated protein kinases from *S. pimpinellifolium*),*SpMPK2*, and *SpMPK3* genes played a crucial role in enhancing the drought tolerance of tomato plants by affecting the production and activity of H_2O_2 via the ABA-H_2O_2 pathway, and thus their inhibition reduced drought tolerance.

However, the modification of the expression of a single gene involved in resistance response such as listed above usually has a limited effect. A better solution seems to be the modification of the expression of transcription factors (TFs). This is an attractive target category for manipulation group, as it activates a cascade of genes that act together in enhancing tolerance towards different stresses. Most of them are classified into several transcription families such as AP2/ERF (APETALA2/Ethylene Responsive Factor), MYC, MYB, NAC, (Cys2His2 zinc finger), bZIP (basic leucine zipper) and WRKY (Shinozaki and Yamaguchi-Shinozaki 2007). Some of them are involved in plants' response to drought. Especially TFs from bZIP (e.g. ABA responsive element binding protein/ABRE binding factor (AREB/ABF)), AP2/EREB (e.g. DRE binding protein/CRT binding factor (DREB/CBF)), NAM (no apical meristem, ATAF1-2, CUC2 (cup shaped cotyledon) (NAC) (e.g. stress-responsive NAC (SNAC)), CCAAT-binding (e.g. C3H2 zinc finger protein ZFP) (Yang et al. 2010). AREB/ABF belong to the bZIP family plant TFs known to function in ABA signaling during dehydration and seed maturation. In response to ABA, an activated AREB/ABF binds to a *cis*-element known as an ABA-responsive element (ABRE) to trigger gene expression (Pandey et al. 2011). Up to now, participation of this kind of TFs in ABA-mediated stress signaling has been described for different plants such as *Arabidopsis thaliana*, rice, wheat and barley (Yang et al. 2010). Research by Yanez et al. (2009) revealed that the expression *SlAREB* in tobacco and tomato leaves was responsible for up-regulation of stress-responsive genes such as RD29B, the LEA genes ERD10B and TAS14 (dehydrin from tomato), the transcription factor PHI-2 and trehalose-6-phosphate phosphatizing gene. These results suggested that this class of bZIPs plays a role in abiotic stress response in the *Solanum* genus. In another study, Hsieh et al. (2010) observed that the overexpression of *Sl*AREB is responsible for increasing tolerance to water and salinity in tomato plants. The sverproduction of *Sl*AREB in transgenic tomato plants regulated genes AtRD29A, AtCOR47, and SlCI-like dehydrin under ABA and abiotic stress treatments. Mishra et al. (2012) inserted the transcription factor gene ATHB-7 (*Arabidopsis thaliana* homeodomain-leucine zipper class I genes) into the tomato genome. *ATHB-7* gene is induced in plants under drought stress via a mechanism that requires the production of ABA and acts as a negative growth regulator in Arabidopsis the expression of *A. thaliana* transcription factor gene ATHB-7 in tomato plants significantly reduced the leaf stomatal density and stomatal pore size, which is probably crucial in preserving higher water potential. However, Mishra et al. (2012) observed in transgenic tomato line (DTL-20) a reduction in plant growth. This characteristic under-soil water deficit is common to many plant species.

Plants have developed several adaptation strategies that allow them to withstand saline stress. Among them we can distinguish sequestration of solutes, limitation of lipid peroxidation and the production of osmoprotectants. Numerous data indicates that there are potential benefits from obtaining transgenic plants overexpressing H+-pyrophosphatase and Na+/H+ antiporter, which increase tolerance to salinity. The data presented by Bhaskaran and Savithramma (2011) and Yarra et al. (2012) support the aforementioned hypothesis. Some research has indicated a significant role for vacuolar H^+-ATPase (V-ATPase) under drought conditions. This multisubunit enzyme is necessary for plant growth because it is responsible for energizing secondary transport in the maintenance of ion homeostasis and in abiotic stress tolerance. Hu et al. (2012) demonstrated that the overexpression of *MdVHA-B* (subunit B of the V-ATPase form apple) in tomato plants resulted in high tolerance to drought stress as well as reduced malondialdehyde (MDA) contents and relative water loss, along with increased levels of free proline and H^+ ATPase activity as compared to the control plants. It should be emphasized that MDA is a widely used marker of oxidative lipid injury whose concentration varies in response to abiotic or biotic stresses. Malondialdehyde accumulation takes place in plants due to membrane lipid peroxidation (Sharma et al.2012).

As previously mentioned, the transfer of gene-coding transcription factors is one of the strategies to increase plant tolerance to drought and salinity. Rai et al. (2013) showed that the overexpression of *BcZAT12* in transgenic tomato plants caused a significant increase in their drought tolerance. Similarly, Mishra et al. (2012) demonstrated that transgenic tomato lines carrying the transcription factor of the *ATHB-7* gene from *A. thaliana,* were highly drought tolerant. On the other hand, Álvarez-Viveros et al. (2013) suggested that the overexpression of two genes, glyoxalase I gene (*GlyI*) and glyoxalase II genes (*GlyII*), might improve the salinity tolerance of tomatoes. It is known that methylygloxal is produced during salt stress, and its detoxification is triggered by glycolases. Thus, transgenic plants subjected to a high concentration of NaCl (800 mM) displayed both reduced lipid peroxidation and the production of H_2O_2. It is not only during drought stress, but also in cases of other stresses, that the scavenging of reactive oxygen species (ROS) is connected with the acting of a range of enzymatic and non-enzymatic antioxidants as well as of organic compounds as polyamines (PAs) (Gill and Tuteja 2010). Polyamines are considered as one of the oldest groups of substances, and include tetramine spermine (Spm), putrescine (Put) and cadaverine (Cad). In plants, polyamines not only play a role in abiotic and biotic stress, but also in many other physiological processes (organogenesis, embryogenesis, floral initiation and development, leaf senescence, fruit development and ripening) (cf. Alcazar et al. 2010). Recent studies have revealed that polyamine signaling is involved in direct interactions with different metabolic pathways and entangled hormonal cross-talks (e.g., abscisc acid involved in the regulation of abiotic stress responses) (Alcazar et al. 2010). Furthermore, many studies using transgenic overexpression or loss-function mutants confirmed protective role PAs in plant response to abiotic stress. As mentioned previously, the example of polyamines is putrescine (Put) in biosynthesis, in which arginine decarboxylase (ADC) is involved. Wang et al. (2011) showed that transgenic tomato lines with an overexpression of the *PtADC*gene isolated from *Poncirus trifoliate* performed better in plant dehydratation and drought stress. As expected, under these stress conditions, ROS accumulation significantly decreased as compared to the control plants.

Numerous data have indicated that the heterologous overexpression of ornithine decarboxylase, ADC, S-adenosyl-l-methionine decarboxylase (SAMDC), spermidine synthase (SPDS) from different animal or plant sources in such plants as tomatoes, rice and tobacco has displayed tolerance traits against different stress conditions (including salt stress, osmotic stress, freezing, heat, drought, etc.) (cf. Alcazar et al. 2010).

In countries with cold climates, tomatoes are grown in greenhouses. Maintaining controlled temperature conditions raises the costs of breeding. Thus, a reasonable solution to manage this problem seems to be obtaining genetically-engineered tomato plants resistant to low temperatures. Generally, cold stress is responsible for, among others, the induction of osmotic disorders. Thus, tolerance to cold activates enzymes responsible for the synthesis of osmoprotectants and antioxidant defense. Osmotin and osmotin-like proteins have been demonstrated to accumulate in response to various biotic and abiotic stresses in plants. Patade et al. (2013) gave clear evidence that the accumulation of both osmotin and proline during cold stress in transgenic tomato lines imparted cold tolerance to them. Several studies on species more tolerant to cold allowed the determination of genes regulated by cold, known as *COR* genes (cold-regulated). Al genes belonging to the*COR* group have two characteristic sequences in promoter: the C-repeat (CRT) and the dehydration responsive element (DRE)-related motifs that interact with the CRT/DRE binding factor (CBF1). When the aforementioned gene from *A. thaliana* was introduced into tomatoes, the transgenic plants revealed higher chilling tolerance (Hsieh et al. 2002). However, transgenic tomato plants showed growth retardation with reduced fruit, seeds and fresh weight numbers. Moreover, transgenic tomato plants contained higher levels of proline than wild-type plants under normal or water-deficient conditions. Singh et al. (2011) obtained similar results by introducing into tomatoes the *AT-CBF1* gene, which is driven by the inducible promoter RD29A (which contained several *cis*-acting elements, including DRE, ABRE). However, Singh et al. (2011) did not observe any morphological disorders. The use of RD29A promoter instead of constitutive promoter in the tomatoes led to the development of cold-tolerant transgenic plants without any phenotypic abnormalities.

On the other hand, in terms of global warming, obtaining transgenic plants resistant to high temperature seems to be fully justified. It is commonly known that the accumulation of polyamines, including betaine, putrescine, spermidine or spermine, under abiotic stresses plays a crucial role in plant defense response to unfavorable conditions. SAMDC is one of the pivotal regulatory enzymes involved in biosynthesis of these compounds. Cheng et al. (2009) reported that transgenic tomatoes carrying the *SAMDC* gene from*Saccharomyces cerevisiae* produced 1.7–2.4 times more polyamines and therefore showed enhanced tolerance to high temperatures as compared to the control plants. Similarly to cold and heat stresses, ultraviolet B (UV-B, 280–320 nm) causes both the production and accumulation of toxic ROS. Furthermore, the interaction of high temperatures and UV-B could trigger sunscald phenomenon (tissue browning and desiccation) among the crop plants' fruit. Under such unfavorable conditions, the plants protect themselves by producing antioxidant enzymes including superoxide dismutase (SOD) and ascorbate peroxidase (APX). The data provided by Wang et al. (2006) clearly indicates that an overexpression of cAPX (cytosolic ascorbate peroxidase) in transgenic tomato plants significantly enhances resistance to high temperature (40 °C) compared to wild-type plants.

Examples of successful genetic engineering of tomatoes for enhanced resistance to abiotic stresses are presented in Table 2.

Table 2: Examples of successful genetic engineering of tomato

Fruit trait	Targeted gene	References
Fruit quality (organoleptic and nutritional)		
Flavor and aroma	Thaumatin, *GES, LeAADC1A, LeAADC2*	Bartoszewski et al. (2003); Davidovich-Rikanati et al. (2007); Mathieu et al. (2009); Tieman et al. (2006)
Size	*fw2.2*	Cong and Tanksley (2006); Liu et al. (2003)
Firmness	β-galactosidase,*EXP1A* (expansine)	Brummell et al. (1999); Smith et al. (2002)
Parthenocarpy	*Arf8*; *IAA9; SIARF7, Sl-IAA27*	Bassa et al. (2012); de Jong et al. (2011); Goetz et al. (2007); Wang et al. (2005)
Soluble solids content	*Lin5* (invertase 5)	Zanor et al. (2009)
Carotenoid content	*Dxs, CrtB, CrtR-b2*(carotene beta hydroxylase), *CrtI, CrtY, PSY-1,Cyc-B, LCY-B,CHY-B, CRY-2, DET-1, COP1LIKE, CUL4*(Cullin4), *FIBRILLIN, spermidine synthase*	Apel and Bock (2009); D'Ambrosio et al. (2011); Davuluri et al. (2005); Dharmapuria et al. (2002); Enfissi et al. (2005); Fraser et al. (2002), (2007); Giliberto et al. (2005); Liu et al. (2004); Neily et al. (2011); Simkin et al. (2007); Wurbs et al. (2007); Wang et al. (2008)
Flavonoid content	*CHI, CHS, CHI, F3H, FLS, STS,CHR, FNSII, MYB12, S1MYB12, Del, Ros, ANT1,AN2*	Adato et al. (2009); Ballester et al. (2010); Bassolino et al. (2013); Butelli et al. (2008); Colliver et al. 2002; Maligeppagol et al. (2013); Muir et al. (2001); Schijlen et al. (2006); Schreiber et al. (2012)

Fruit trait	Targeted gene	References
Carboxylic acids	*SlAco3b*	Morgan et al. (2013)
Ascorbic acid content	*GalLDH, GME, GCHI, ADCS*	de la Garza et al. (2004), (2007); Garcia et al. (2009); Gilbert et al. (2009); Zhang et al. (2011); Waller et al. (2010)
Abiotic stress		
	GlyI, GlyII, cAPX, SpMPK1, SpMPK2,SpMPK3,Osmotin, HMA4 (P1B-ATPase), *SAMDC, mtlD, codA, AVP1, PgNHX1,* BcZ*AT12, TaNHX2,tas14, PtADC, MdVHA-B*	Álvarez-Viveros et al. (2013); Barabasz et al. (2012); Bhaskaran and Savithramma (2011); Chen et al. (2009a); Goel et al. (2011); Hu et al. (2012); Khare et al. (2010); Li et al. (2013); Mishra et al. (2012); Muñoz-Mayor et al. (2012); Park et al. (2005); Patade et al. (2013); Rai et al. 2013; Wang et al. (2006), (2011); Yarra et al. (2012)
Biotic stress		
	AFP,amiR-AV1-3,*hCAP18/LL-37, Bs2,CHI, alfAFP,ech42,Cry 2Ab*, *LF, Cry1Ac*	Chen et al. (2009b); El-Siddig et al. (2011); Herbette et al. (2011); Horvath et al. (2012); Jung (2013); Lee et al. (2002); Ma et al. (2011); Rashid and Bal (2011); Saker et al. (2008), (2011); Shah et al. (2010); Vu et al. (2013)
Pharmaceuticals protein		
	PfCP-2.9, BACE1, IL-12; F1-V, sDSP, hIgA_2A1, Tα1, miraculin, *hFIX*, AGAP, Hiv-1 Tat,*HBsAg*	Alvarez et al. (2006); Alvarez and Cardineau (2010); Baesi et al. (2011); Biswas et al. (2012); Chen et al. (2009a); Cueno et al. (2010); Elías-López et al. (2008); Hirai et al. (2010); Juárez et al. (2012); Kantor et al. (2013); Kato et al. (2011); Kim et al. (2012); Kurokawa et al. (2013); Lai et al. (2009); Li et al. (2011); Lou et al. (2007);Ramirez et al. (2007); Soria-Guerra et al. (2007), (2011); Zhang et al. (2007); Youm et al. (2008)

Resistance to Biotic Stresses

The enormous economic success of crop plants, including the tomato, is due to the application of pesticides and control of bacterial and viral diseases. Currently, pests and diseases are controlled by pesticides, but plants' acquisition of resistance to pesticides as well as the appearance of new diseases may be side effects. Therefore, genetic engineering seems to be a reasonable solution to limit the usage of pesticides. Molecular studies on plant resistance mechanisms allowed the identification of genes whose manipulation could improve resistance to pathogens. Furthermore, an examination of plants susceptible to different pathogens allowed the identification of genes which are crucial for plant susceptibility to pathogens and could be potential targets for RNA interference (RNAi) strategy. Below, we discuss some of the achievements in this field.

Diseases caused by geminiviruses strongly affect the infected crops' yield, leading to significant economic losses. Aerial spraying is traditional method of eliminating virus infections in crop plants. This approach is extremely effective, but unfavorable from environmental protection point of view. Many studies have shown that virus-encoded RNAi suppressors are responsible for pathogenesis in host plants. For this reason, they are very important targets for antiviral strategies. Vu et al. (2013) generated tomato plants overproducing amiRs (artificial micro RNAs) to silence viral AV2/AV1 (coat proteins) transcripts. Since this study revealed that some of the transformants displayed tolerance/resistance against Tomato Leaf Curl New Delhi Virus, it is believed that the amiR strategy could be effectively employed to protect crop plants against viruses.

In the course of their evolution, plants have developed specific intracellular immune receptors encoded by disease resistance (R) genes. These genes recognize gene products originating from different pathogen species. For example AvrBs2 (an effector that is highly conserved in a number of *Xanthomonas* species) is recognized by Bs2 R protein in pepper (a close relative of tomato). Horvath et al. (2012) investigated whether the obtained Bs2 transgenic tomato lines are resistant to different bacterial strains. Their results showed that all tested genotypes from the Bs2 lines had developed high resistance to bacteria as compared to the control group. Another strategy to improve plant defense against different pathogens (including fungi and bacteria) consists of constructing genetically-engineered plants that express antimicrobial peptides. Jung (2013) obtained transgenic tomato lines by producing a human antimicrobial peptide (hCAP18/ ll-37)—cathelicidin; these lines show high expressions of PR protein (Pathogenesis-related), lipid transfer protein, and antifungal protein and exhibited significant resistance to bacterial soft rot and bacterial spot diseases. Mitochondrial alternative oxidases (AOXs) and important components of the alternative respiratory pathway in plants. This particular AOX pathway can be induced by the pathogens ROS, salicylic acid (SA) and high light intensity. It is known that the induction of SA is strictly linked to defense responses in plants, therefore it has been speculated that the alternative pathway might be connected with plant resistance to pathogens. This hypothesis was supported by the results obtained by Ma et al. (2011). They provided evidence that all plants with modified AOX expression levels could cope with Tomato Spotted Wilt Virus in contrast to non-transformed, wild-type plants.

Early defense responses include changes in the plant cell membrane. Potassium and chloride ions leak from a cell and are replaced by calcium ions. This leads to increased synthesis of hydrogen peroxide (H_2O_2), which is a toxic to plants as it inactivates enzymes strictly associated with ROS such as APX, SOD, catalase and glutathione peroxidases (GPx). Interesting results were obtained by Herbette et al. (2011), who investigated the role of selenium-independent glutathione peroxidase in the response to abiotic (mechanical damage) and biotic (exposition on *Oidium neolycopersici* and *Botrytis cinerea*) stresses in transgenic tomato lines overexpressing GPx. They reported that in the case of mechanical damage, GPx overexpression alleviated stress, while when plants were challenged by biotic stress GPx overexpression abolished plant defense response and increased formation of necrotic lesions. The authors concluded that GPx helped cells overcome elevated ROS generation in abiotic stress response, whereas in biotic stresses GPx activity clashed with ROS-mediated signaling events. In the light of the aforementioned data expression of δ-endotoxins in transgenic plants, this could prevent lepidopterus insect-caused damage. Such research was conducted by Saker et al. (2011), who evaluated resistance of transgenic tomato plants expressing the *Cry2Ab* (δ-endotoxins from *Bacillus thuringiensis*) gene to *Helicoverpa armigera* (Hűbner) and*Phthorimaea operculella* (Zeller). This study showed that the mortality of larvae fed with transgenic plants was 100 %, as compared to only 8 % for the control plants.

Tomato plants are exposed to attack of a broad spectrum of pathogens. In the case of tomatoes, Fusarium wilt (*Fusarium oxysporum* f. sp. *lycopersici*), Verticillium wilt (*Verticillium dahliae*), early blight (*Alternaria solani*) and late blight (*Phytophthora infestans*) are major fungal diseases. Several attempts to generate transgenic tomato plants showing tolerance to fungal pathogens were made. For example, Shah et al. (2010) obtained transgenic tomato lines expressing endochitinase (*ech42*) gene from*Trichoderma virens*. These transgenic lines revealed enhanced resistance to fungal pathogens as compared to control plants. On the other hand, Chen et al. (2009b) generated tomato plants resistant to *B. cinerea* and the resistance levels were related to the expression levels of the transgene, displaying the gene-dosage effect. The highest resistance was noticed in the case of plants containing *CHI-AFP* (bivalent gene chitinase and alfalfa defensin).

Examples of successful genetic engineering for enhanced tolerance to biotic stresses and production of biopharmaceutical in tomato are presented in Table 2.

Improvement of Fruit Quality

Tomato fruits have two categories of intrinsic qualities, organoleptic properties and nutritional value. Organoleptic qualities include the texture of the fruit, their taste and aroma. With regards to nutritional values, tomato fruits are a low-fat, high-fiber, low-calorie source of many vitamins and minerals and many other substances such as sugars, flavonoids, ascorbic acids and folate, and carotenoids. Other very important qualities of tomato fruits are their color, shape, fruit firmness and shelf life. Examples of successful genetic engineering to enhance fruit quality traits have been presented in Table 2.

From the point of view of a potential consumer, the fruit's taste appears as a very important organoleptic trait. Until now, only two successful attempts to change tomato flavor have been reported. Bartoszewski et al. (2003) generated transgenic lines of tomatoes expressing biologically active thaumatin, a sweet-tasting, flavour-enhancing protein produced by the fruits of the African plant *Thaumatococcus daniellii* Benth. Their results revealed that fruits from the T_2 plants were sweeter as compared to the control, wild-type fruits and possessed a specific aftertaste. This achievement demonstrated that it was possible to overcome poor fruit taste in breeding tomato lines such as those bearing a non-ripening mutation. Furthermore Davidovich-Rikanati et al. (2007) reported simultaneous modification of fruits' flavor and aroma. They obtained transgenic lines with modified both flavor and aroma by expressing *Ocium basillicum* geraniol synthase (*GES*) under the control of the tomato ripening-specific polygalacturonase promoter (PG). Geraniol synthase belongs to the monoterpenes group, which are important contributors to many fruit scents and also intermediate in carotenoid biosynthesis. Generally, they are synthesized from geranyl diphosphate (GDP) where GES catalyzes the conversion of GDP to geraniol, which is a pivotal precursor of the isomeric monoterpenes aldehydes having different aroma (e.g. lemon, rose-like aroma). While the tomato ripens, carotenoid biosynthesis is highly active, but the ripened fruits contains small amounts of monoterpenes (Iijima et al. 2004). Davidovich-Rikanati et al. (2007) indicated that this modification could be applicable to other carotenoid-accumulating fruit species important from agricultural and horticultural point of view. Furthermore, because volatile terpenoids have some advantages in antimicrobial, antifungal or pesticidal activities, this modification can be very useful for improving the fruits' shelf life or reducing pesticide application.

The unique flavor of the tomato fruit is a combination of different components, including sugars, amino acids, lipids and carotenoids. It is well known that the flavor of commercially-produced tomatoes is unsatisfactory, so flavor improvement appears as one of major challenges for scientists. Unfortunately, until now very few genes involved in biosynthesis of volatile compounds have been identified (Mathieu et al. 2009). Tieman et al. (2006) showed that overexpression of *LeAADC1* (carotenoid cleavage dioxygenase gene) or *LeAADC2* in fruits brought about a nearly tenfold increased emission of pathway products involving 2-phenylacetaldehyde, 2-phenylethanol and 1-nitro-2-phenylethane. On the other, hand antisense silencing of tomato genes (*LeAADC2*) reduced the emission of these volatiles by 50 %. Interestingly, Mathieu et al. (2009) described new quantitative trait loci (QTLs) that affected the volatile emission of red-ripe tomato fruits. These results suggested that QTLs could be used as a tool to identify genes responsible for changes in volatile levels.

Genetic modification of tomato fruit firmness has been achieved by engineering genes involved in the regulation of a single enzymatic step in the cell wall-formation pathways. Brummell et al. (1999) and Smith et al. (2002) provided evidence that expansin or β-galactosidase controlled fruit softening and firmness in the case of genetically modified tomatoes.

The flavor and firmness of tomato fruit not the only important factors for potential consumers, fruit size is also of importance. Therefore, some research has been conducted regarding to this fruit characteristic. Detailed studies on dosage series of the *fw2.2* gene encoding plant-specific protein regulating cell divisions, particularly in fruits, were performed on transgenic tomatoes, allowing for fruit size to be altered (Liu et al. 2003; Cong and Tanksley 2006). Since the detailed mechanism by which *fw2.2* participates in fruit development is unknown, deciphering this enigma is one of the keys to understanding the phenomenon of fruit development. Cong and Tanksley (2006) suggested that *fw2.2*might have mediated this process through gene co-option and recruitment of a cell cycle control pathway. As it turns out, *fw2.2* interacted with the highly conserved regulatory unit of CKII b ((beta) subunit of a CKII kinase) and thus may affect regulation of cell division via CKII mediated pathways.

Parthenocarpic fruits can be obtained in two ways: natural or artificial without ovule fertilization. This phenomenon is very desirable not only in the case of edible fruits, but also in the case of fruit crops that may be difficult to pollinate or fertilize such as tomatoes. It is noteworthy that parthenocarpy resulted in increasing the content of the soluble solid. In light of numerous publications, in order to induce and develop parthenocarpic tomato fruit, genetic transformation targeting a single TF (transcription factor) has been used. For example, Wang et al. (2005) reported that the downregulation of Aux/*IAA9* (the auxin/indole-3-acetic acid (Aux/IAA) and auxin response factor (ARF) familie) TF triggered parthenocarpic fruit development. On the other hand, Goetz et al. (2007) obtained a similar effect by the overexpression of auxin response factor 8 (*ARF8*) and Bassa et al. (2012) by the overexpression of *Sl-IAA27*. Moreover, de Jong et al. (2011) revealed that transgenic tomato lines with decreased *SlARF7* mRNA levels promoted fruit parthenocarpic development, indicating that this TF might act as a negative regulator of the fruit set. Furthermore, it turned out that *SlARF7* played a pivotal role in the modulation of the GA response during the early stages of tomato fruit development.

Not only organoleptic attributes, but also the nutritional properties of tomato fruits have recently attracted attention of scientists. To prove involvement of cell wall invertase (*LIN5*) in controlling the content of soluble solid, Zanor et al. (2009) exploited the RNAi approach in transgenic tomato plants. As a result, the transgenic plants displayed several changes in morphology (including flower architecture, reduced fruit size and reduced seed amount) and metabolic pathways (particularly sugar metabolism and hormones). Numerous attempts have been made to obtain transgenic lines with increased levels of lycopene, xantophylls and β-carotene (Fraser et al. 2007). Neily et al. (2011) reported obtaining transgenic tomato lines that showed 1.5–2fold increase in polyamine content by the overexpression of the SPDS gene, an enzyme crucial for polyamine biosynthesis. It has to be emphasized that the constitutive expression of the *SPDS* gene enhanced the accumulation not only of spermidine but also putrescine. Remarkably, the transgenic tomato fruits also revealed an increase in carotenoid accumulation, especially of lycopene (1.3- to 2.2-fold), and increased ethylene production (1.2- to 1.6-fold) as compared to wild-type fruits. Based on these results, it is believed that a high level of accumulation of polyamines in tomatoes regulates the steady-state level of transcription of the genes responsible for the lycopene metabolic pathway, resulting in a higher accumulation of lycopene in the fruit. The research conducted by D'Ambrosio et al. (2011) is another example of engineering high carotenoid content in transgenic tomato fruit. They found that plants of transgenic tomato lines carrying tomato carotene beta hydroxylase 2 transgene showed statistically higher content of total carotenoids (including β-xantophylls, violaxanthin, neoxanthin) as compared to the control. Flavonoids, another group of compounds that occurring in tomato fruit, cause great interest among scientists because of their anti-inflammatory and antioxidant properties. Several strategies are used to achieve an enhanced level of flavonoids in tomato fruits. The first approach is strictly connected with engineering of single structural genes involved in crucial steps in the flavonoid biosynthesis pathway such as chalcone isomerase (*CHI*) or chalcone synthase (*CHS*) (Muir et al. 2001; Colliver et al. 2002). However, more spectacular results were obtained when multiple genes were targeted within the flavonoid pathway. For example, Colliver et al. (2002) reported that the ectopic expression of *CHS*, *F3H* (flavonole hydroxylase), and flavonole synthase (*FLS*) from the *Petunia hybrida* in tomato fruits resulted in enhanced levels of flavonoids and in peel tissue. Furthermore, they reported that *CHS* and *FLS*transgenes had a synergistic effect on flavonoid biosynthesis in tomato flesh tissues when they acted together. Another strategy demonstrated the possibility of targeting the flavonoid pathway in tomatoes towards synthesizing atypical flavonoids, e.g. grape stilbene synthase (STS), that normally are not present in tomato fruit. The results obtained by Schijlen et al. (2006) revealed that *STS* overexpressing tomatoes showed an increased accumulation of resveratrol aglycone in the tomato fruit peel. Another strategy consists of engineering transcription factors in order to increase the content of the flavonoid compounds. Such attempts were undertaken by several research teams (Adato et al. 2009; Ballester et al. 2010; Schreiber et al. 2012; Bassolino et al. 2013; Maligeppagol et al. 2013). Maligeppagol et al. (2013), with fruit-specific expressions of two transcription factors*Delila* and *Rosea*1, isolated from *Antirrhinum majus* generated transgenic tomato plants accumulating a 70–100-fold higher amount of anthocyanin in the fruit. The transgenic tomato plants were identical to the control plants, except for the accumulation of high levels of anthocyanin pigments in the mature fruit. Insightful analysis confirmed the elevated expression of the downstream genes of the anthocyanin pathway due to the expression of the aforementioned TFs, and that the anthocyanin expression levels coincided with fruit ripening stages, with the highest expression occurring at the breaker stage. Similar studies were conducted by Butelli et al. (2008) and Orzaez et al. (2009) on tomato line (cv. MicroTom) expressing *Delia* and *Rosea*1 TFs under the control E8 promoter (tomato fruit-specific E8 promote) where the fruit displayed the purple-fruited phenotype. In their study, Ballester et al. (2010) exploited VIGS to down-regulate*SlMYB12* (TFs family MYB form *S. lycopersicum*) gene expression in tomato fruit to demonstrate direct involvement of *SlMYB12* in the establishment of a pink phenotype. The research of Schreiber et al. (2012) confirmed the hypothesis that *ANT1* (gene encoding homologous R2R3) was the gene responsible for anthocyanin accumulation in tomato fruit peels.

In contrast to above-mentioned achievements in flavonoid pathway engineering, relatively less research has been conducted regarding the enhanced level of carboxylic or ascorbic acids in tomato fruit. It is believed that high organic acid content is a very important attribute of fresh tomato fruits. Notwithstanding, the complexity of their metabolism makes it difficult to choose the best way to influence carboxylic acid levels. Morgan et al. (2013) analyzed a tomato introgression line with increased levels of fruit citrate and malate at the breaker stage to identify a metabolic engineering target that was afterward investigated in transgenic plants. Morgan et al. (2013) analyzed transgenic lines of tomato fruit expressing an antisense construct against SlAco3b (one of the two tomato genes encoding aconitase). They indicated that in transgenic tomato lines, both the aconitase transcript level and aconitase activity were reduced. Increased levels of both citrate and malate were noticed in the ripe fruit, and as a result the total carboxylic acid content was raised by 50 % at maturity. It is known that ascorbic acid plays an essential role as an antioxidant in defense against various abiotic stresses, so several scientists have investigated the effect of increased ascorbate accumulation in transgenic plants on their tolerance of oxidation, cold and salt stresses. Zhang et al. (2011) revealed that the overexpression of GDP-D- mannose-3′,5′-epimerase genes (*SlGME1* and *SlGME2*) resulted in increased ascorbate accumulation in tomato fruits, improving their tolerance to abiotic stresses. Interactions between ascorbate levels and fruit metabolism were studied using RNAi technology. Garcia et al. (2009) generated transgenic tomato lines with silenced l-galactono-1,4-lactone dehydrogenase, producing fruits with improved ascorbic acid content. In contrast to the studies of Zhang et al. (2011), they silenced GDP-d-mannose-3′,5′-epimerase, which resulted in a reduced level of ascorbic acid in the fruit (Gilbert et al. 2009). Recently, Garchery and Gest (2013), using the RNAi approach, obtained transgenic lines with decreased ascorbate oxidase activity and thus the plants showed increased levels of total ascorbic acid.

Folate (all forms of vitamin B) is essential for numerous bodily functions, such as regulating cell growth and functioning, and also has positive effect on the nervous system and brain. Moreover, it participates in maintaining the genetic material in the transmission of hereditary cell characteristics, regulating their distribution; it improves the digestive system, and is involved in the formation of gastric juice for the efficient operation of the liver, stomach and intestines; it stimulates hematopoiesis, i.e. the formation of red blood cells; and protects the body against cancer (particularly cancer of the uterus). Folate deficiency leads to neural tube defects in developing embryos and other human diseases including diarrhea, macrocytic anemia, neuropathy, mental confusion, pregnancy complications, different kind of cancers, etc. (Bailey 2010). Because the human body cannot synthesize folate *de novo*, it has to be supplied through diet. A potential source of folate are leafy vegetables (e.g. spinach, broccoli, cabbage), but in folate occurs slightly smaller quantities in tomatoes, lentils, beets, sunflowers, etc. In plants, folates are synthesized from pteridine p- aminobenzoate (PABA) and glutamate precursors (de la Garza et al. 2007). Because folate is very important for human health and plants are known as one of the major source of folate, some attempts to enhance plants' folate levels have been undertaken (Raiola et al. 2014).

As is the case with ascorbic acid, there is very little research related to modification of folate content in tomato fruit. de la Garza et al. (2004) reported a twofold increase in folate content in tomato fruits overexpressing GTP cyclohydrolase I. 3 years later, the same research team reported a 25-fold increase in folate accumulation in transgenic tomato fruits (de la Garza et al. 2007). On the other hand, Waller et al. (2010) showed that in engineered fruit overexpressing foreign GTP cyclohydrolase I and aminodeoxychorismate synthase genes, the expression of endogenous genes was not changed, but those of three downstream pathway genes, aminodeoxychorismate lyase, dihydroneopterin aldolase and mitochondrial folylpolyglutamate synthase, increased by up to 7.8-, 2.8-, and 1.7-fold respectively, apparently in response to the build-up of specific folate pathway metabolites.

Molecular Farming of Tomatoes

Molecular farming is technology involving the use of plants, and potentially also animals, as the means of producing compounds that are of therapeutic value (safely and inexpensively). Until recently, the production of recombinant proteins was based on bacteria, mammalian or insect in vitro cultures and fungal cell cultures, which are insufficient to producing more complex polypeptides. Moreover, these technologies as applied now have some limitations resulting from differences in metabolic pathways and translation processes and refer mainly to expression systems based on bacterial, insect or fungal cell cultures. This may generate changes in the molecular structure of recombinant proteins, resulting in the limitation or total loss of their desirable activities. Plants offer many advantages over these systems in terms of safety, cost, time involved, protein complexity, storage and distribution issues. In this system, the desired foreign protein can be produced e.g. at 2–10 % of the cost of a microbial fermentation system and at 0.1 % of mammalian cell cultures, although it depends on the protein of interest, product field and the plant used. Additionally, plants have a higher eukaryote protein synthesis pathway very similar to animal cells with only minor differences in protein glycosylation. Therefore, plant biosynthesis pathway ensures correct structures even in the case of highly complex proteins. Furthermore, the use of plants avoids the risk of contamination with animal pathogens, such as viruses, that could be harmful to humans (e.g. HIV, hepatitis viruses, prions). It should be emphasized that no plant viruses have been found to be pathogenic to humans. Purification of the desired polypeptide from plants is often easier than from bacteria. Moreover, in some cases (e.g. edible vaccines), the purification process can be omitted. Whereas transgenic plants or virus-infected plants can be grown on field requiring only water, minerals and sunlight, mammalian cell cultivation is a very expensive process, requiring bioreactors that cost several hundred million dollars when production is scaled up to commercial levels So far, more than 100 diagnostic and therapeutic recombinant proteins as well as vaccines have been produced in various plants including tobacco, potato, tomato, lettuce, carrot, cereals, legumes (Wiktorek-Smagur et al. 2012). Here, we present some examples of the production of recombinant pharmaceutical proteins in tomatoes.

Malaria is a potentially fatal tropical disease that is caused by protozoan parasites of the genus *Plasmodium*. Vaccines are a cost-effective way to overcome infectious diseases of this magnitude. Notwithstanding, the available malaria vaccines produced by conventional methods are still unaffordable for most people due to their high price. Thus, Kantor et al. (2013) initiated the production of antigen gene *PfCP-2.9* of *Plasmodium falciparum* in tomato fruits. This research is the first report of a successful transformation with the expression of a malaria antigen (*PfCP-2.9*) in transgenic tomato plants of the T_0 and T_1generations. Kantor et al. (2013) found that transgenic tomatoes produced 35 mg per gram of fresh weight of leaves of malaria antigen protein. The TSP (Total Soluble Protein) extracted from the tomato leaves was similar to that obtained by other researchers (Biswas et al. 2012).

Alzheimer's disease (AD) is the most common cause of dementia in older adults. The cause and progression of Alzheimer's disease are not well understood. AD is a neurological disorder in which the death of brain cells causes memory loss and cognitive decline. Research indicates that the disease is associated with plaques and tangles in the brain. In patients with Alzheimer's disease, the presence of insoluble protein deposits (Aβ amyloid) in the brain was revealed. It seems that inhibiting the formation of Aβ by using vaccines directed against Aβ would be one of the most promising approaches towards the treatment of AD. Such attempts were undertaken by Youm et al. (2008). They obtained tomato plants with a satisfactory level of Aβ protein expression to be used in an oral immunization assay on mice. Elías-López et al. (2008) explored the use of transgenic tomato (TT) expressing IL-12 (interleukin) (TT–IL-12, transgenic tomato interleukine) as a single polypeptide as a possible strategy to produce constant and therapeutic IL-12 levels when administered through the oral route in a well-characterized murine model of progressive pulmonary tuberculosis. They demonstrated that oral administration of TT-IL12 crude fruit extracts ameliorated protective immunity and reduced lung tissue damage during early and late drug-sensitive and drug-resistant mycobacterial infection in an albino, laboratory-bred strain of the house mouse (BALB/c).

On the other hand, Kim et al. (2012) initiated the production of human β-secretase (BACE1) in transgenic tomato fruits, which serves as a vaccine antigen that would promote immune response. Furthermore, the proteolytic activity of the tomato-derived rBACE1 was similar to that of a commercial sample of *Escherichia coli*-derived BACE1. In 2006, Alvarez et al. received transgenic tomato plants with the *Yersinia pestis f1-v* fusion gene encoding for F1-V, an antigen fusion protein. The immunogenicity of F1-V against a challenge with subcutaneous *Y. pestis* was confirmed in mice that had been vaccinated orally with freeze-dried fruits (Alvarez and Cardineau, 2010). In 2007, Soria-Guerra et al. obtained transgenic tomato plants (cv AC) expressing a synthetic gene encoding a novel synthetic recombinant polypeptide, sDTP (Diphtheria–Pertussis–Tetanus) containing two adjuvant and six DPT immunoprotective exotoxin epitopes. In the course of subsequent studies, Soria-Guerra et al. (2011) examined whether the ingestion of tomato-derived sDPT protein induced specific antibodies in mice. The results showed that the sera of the immunized mice tested for IgG antibodies, the response to pertussis, tetanus and diphtheria toxin, and showed responses to the foreign antigens. Furthermore, the high response of IgA against tetanus toxin was apparent in the gut.

There are many causes of infectious diarrhea (viruses, bacteria and parasites). *Norovirus*is the most common cause of viral diarrhea in adults, but rotavirus is the most common cause in children under 5 years of age. Recombinant production of rotavirus antigens in plants has been proposed as an alternative to traditional production platforms. Juárez et al. (2012) obtained transgenic tomato plants expressing a recombinant human immunoglobulin A (hIgA_2A1) selected against the VP8* peptide of rotavirus SA11 strain. The amount of hIgA_2A1 protein reached 3.6 ± 0.8 % of the TSP in the fruit of the transformed plants. Fruit-derived products suitable for oral intake showed anti-VP8* binding activity and strongly inhibited virus infection in an in vitro virus neutralization assay.

Thymosine (Tα1) plays a crucial role in the treatment of diseases induced by viral infections (e.g. hepatitis B and C) and also cancers as an immune booster. For clinical use, Tα1 is mainly derived from animal thymus extraction or chemical synthesis. However, Chen et al. (2009b) reported the possible production the aforementioned protein in tomato plants. They revealed that Tα1 protein reached a maximum of 6.098 µg/g fresh weight in mature tomato fruit. Moreover, the specific activity of Tα1 protein produced by tomato plants was higher than that from the synthetic *E. coli* system. Some research demonstrated that tomato plants can be exploited for the production of hepatitis B surface antigen (Lou et al. 2007; Baesi et al. 2011; Li et al. 2011). All of the above findings support the concept of using transgenic tomato plants as a model for edible vaccines or producing antibodies. However, there are some disadvantages of this technology such as the short shelf life of fresh tomato fruits. To overcome this problem, food-processing techniques such as freeze-drying could be applied. Plant material prepared in this way can be stored for a long period of time and directly consumed without cooking. Notwithstanding, this procedure allows the plant-made vaccine to equalize and concentrate. To date, several studies have demonstrated the use of this technique to produce vaccines in transgenic tomatoes (Alvarez et al. 2006; Elías-López et al. 2008; Alvarez and Cardineau 2010, Soria-Guerra et al. 2011).

Tomato plants are used not only for production of vaccines or antibodies, but also for the production of other recombinant proteins such as miraculin. This protein is a taste-modifying glycoprotein extracted from a miracle fruit (*Richadella dulcifica*) and changes sour taste into sweet taste (Kato et al. 2011). In transgenic tomato plants, the recombinant miraculin content reached a concentration of up to 90 µg per g fresh weight (FW) of tomatoes (Hirai et al. 2010). Further studies by Kurokawa et al. (2013) revealed the miraculin accumulation levels in red fruits varied among the lines. Miraculin gene expression was driven by the E8 promoter and HSP terminator cassette (E8–MIR–HSP) in transgenic tomato plants, and the miraculin concentration was the highest in ripening fruits, 30–630 µg per gram of FW. The results achieved by Kurokawa et al. (2013) confirmed that combination of the appropriate promoter and terminator cassettes was important for significantly increasing the accumulation of recombinant proteins in a ripening fruit.

A number of studies over the past decades have proved that transgenic plants (including tomatoes) can be used as bioreactors for the production of recombinant therapeutic proteins (Wiktorek-Smagur et al. 2012). Zhang et al. (2007), using *Agrobacterium*-mediated transformation, demonstrated that the *hFIX* (human coagulation Factor IX) gene was expressed specifically in tomato fruits. The highest expression level was 15.84 ng/g FW (approx. 0.016 % of total soluble protein) and found in mature fruit. The analgesic–antitumor peptide (AGAP) from the venom of *Buthus martensii* Karsch is another therapeutic protein produced in transgenic tomato plants is (Lai et al. 2009). Earlier studies showed that AGAP would be useful in clinical therapy as an antitumor drug.

Examples of successful genetic engineering for biopharmaceutical in tomatoes are presented in Table 2.

CONCLUSION

Since the tomato (*S. lycopersicum* L.) was imported to Europe in the 16th century, it has become one of the most important vegetables around the world. Recently, interest in the tomato has significantly increased because of its nutritional values as well as its anti-cancer and anti-oxidative properties. In this review, we looked into new insights from recent developments in tomato biotechnology. Generally, it is known that traditional methods for improving tomatoes are time-consuming and troublesome due to breeding times. For this reason, it is necessary to develop efficient methods for the in vitro regeneration of different varieties of tomato. This would make a prerequisite step for further modification of tomato genome. Since more than 10,000 tomato varieties exist, it seems obvious that establishment of one universal protocol for regeneration is rather impossible since it would require very extensive analytical research on the physiological and genetic background of tomatoes' regeneration capacity. At present, it seems more likely to establish a tissue culture protocol for select commercially important tomato cultivars preceded by wide screening of their regeneration potential. According to numerous data outlined in this review, the in vitro culture of tomatoes has been successfully used in different biotechnological applications. It should be pointed out that different genotypes of tomato are characterized by diverse morphogenic potential, and unfortunately there are some reports describing their partial recalcitrance or total inability to respond to in vitro cultures. Therefore, improvement of existing regeneration protocols is still required. Despite various difficulties, currently a procedure of successful stable*Agrobacterium*-mediated transformation of tomato plants has been achieved. In the light of the numerous data presented here, genetic engineering has opened amazing opportunities for tomato plant improvement. So far, transgenic tomato lines have been generated with enhanced resistance for wide range of stresses, including abiotic and biotic ones. This has become possible through the overexpression several genes or TFs. Additionally, understanding the underlying physiological process in response to different stresses could help in determining what promoter or TFs would be appropriate to use for transformation. It should be pointed out that constantly expending knowledge regarding the physiological and genetics basis of stress tolerance, along with genetic transformation technologies, could allow for essential progress in the development of tomato cultivars with improving stress tolerance. Moreover, using GM technology, researchers are able to obtain tomato fruits with improved nutritional and organoleptic values. Finally, the credibility of the use of tomatoes in molecular farming has been proven beyond all doubt. Although promising achievements in tomato engineering, the culture of GM tomato face serious problems in most leading producer countries. The cultivation of GM tomatoes was stopped in the USA in 2002, so only China remains a producer of GM tomatoes. The main reason for this seems to be a negative opinion of the public towards GM plants. There is a general belief that GM crops are harmful for human health as well as the environment. Therefore, one of the tasks of the scientific community is not only the production of GM crops, but also educate the public about the benefits they bring to us. It should be pointed out that broad research has provided no evidence that transgenic crops cause a greater risk to human or animal health than stereotyped crops. The Federal Office of Consumer Protection and Food Safety of Germany and partners published the BEETLE (Biological and Ecological Evaluation towards Long-term Effects) report to provide scientific data (reviewed over 100 publications) to the European Commission (FOCPFS 2009). The BEETLE report gave clear evidence that, so far, no adverse effect to human health from eating GM plants have been found. Furthermore, although unexpected harmful effects are known, none have appeared in GM plants. Additionally, to convince consumers about GM plants, the use of marker-free transgenic plants (e.g. deprived of resistance of herbicides or antibiotic) could be a good argument. The continuously expanding knowledge of genomics of tomatoes' wild relative species, including knowledge about e.g. introgression of genetic information from related species into cultivated tomato, would significantly limit the risk of harmful effects on human or animal health or on the environment.

Although GM tomatoes are promising for improving the quality of human life, their potential has been seldom validated in field trials. Such trials as well as BEETLE report have to be expanded and their results have to be provided to society in order to raise awareness. It is only if the safety of GM crops and the benefits they bring to breeders and consumers, that biotechnology-derived plants will contribute to the success of their development.

REFERENCES

1. Adato A, Mandel T, Mintz-Oron S, Venger I, Levy D, Yativ M, Domınguez E, Wang Z, De Vos RC, Jetter R, Schreiber L, Heredia A, Rogachev I, Aharoni A (2009) Fruit-surface flavonoid accumulation in tomato is controlled by a SlMYB12-regulated transcriptional network. PLoS Genet 512:e1000777. doi:10.1371/journal.pgen.1000777
2. Afroz A, Chaudry Z, Rashid U, Khan MR, Ghulam MA (2010) Enhanced regeneration in explants of tomato (*Lycopersicon esculentum* L.) with the treatment of coconut water. Afr J Biotechnol 24:3634–3644. doi:10.5897/AJB2010.000-3228
3. Ajenifujah-Solebo SOA, Isu NA, Olorode O, Ingelbrecht I, Abiade OO (2012) Tissue culture regeneration of three Nigerian cultivars of tomatoes. Afr J Plant Sci 14:370–375. doi:10.5539/sar.v2n3p58
4. Alćazar R, Altabella T, Marco F, Bortolotti C, Reymond M, Koncz C, Tiburcio AF (2010) Polyamines: molecules with regulatory functions in plant abiotic stress tolerance. Planta 231:1237–1249. doi:10.10007/s00425-010-1130-0PubMed
5. Ali AA, Yossef TR, El-Banna A (2012) Cytokinin-cytokinin interaction ameliorates the callus induction and plant regeneration of tomato (*Solanum lycopersicon* Mill.). Acta Agron Hung 1:47–55. doi:10.1556/AAgr.60.2012.1.6
6. Alvarez ML, Cardineau GA (2010) Prevention of bubonic and pneumonic plague using plant-derived vaccines. Biotechnol Adv 28:184–196. doi:10.1016/j.biotechadv.2009.11.006PubMed
7. Alvarez ML, Pinyerd HL, Crisantes JD, Rigano MM, Pinkhasov J, Walmsley AM, Mason HS, Cardineau GA (2006) Plant-made subunit vaccine against pneumonic and bubonic plague is orally immunogenic in mice. Vaccine 24:2477–2490. doi:10.1016/j.vaccine.2005.12.057PubMed
8. Álvarez-Viveros MF, Inostroza-Blancheteau C, Timmermann T, González M, Arce-Johnson P (2013) Overexpression of *GlyI* and *GlyII* genes in transgenic tomato (*Solanum lycopersicum* Mill.) plants confers salt tolerance by decreasing oxidative stress. Mol Biol Rep 4:3281–3290. doi:10.1007/s11033-012-2403-4
9. Apel W, Bock R (2009) Enhancement of carotenoid biosynthesis in transplastomic tomatoes by induced lycopene-to-provitamin A conversion. Plant Physiol 1:59–66. doi:10.1104/pp.109.140533
10. Ashakiran K, Sivankalyani V, Jayanthi M, Govindasamy V, Girija S (2011) Genotype specific shoots regeneration form different explants of tomato (*Solanum lycopersicum* L.) using TDZ. Asian J Plant Sci Res 2:107–113
11. Baesi M, Nabati Ahmadi D, Rajabi-Memari H, Siahpoosh MR, Abdollahi MR, Jaberolansar N (2011) Cloning and transformation of hepatitis B surface antigen (HBsAg) gene to tomato (*Lycopersicon esculentum* Mill.). Jundishapur J Nat Pharm Prod 1:32–41
12. Bahurpe JV, Patil SC, Pawar BD, Chimote VP, Kale AA (2013) Callus induction and plantlet regeneration in tomato (*Solanum lycopersicum* L.). J Cell Tissue Res 2:3765–3768
13. Bai Y, Lindhout P (2007) Domestication and breeding of tomatoes: what have we gained and what can we gain in the future? Ann Bot 100:1085–1094. doi:10.1093/aob/mcm150PubMedCentralPubMed
14. Bailey LB (2010) Folate in health and disease, second edition. CRC Press Taylor & Francis group 6000 Broken Sound Parkway NW, Suite 300 Boca Rarton, FL 33487-2742
15. Ballester AR, Molthoff J, de Vos R, te Lintel Hekkert B, Orzaez D, Fernandez-Moreno JP, Tripodi P, Grandillo S, Martin C, Heldens J, Ykema M, Granell A, Bovy A (2010) Biochemical and molecular analysis of pink tomatoes: deregulated expression of the gene encoding transcription factor SlMYB12 leads to pink tomato fruit colour. Plant Physiol 1:71–84. doi:10.1104/pp.109.147322
16. Barabasz A, Wilkowska A, Ruszczyńska A, Bulska E, Hanikenne M, Czarny M, Krämer U, Antosiewicz

DM (2012) Metal response of transgenic tomato plants expressing P1B-ATPase. Physiol Plant 145:315–331. doi:10.1111/j.1399-3054.2012.01584.xPubMed

17. Bartoszewski G, Niedziela A, Szwacka M, Niemirowicz-Szczyt K (2003) Modification of tomato taste in transgenic plants carrying a thaumatine gene from *Thaumatococcus daniellii* benth. Plant Breed 4:347–351. doi:10.1046/j.1439-0523.2003.00864.x

18. Bassa C, Mila I, Bouzayen M, Audran-Delalande C (2012) Phenotypes associated with down-regulation of Sl-IAA27 support functional diversity among Aux/IAA family members in tomato. Plant Cell Physiol 9:1583–1595. doi:10.1093/pcp/pcs101

19. Bassolino L, Zhang Y, Schoonbeek HJ, Kiferle C, Perata P, Martin C (2013) Accumulation of anthocyanins in tomato skin extends shelf life. New Phytol 3:650–655. doi:10.1111/nph.12524

20. Bhaskaran S, Savithramma DL (2011) Co-expression of *Pennisetum glaucum* vacuolar Na+/H+ antiporter and *Arabidopsis* H+ -pyrophosphatase enhances salt tolerance in transgenic tomato. J Exp Bot 15:5561–5570. doi:10.1093/jxb/err237

21. Bhatia P, Ashwath N (2005) Effect of duration of light: dark cycles on in vitro shoot regeneration of tomato. Asian J Plant Sci 3:255–260. doi:10.3923/ajps.2005.255.260

22. Bhatia P, Ashwath N (2008) Improving the quality of in vitro cultured shoots of tomato (*Lycopersicon esculentum* Mill.) cv. Red Coat. Biotechnology 2:188–193. doi:10.3923/biotech.2008.188.193

23. Biswas SK, Pandey NK, Rajik M (2012) Inductions of defense response in tomato against Fusarium Wilt through inorganic chemicals as inducers. J Plant Pathol Microbiol 3:128. doi:10.4172/2157-7471.1000128

24. Brummell AA, Harpster MH, Civello PC, Palys JM, Bennett AB, Dunsmuira P (1999) Modification of expansin protein abundance in tomato fruit alters softening and cell wall polymer metabolism during ripening. Plant Cell 11:2203–2216. doi:10.1105/tpc.11.11.2203PubMedCentralPubMed

25. Butelli E, Titta L, Giorgio M, Mock HP, Matros A, Peterek S, Schijlen EGM, Hall RD, Bovy AG, Luo J, Martin C (2008) Enrichment of tomato fruit with health-promoting anthocyanins by expression of select transcription factors. Nat Biotechnol 26:1301–1308. doi:10.1038/nbt.1506PubMed

26. Chaudry A, Abbas S, Yasmin A, Rashid H, Ahmed H, Anjum MA (2010) Tissue culture studies in tomato (*Lycopesricon esculentum*) var. Moneymaker. Pak J Bot 1:155–163

27. Chen SC, Liu AR, Wang FH, Ahammed G (2009a) Combined overexpression of chitinase and defensine genes transgenic tomato enhances resistance to *Botrytis cinerea*. Afr J Biotechnol 20:5182–5188. doi:10.5897/AJB09.704

28. Chen Y, Wang A, Zhao L, Shen G, Cui L, Tang K (2009b) Expression of thymosin $\alpha 1$ concatemer in transgenic tomato (*Solanum lycopersicum*) fruits. Biotechnol Appl Biochem 52:303–312. doi:10.1042/BA20080054PubMed

29. Cheng L, Zou Y, Ding S, Zhang J, Yu X, Cao J, Lu G (2009) Polyamine accumulation in transgenic tomato enhances the tolerance to high temperature stress. J Integr Plant Biol 5:489–499. doi:10.1111/j.1744-7909.2009.00816.x

30. Chetty VJ, Ceballos N, Garcia D, Narvaez-Vasquez J, Lopez W, Orozco-Cardenas ML (2013) Evaluation of four *Agrobacterium tumefaciens* strains for the genetic transformation of tomato (*Solanum lycopersicum* L.) cultivar Micro-Tom. Plant Cell Rep 32:239–247. doi:10.1007/s00299-012-1358-1PubMed

31. Colliver S, Bovy A, Collins G, Muir S, Robinson S, de Vos CHR, Verhoeyen ME (2002) Improving the nutritional content of tomatoes through reprogramming their flavonoid biosynthetic pathway. Phytochem Rev 1:113–123

32. Cong B, Tanksley SD (2006) FW2.2 and cell cycle control in developing tomato fruit: a possible example of gene co-option in the evolution of a novel organ. Plant Mol Biol 62:867–880. doi:10.1007/s11103-006-9062-6PubMed

33. Cueno ME, Hibi Y, Karamatsu K, Yasutomi Y, Imai K, Laurena AC, Okamoto T (2010) Preferential expression and immunogenicity of HIV-1 Tat fusion protein expressed in tomato plant. Transgenic Res 5:889–895. doi:10.1007/s11248-009-9358-9

34. D'Ambrosio C, Stigliani AL, Giorio G (2011) Overexpression of CrtR-b2 (carotene beta hydroxylase 2) from *S. lycopersicum* L. differentially affects xanthophylls synthesis and accumulation in transgenic tomato plants. Transgenic Res 20:47–60. doi:10.1007/s11248-010-9387-4PubMed

35. Dai CX, Mertz D, Lambeth VN (1988) Effect of seedling age, orientation and genotype of hypocotyl and cotyledon explants of tomato on shoot and root regeneration. Genet Manip Crops Newslett 4:26–35. doi:10.3103/S1068367413030178

36. Davidovich-Rikanati R, Sitrit Y, Tadmor Y, Iijima Y, Bilenko N, Bar E, Carmona B, Fallik E, Dudai NE, Simon JE, Pichersky E, Lewinsohn E (2007) Enrichment of tomato flavour by diversion of the early plastidial terpenoid pathway. Nat Biotechnol 25:899–901. doi:10.1038/nbt1312PubMed

37. Davuluri GR, van Tuinen A, Fraser PD, Manfredonia A, Newman R, Burgess D, Brummell DA, King SR, Palys J, Uhlig J, Bramley PM, Pennings HM, Bowler C (2005) Fruit-specific RNAi-mediated suppression of *DET1* enhances carotenoid and flavonoid content in tomatoes. Nat Biotechnol 7:890–895. doi:10.1038/nbt1108

38. de Jong M, Wolters-Arts M, Garcıa-Martınez JL, Mariani C, Vriezen WH (2011) The*Solanum lycopersicum* AUXIN RESPONSE FACTOR 7 (SlARF7) mediates cross-talk between auxin and gibberellins signalling during tomato fruit set and development. J Exp Bot 2:617–626. doi:10.1093/jxb/erq293

39. de la Garza RID, Quinlivan PE, Klaus SMJ, Basset GJC, Gregory JF, Hanson AD (2004) Folate biofortification in tomatoes by engineering the pteridine branch of folate

synthesis. Proc Natl Acad Sci USA 38:13720–13725. doi:10.1073/pnas.0404208101

40. de la Garza RID, Gregory JF, Hanson AD (2007) Folate biofortification of tomato fruit. Proc Natl Acad Sci USA 10:4218–4222. doi:10.1073/pnas.0700409104
41. Dharmapuria S, Rosatia C, Pallara P, Aquilani R, Bouvier F, Camara B, Giuliano G (2002) Metabolic engineering of xanthophyll content in tomato fruits. FEBS Lett 519:30–34. doi:10.1016/S0014-5793(02)02699-6
42. Elías-López AL, Marquina B, Gutiérrez-Ortega A, Aguilar D, Gomez-Lim M, Hernández-Pando R (2008) Transgenic tomato expressing interleukin-12 has a therapeutic effect in a murine model of progressive pulmonary tuberculosis. Clin Exp Immunol 154:123–133. doi:10.1111/j.1365-2249.2008.03723.xPubMedCentralPubMed
43. El-Siddig MA, El-Hussein AA, Saker MM (2011) *Agrobacterium*-mediated transformation of tomato plants expressing defensin gene. Int J Agric Res 4:323–334. doi:10.3923/ijar.2011.323.334
44. Enfissi EMA, Fraser PD, Lois LM, Boronat A, Schuch W, Bramley PM (2005) Metabolic engineering of the mevalonate and nonmevalonate isopentenyl diphosphate-forming pathways for the production of health-promoting isoprenoids in tomato. Plant Biotechnol J 3:17–27. doi:10.1111/j.1467-7652.2004.00091.xPubMed
45. FAOSTAT (2011). http://faostat3.fao.org/faostat-gateway/go/to/download/Q/QC/E
46. Federal Office of Consumer Protection and Food Safety (German) and Partners (2009) Long-term effects of genetically modified (GM) crops on helath and the environment (including biodiversity): Prioritisation of potential risks and delimitation of uncertainties. Federal Office of Consumer of Food Safety, Berlin. http://bch.cbd.int/database/record-v4-sthtml?documentid=101007
47. Fernandez-Moreno JP, Orzaez D, Granell A (2013) VIGS: a tool to study fruit development in *Solanum lycopersicum*. Methods Mol Biol 975:183–196. doi:10.1007/978-1-62703-278-0_14PubMed
48. Foolad MR (2007) Genome mapping and molecular breeding of tomato. Int J Plant Genomics 64358:52. doi:10.1155/2007/64358
49. Fraser PD, Romer S, Shipton CA, Mills PB, Kiano JW, Misawa N, Drake RG, Schuch W, Bramley PM (2002) Evaluation of transgenic tomato plants expressing an additional phytoene synthase in a fruit-specific manner. Proc Natl Acad Sci USA 2:1092–1109. doi:10.1073/pnas.241374598
50. Fraser PD, Enfissi EMA, Halket JM, Truesdale MR, Yu D, Gerrish C, Bramleya PM (2007) Manipulation of phytoene levels in tomato fruit: effects on isoprenoids, plastids, and intermediary metabolism. Plant Cell 19:3194–3211. doi:10.1105/tpc.106.049817PubMedCentralPubMed
51. Fuentes AD, Ramos PL, Sanchez Y, Callard D, Ferreira A, Tiel K, Cobas K, Rodriguez R, Borroto C, Doreste V, Pujol M (2008) A transformation procedure for recalcitrant tomato by addressing transgenic plant-recovery limiting factors. Biotechnol J 3:1088–1093. doi:10.1002/biot.200700187PubMed
52. Fukkuda-Parr S (2012) The green revolution: GM crops and unequal development. UK Bath Press, Bath
53. Gamborg OL, Miller RA, Ojima K (1968) Nutrient requirements of suspension cultures of soybean root cells. Exp Cell Res 50:151–158. doi:10.1016/0014-4827(68)90403-5PubMed
54. Garchery C, Gest N, Do PT, Alhagdow M, Baldet P, Menard G, Rothan C, Massot C, Gautier H, Aarrouf J, Fernie AR, Stevens R (2013) A diminution in ascorbate oxidase activity affects carbon allocation and improves yield in tomato under water deficyt. Plant Cell Environ 36:159–175. doi:10.1111/j.1365-3040.2012.02564.xPubMed
55. Garcia V, Stevens R, Gil L, Gilbert L, Gest N, Petit J, Faurobert M, Maucourt M, Deborde C, Moing A, Poessel JL, Jacob D, Bouchet JP, Giraudel JL, Gouble B, Page D, Alhagdow M, Massot C, Gautier H, Lemaire-Chamley M, Rolin D, Usadel B, Lahaye M, Causse M, Baldet P, Rothan C (2009) An integrative genomics approach for deciphering the complex interactions between ascorbate metabolism and fruit growth and composition in tomato. C R Biol 11:1007–1021. doi:10.1016/j.crvi.2009.09.013
56. Garcia-Hurtado N, Carrera E, Ruiz-Rivero O, López-Gresa MP, Hedden P, Gong F, García-Martínez JL (2012) The characterization of transgenic tomato overexpressing *gibberellin 20-oxidase* reveals induction of parthenocarpic fruit growth, higher yield, and alteration of the gibberellins biosynthetic pathway. J Exp Bot 16:5803–5813. doi:10.1093/jxb/ers229
57. Gilbert L, Alhagdow M, Nunes-Nesi A, Quemener B, Guillon F, Bouchet B, Faurobert M, Gouble B, Page D, Garcia V, Peti J, Stevens R, Causse M, Fernie AR, Lahaye M, Rothan C, Baldet P (2009) GDP-d-mannose 3,5-epimerase (GME) plays a key role at the intersection of ascorbate and non-cellulosic cell-wall biosynthesis in tomato. Plant J 3:499–508. doi:10.1111/j.1365-313X.2009.03972.x
58. Giliberto L, Perrotta G, Pallara P, Weller JL, Fraser PD, Bramley PM, Fiore A, Tavazza M, Giuliano G (2005) Manipulation of the blue light photoreceptor cryptochrome 2 in tomato affects vegetative development, flowering time, and fruit antioxidant content. Plant Physiol 137:199–208. doi:10.1111/j.1365-313X.2009.03972.xPubMedCentralPubMed
59. Gill SS, Tuteja N (2010) Reactive oxygen species and antioxidant machinery in abiotic stress tolerance in crop plants. Plant Physiol Biochem 48:909–930. doi:10.1016/j.plphy.2010.08.016PubMed
60. Głowacka B (2004) Influence of light colour on micropropagation of tomato (*Lycopersicon esculentum* Mill.). Biotechnologia 2:168–175
61. Godishala V, Mangamoori L, Nanna R (2011) Plant regeneration via somatic embryogenesis in cultivated tomato (*Solanum lycopersicum* L.). J Cell Tissue Res 1:2521–2528
62. Goel D, Singh AK, Yadav V, Babbar SB, Murata N, Bansal KC (2011) Transformation of tomato with a

bacterial *coda* gene enhances tolerance to salt and water stresses. J Plant Physiol 11:1286–1294. doi:10.1016/j.jplph.2011.01.010

63. Goetz M, Hooper LC, Johnson SD, Rodrigues JC, Vivian-Smith A, Koltunov AM (2007) Expression of aberrant forms of auxin response factor 8 stimulates parthenocarpy in*Arabidopsis* and tomato. Plant Physiol 2:336–351. doi:10.1104/pp.107.104174
64. Guan ZJ, Guo B, Huo YL, Dai JK, Wei YH (2012) Histocytological examination on organogenesis and somatic embryogenesis of HBsAg-transgenic cherry tomato mutant. Int J Exp Bot 81:51–58
65. Guo M, Zhang YL, Meng ZJ, Jiang J (2012) Optimization of factors affecting*Agrobacterium*-mediated transformation of Micro-Tom tomatoes. Genet Mol Res 1:661–671. doi:10.4238/2012
66. Hanus-Fajerska E (2006) Variation in tomato plants regenerated from Cucumber Mosaic Viruse infected tissue. ISHS Acta Hort789: XV Meeting of the EUCARPIA Tomato Working Group. http://www.actahort.org/books/789/789_40.htm
67. Harish MC, Rajeevkumar S, Sathishkumar R (2010) Efficient in vitro callus induction and regeneration of different tomato cultivars of India. Asian J Biotechnol 3:178–184. doi:10.3923/ajbkr.2010.178.184
68. Hasan M, Khan AJ, Khan S, Shah AH, Khan AR, Mirza B (2008) Transformation of tomato (*Lycopesricon esculentum* Mill.) with *Arabidopsis* early flowering gene *APETALI* (*API*) Through *Agrobacterium* infiltration of ripened fruits. Pak J Bot 1:161 173
69. Herbette S, Tourvielle de Labrouheb D, Drevetc JR, Roeckel-Dreveta P (2011) Transgenic tomatoes showing higher glutathione peroxydase antioxidant activityare more resistant to an abiotic stress but more susceptible to biotic stresses. Plant Sci 180:548–553. doi:10.1016/j.plantsci.2010.12.002PubMed
70. Hirai T, Fukukawa G, Kakuta H, Fukuda N, Ezura H (2010) Production of recombinant miraculin using transgenic tomatoes in a closed cultivation system. J Agric Food Chem 58:6096–6101. doi:10.1021/jf100414vPubMed
71. Horvath DM, Stall RE, Jones JB, Pauly MH, Vallad GV, Dahlbeck D, Staskawicz BJ, Scott JW (2012) Transgenic resistance confers effective field level control of bacterial spot disease in tomato. PLoS ONE 7(8):e42036. doi:10.1371/journal.pone.0042036PubMedCentralPubMed
72. Hsieh TH, Lee JT, Yang PT, Chiu LH, Charng YY, Wang YC, Chan MT (2002) Heterology expression of the *Arabidopsis C-Repeat/Dehydration Response Element Binding Factor 1*gene confers elevated tolerance to chilling and oxidative stresses in transgenic tomato. Plant Physiol 129:1086–1094PubMedCentralPubMed
73. Hsieh TH, Li CW, Su RC, Cheng CP, Tsai YC, Chan MT (2010) A tomato bZIP transcription factor, SlAREB, is involved in water deficit and salt stress response. Planta 231:1459–1473. doi:10.1007/s00425-010-1147-4PubMed
74. Hu DG, Wang SH, Luo H, Ma QJ, Yao YX, You CX, Hao YJ (2012) Overexpression of*MdVHA-B*, *V-ATPase* gene from apple, confers tolerance to drought in transgenic tomato. Sci Hortic 145:94–101. doi:10.1016/j.scienta.2012.08.010
75. Iijima Y, Gang DR, Fridman E, Lewinson E, Pichersky E (2004) Characterization of geraniol synthase from the peltate glands of sweet basil. Plant Physiol 134:370–379. doi:10.1104/pp.103.032946PubMedCentralPubMed
76. Ishag S, Osman MG, Khalafalla MM (2009) Effects of growth regulators and genotype on shoot regeneration in tomato (*Lycopresicon esculentum* c.v. Omdurman). Int J Sustain Crop Prod 6:7–13
77. Jabeen N, Mirza B, Chaudhary Z, Rashid H, Gulfraz M (2009) Study of the factors affecting *Agrobacterium* mediated gene transformation in tomato (*Lycopersicon esculentum* Mill.) cv. Riogrande using rice chitinase (*CHT-3*) gene. Pak J Bot 5:2605–2614
78. Jaberolansar N, Hayati J, Rajabi-Memari H, Hosseini-Tafreshi SA, Nabati-Ahmadi D (2010) Tomato and tobacco phytoene desaturase gene silencing by virus-induced gene silencing (VIGS) technique. Iran J Virol 1:7–11
79. Jehan S, Hassanein AM (2013) Hormonal requirements trigger different organogenic pathways on tomato nodal explants. Am J Plant Sci 4:2118–2125. doi:10.4236/ajps.2013.411263
80. Juárez P, Presa S, Espí J, Pineda B, Antón MT, Moreno V, Buesa J, Granell A, Orzaez D (2012) Neutralizing antibodies against rotavirus produced in transgenically labelled purple tomatoes. Plant Biotechnol J 3:341–352. doi:10.1111/j.1467-7652.2011.00666.x
81. Jung YJ (2013) Enhanced resistance to bacterial pathogen in transgenic tomato plants expressing cathelicidin antimicrobial peptide. Biotechnol Bioprocess Eng 18:615–624. doi:10.1007/s12257-013-0392-3
82. Kantor M, Sestras R, Chowudhury K (2010) Identification of the most organogenic-responsive variety of tomato using the variety X medium interaction. Rom Biotechnol Lett 5:5640–5645
83. Kantor M, Sestras R, Chowdhury K (2013) Transgenic tomato plants expressing the antigen gene PfCP-2.9 of Plasmodium falciparum. Pesquisa Agropecuária Brasileira 1:73–79. doi:10.1590/S0100-204X2013000100010
84. Kato K, Maruyama S, Hirai T, Hiwasa-Tanase K, Mizoguchi T, Goto E, Ezura H (2011) A trial of production of the plant-derived high-value protein in a plant factory. Photosynthetic photon fluxes affect the accumulationof recombinant miraculin in transgenic tomato fruits. Plant Signal Behav 8:1172–1179. doi:10.4161/psb.6.8.16373
85. Khare N, Goyary D, Singh NK, Shah P, Rathore M, Anandhan S, Sharma D, Arif M, Ahmed Z (2010) Transgenic tomato cv. Pusa Uphar expressing a bacterial mannitol-1-phosphate dehydrogenase gene confers abiotic stress tolerance. Plant Cell Tissue Organ Cult 103:267–277. doi:10.1007/s11240-010-9776-7
86. Khoudi H, Nouri-Khemakhem A, Gouiaa S, Masmoudi K (2009) Optimization of regeneration and transformation

parameters in tomato and improvement of its salinity and drought tolerance. Afr J Biotechnol 22:6068–6076. doi:10.5897/AJB09.057

87. Khuong TTH, Crété P, Robaglia C, Caffarri S (2013) Optimisation of tomato Micro-tom regeneration and selection on glufosine/Basta and dependency of gene silencing on transgene copy number. Plant Cell Rep 32:1441–1454. doi:10.1007/s00299-013-1456-8PubMed

88. Kim HS, Youma JW, Moona KB, Ha JH, Kim YH, Joung H, Jeon JH (2012) Expression analysis of human β-secretase in transgenic tomato fruits. Protein Expr Purif 82:125–131. doi:10.1016/j.pep.2011.11.012PubMed

89. Kobayashi M, Nagasaki H, Garcia V, Just D, Bres C, Mauxion JP, Paslier MCL, Brunel D, Suda K, Minakuchi Y, Toyoda A, Fujiyama A, Toyoshima H, Suzuki T, Igarashi K, Rothan C, Kaminuma E, Nakamura Y, Yano K, Aoki K (2013) Genome-wide analysis of intraspecific DNA polymorphism in 'Micro-Tom', a model cultivar of tomato (*Solanum lycopersicum*). Plant Cell Physiol 2:445–454. doi:10.1093/pcp/pct181

90. Koenig D, Jimenez-Gomez JM, Kimura S, Fulop D, Chitwood DH, Hedland LR, Kumar R, Covington MF, Devisetty UK, Tat AV, Toghe T, Bolger A, Schneeberger K, Ossowski S, Lanz Ch, Xiong G, Taylor-Teeples M, Rady SM, Pauly M, Weigel D, Usadel B, Fernie AF, Peng J, Sinnha NR, Maloof JN (2013) Comparative transcriptomics reveals patterns of selection in domesticated and wild tomato. Proc Natl Acad Sci USA 28:E2655–E2662. doi:10.1073/pnas.1309606110

91. Koleva Gudeva L, Dedejski G (2012) *In vivo* and in vitro production of some genotypes of cherry tomato *Solanum lycopersicum* var. *Cerasiforme* (DUNAL). Int J Farm Allied Sci 4: 91–96. URL: http://ijfas.com/2012-1-4/

92. Koul B, Sirivastava S, VijayAmla D, Sanyal I (2014) Establishment and optimization of*Agrobacterium*-mediated transformation and regeneration of tomato (*Solanum lycopersicum* L.) Int. J Biosci 10:51–69. doi:10.12692/ijb/4.10.51-69

93. Kurokawa N, Hirai T, Takayama M, Hiwasa-Tanase K, Ezura H (2013) An E8 promoter–HSP terminator cassette promotes the high-level accumulation of recombinant protein predominantly in transgenic tomato fruits: a case study of miraculin. Plant Cell Rep 32:529–536. doi:10.1007/s00299-013-1384-7PubMed

94. Lai L, Huang T, Wang Y, Liu Y, Zhang J, Song Y (2009) The expression of analgesic-antitumor peptide (*AGAP*) from Chinese *Buthus martensii* Karsch in transgenic tobacco and tomato. Mol Biol Rep 36:1033–1039. doi:10.1007/s11033-008-9277-5PubMed

95. Lee TJ, Coyne DP, Clemente TE, Mitra A (2002) Partial resistance to bacterial wilt in transgenic tomato plants expressing antibacterial *lactoferrin* gene. J Am Soc Hortic Sci 2:150–164

96. Li T, Sun JK, Lu ZH, Liu Q (2011) Transformation of HBsAg (Hepatitis B Surface Antigen) gene into tomato mediated by *Agrobacterium tumefaciens*. Czech J Genet Plant Breed 2:69–77

97. Li C, Yan JM, Li YZ, Zhang ZC, Wang QL, Liang Y (2013) Silencing the *SpMPK1*, *SpMPK2*, and *SpMPK3* genes in tomato reduces abscisic acid—mediated drought tolerance. Int J Mol Sci 14:21983–21996. doi:10.3390/ijms141121983PubMedCentralPubMed

98. Liu J, Cong B, Tanksley SD (2003) Generation and analysis of an artificial gene dosage series in tomato to study the mechanism by which the cloned quantitative trait locus fw2.2 controls fruit size. Plant Physiol 1:292–299. doi:10.1104/pp.102.018143

99. Liu Y, Roof S, Ye Z, Barry C, van Tuinen A, Vrebalov J, Bowler C, Giovannoni J (2004) Manipulation of light signal transduction as a means of modifying fruit nutritional quality in tomato. Proc Natl Acad Sci USA 26:9897–9902. doi:10.1073/pnas.0400935101

100. Lou XM, Yao QH, Zhang Z, Peng RH, Xiong AS, Wang HK (2007) Expression of the human hepatitis B virus large surface antigen gene in transgenic tomato plants. Clin Vaccine Immunol 4:464–469. doi:10.1128/CVI.00321-06

101. Lozano R, Gimenez E, Cara B, Capel J, Angosto T (2009) Genetic analysis of reproductive development in tomato. Int J Dev Biol 53:1635–1648. doi:10.1387/ijdb.072440rlPubMed

102. Ma H, Song C, Borth W, Sether D, Melzer M, Hu J (2011) Modified expression of alternative oxidase in transgenic tomato and petunia affects the level of tomato spotted wilt virus resistance. BMC Biotechnol 11:96. doi:10.1186/1472-6750-11-96PubMedCentralPubMed

103. Maligeppagol M, Chandra GS, Prakash M, Navale PM, Deepa H, Rajeev PR, Asokan R, Babu KP, Babu CCB, Rao VK, Kumar KNK (2013) Anthocyanin enrichment of tomato (*Solanum lycopersicum* L.) fruit by metabolic engineering. Curr Sci 1:72–80

104. Mamidala P, Nanna RS (2011) Effect of genotype, explants source and medium on in vitro regeneration of tomato. Int J Genet Mol Biol 3:45–50

105. Marti E, Gisbert C, Bishop GJ, Dixon MS, Garcia-Martinez JL (2006) Genetic and physiological characterization of tomato cv. Micro-Tom. J Exp Bot 9:2037–2047. doi:10.1093/jxb/erj154

106. Mathieu S, Dal Cin V, Fei Z, Li H, Bliss P, Taylor MG, Klee HJ, Tieman DM (2009) Flavour compounds in tomato fruits: identification of loci and potential pathways affecting volatile composition. J Exp Bot 1:325–337. doi:10.1093/jxb/ern294

107. McCormick S, Niedermeyer J, Fry J, Barnason A, Horsch R, Fraley R (1986) Leaf disc transformation of cultivated tomato (*L. esculentum*) using *Agrobacterium tumefaciens*. Plant Cell Rep 2:81–84

108. Mensuali-Sodi A, Panizza M, Tognoni F (1995) Endogenous ethylene requirement for adventitious root induction and growth in tomato cotyledons and lavandin microcuttings in vitro. Plant Growth Regul 17:205–212. doi:10.1007/BF00024727

109. Minoia S, Petrozza A, D'Onofrio O, Piron F, Mosca G, Sozio G, Cellini F, Bendahmane A, Carriero F (2010) A new mutant genetic resource for tomato crop improvement

by TILLING technology. BMC Res Notes 3:69–76. doi:10.1186/1756-0500-3-69PubMedCentralPubMed

110. Mishra KB, Iannacone R, Petrozza A, Mishra A, Armentano N, La Vecchia G, Trtilek M, Cellini F, Nedbal L (2012) Engineering drought tolerance in tomato plants is reflected in chlorophyll fluorescence emission. Plant Sci 182:79–86. doi:10.1016/j.phytochem.2012.09.007PubMed

111. Morgan MJ, Osorio S, Gehl B, Baxter CJ, Kruger NJ, Ratcliffe RG, Fernie AR, Sweetlove LJ (2013) Metabolic engineering of tomato fruit organic acid content guided by biochemical analysis of an introgression line. Plant Physiol 1:397–407. doi:10.1104/pp.112.209619

112. Muir SR, Collins GJ, Robinson S, Hughes SG, Bovy AG, de Vos CH, van Tunen AJ, Verhoyen ME (2001) Overexpression of petunia chalcone isomerase in tomato results in fruit containing increased levels of flavonols. Nat Biotechnol 5:470–474. doi:10.1038/88150

113. Muñoz-Mayor A, Pineda B, Garcia-Abellán JO, Antón T, Garcia-Sogo B, Sanchez-Bel P, Flores FB, Atarés A, Angosto T, Pintor-Toro JA, Moreno V, Bolarin MC (2012) Overexpression of dehydrin *tas14* gene improves the osmotic stress imposed by drought and salinity in tomato. J Plant Physiol 169:459–468. doi:10.1016/j.jplph.2011.11.018PubMed

114. Murashige T, Skoog F (1962) A revised medium for rapid growth and bioassays with tobacco tissue cultures. Physiol Plant 15:473–497. doi:10.1111/j.1399-3054.1962.tb08052.x

115. Namitha KK, Negi P (2013) Morphogenetic potential of tomato (*Lycopersicon esculentum,*) cv. Arka Ahuti to plant growth regulators. Notulae Scientia Biologicae 2:220–225

116. Neily MH, Matsukura Ch, Maucourt M, Bernillon S, Deborde C, Moing A, Yin YG, Saito T, Mori K, Asamizua E, Rolin D, Moriguchi T, Ezura H (2011) Enhanced polyamine accumulation alters carotenoid metabolism at the transcriptional level in tomato fruit over-expressing spermidine synthase. J Plant Physiol 168:242–252. doi:10.1016/j.jplph.2010.07.003PubMed

117. Orzaez D, Granell A (2009) Reverse genetics and transient gene expression in fleshy fruits. Plant Signal Behav 9:864–867. doi:10.1104/pp.109.13900

118. Orzaez D, Mirabel S, Wieland WH, Granell A (2006) Agroinjection of tomato fruits. A tool for rapid functional analysis of transgenes directly in fruit. Plant Physiol 1:3–11. doi:10.1104/pp.105.068221

119. Orzaez D, Medina A, Torre S, Fernandez-Moreno JP, Rambla JL, Fernandez-del-Carmen A, Butelli E, Martin C, Granell A (2009) A visual reporter system for virus-induced gene silencing in tomato fruit based on anthocyanin accumulation. Plant Physiol 3:1122–1134. doi:10.1104/pp.109.139006

120. Paduchuri P, Gohokar S, Thamke B, Subhas M (2010) Transgenic tomatoes. Int J Adv Biotechnol Res 2:69–72. http://www.bipublication.com

121. Pandey SK, Nookaraju A, Upadhayaya CP, Gururani MA, Venkatesh J, Kim DH, Park SW (2011) An update biotechnological approaches for improving abiotic tolerance stress in tomato. Crop Sci 51:1–22. doi:10.2135/cropsci2010.10.0579

122. Park S, Jinsheng L, Pittman JK, Berkowitz GA, Yang H, Undurraga S, Morris J, Hirsch KD, Gaxiola RA (2005) Up-regulation of H^+-pyrophosphatase (H^+-PPase) as a strategy to engineer drought-resistant crop plants. Proc Natl Acad Sci USA 52:18830–18835. doi:10.1073/pnas.0509512102

123. Patade VY, Khatri D, Kumari M, Grover A, Gupta SM, Ahmed Z (2013) Cold tolerance in Osmotin transgenic tomato (*Solanum lycopersicum L.*) is associated with modulation in transcript abundance of stress responsive genes. SpringerPlus 2:117. doi:10.1186/2193-1801-2-117PubMedCentralPubMed

124. Perlata IE, Spooner DM (2007) History, origin and early cultivation of tomato (Solanaceae). In: Rozdan MK, Matto AK (eds) Genetic improvement of solanaceous crops: tomato, vol 2. Science Publishers, Enfield, NH, pp 1–27

125. Plana D, Fuentes A, Alvarez M, Lara RM, Alvarez F, Pujol M (2006) A new approach for in vitro regeneration of tomato plants devoid of exogenous plant growth hormones. Biotechnol J 1:1153–1157. doi:10.1002/biot.200500042PubMed

126. Rai AC, Singh M, Shah K (2013) Engineering drought tolerant tomato plants over-expressing Bc*ZAT12* gene encoding a C_2H_2 zinc finger transcription factor. Phytochemistry 85:44–50. doi:10.1016/j.phytochem.2012.09.007PubMed

127. Raiola A, Rigano MM, Calafiore R, Frusciante L, Barone A (2014). Enhancing the human-promoting effects of tomato fruit for bofortified food. Hindawi Publishing Corporation Mediators of Inflammation. doi:10.1155/2014/139873

128. Ramirez YJP, Tasciotti E, Gutierrez-Ortega A, Donayre Torres AJ, Olivera Flores MT, Giacca M, Gomez Lim MA (2007) Fruit-specific expression of the human immunodeficiency virus type 1 tat gene in tomato plants and its immunogenic potential in mice. Clin Vaccine Immunol 6:685–692. doi:10.1128/CVI.00028-07

129. Rashid R, Bal SS (2010) Effect of hormones on direct shoot regeneration in hypocotyl explants of tomato. Notulae Scientia Biologicae 1:70–73

130. Rashid R, Bal SS (2011) *Agrobacterium* -mediated genetic transformation of tomato (*Solanum lycopesricum L)* with *Cry1Ac* gene for resistance against fruit borer. J Trop Agric 49(1–2):110–113

131. Romero I, Tikunov Y, Bovy A (2011) Virus-induced gene silencing in detached tomatoes and biochemical effects of phytoene desaturase gene silencing. J Plant Physiol 168:1129–1135. doi:10.1016/j.jplph.2010.12.020PubMed

132. Ruf S, Hermann M, Berger IJ, Carrer H, Bock R (2001) Stable genetic transformation of tomato plastids and expression of a foreign protein in fruit. Nat Biotechnol 9:870–875. doi:10.1038/nbt0901-870

133. Saito T, Ariizumi T, Okabe Y, Asamizu E, Hiwasa-Tanase K, Fukuda N, Mizoguchi T, Yamazaki Y, Aoki K, Ezura H (2011) TOMATOMA: a novel tomato mutant database

distributing Micro-Tom mutant collections. Plant Cell Physiol 2:283–296. doi:10.1093/pcp/pcr004

134. Saker MM, Hussein HA, Osman NH, Soliman MH (2008) *In vitro* production of transgenic tomatoes expressing defensin gene using newly developed regeneration and transformation system. Arab J Biotechnol 1:59–70

135. Saker MM, Salama HS, Salama M, El-Banna A, AbdelGhany NM (2011) Production of transgenic tomato plants expressing *Cry* 2Ab gene for the control of some lepidopterous insects endemic in Egypt. J Genet Eng Biotechnol 9:149–155. doi:10.1016/j.jgeb.2011.08.001

136. Schijlen E, de Vos CHR, Jonker H, van den Broeck H, Molthoff J, van Tunen A, Martens S, Bovy A (2006) Pathway engineering for healthy phytochemicals leading to the production of novel flavonoids in tomato fruit. Plant Biotechnol J 4:433–444. doi:10.1111 /j.1467-7652.2006.00192.xPubMed

137. Schreiber G, Reuveni M, Evenor D, Oren-Shamir M, Ovadia R, Sapir-Mir M, Bootbool-Man A, Nahon S, Shlomo H, Chen L, Levin I (2012) ANTHOCYANIN1 from *Solanum chilense* is more efficient in accumulating anthocyanin metabolites than its *Solanum lycopersicum* counterpart in association with the ANTHOCYANIN FRUIT phenotype of tomato. Theor Appl Genet 124:295–307. doi:10.1007/s00122-011-1705-6PubMed

138. Shah MR, Mukherjee PK, Eapen S (2010) Expression of a fungal endochitinase gene in transgenic tomato and tobacco results in enhanced tolerance to fungal pathogens. Physiol Mol Biol Plants 1:39–51. doi:10.1007/s12298-010-0006-x

139. Sharma MK, Solanke AU, Jani D, Singh Y, Sharma AK (2009) A simple and efficient*Agrobacterium*-mediated procedure for transformation of tomato. J Biosci 3:423–433

140. Sharma P, Jha AB, Dubey RS, Pessarakli M (2012) Reactive oxygen species, oxidative damage, and antioxidative defense mechanism in plants under stressful conditions. J Bot 2012:1–26. doi:10.1155/2012/217037

141. Sherkar HD, Chavan AM (2014) Studies on callus induction and shoot regeneration in tomato. Sci Res Rep 1:89–93

142. Shinozaki K, Yamaguchi-Shinozaki K (2007) Gene networks involved in drought stress response and tolerance. J Exp Bot 58:221–227. doi:10.1093/jxb/erl164PubMed

143. Simkin AJ, Gaffe J, Alcaraz JP, Carde JP, Bramley PM, Fraser PD, Kuntz M (2007) Fibrillin influence on plastid ultrastructure and pigment content in tomato fruit. Phytochemistry 68:1545–1556. doi:10.1016/j.phytochem.2007.03.014PubMed

144. Singh S, Rathore M, Goyar D, Singh RK, Anandhan S, Sharma DK, Ahmed Z (2011) Induced ectopic expression of At-CBF1 in marker-free transgenic tomatoes confers enhanced chilling tolerance. Plant Cell Rep 30:1019–1028. doi:10.1007/s00299-011-1007-0PubMed

145. Smith DL, Abbott AA, Gross KC (2002) Down-regulation of tomato β-galactosidase 4 results in decreased fruit softening. Plant Physiol 4:1755–1762. doi:10.1104/pp.011025

146. Soria-Guerra RE, Rosales-Mendoza S, Marquez-Mercado C, Lopez-Revilla R, Castillo-Collazo R, Alpuche-Solıs AG (2007) Transgenic tomatoes express an antigenic polypeptide containing epitopes of the diphtheria, pertussis and tetanus exotoxins, encoded by a synthetic gene. Plant Cell Rep 26:961–968. doi:10.1007/s00299-007-0306-yPubMed

147. Soria-Guerra RE, Rosales-Mendoza S, Moreno-Fierros L, Lopez-Revilla R, Alpuche-Solıs AG (2011) Oral immunogenicity of tomato-derived sDPT polypeptide containing*Corynebacterium diphtheriae*, *Bordetella pertussis* and *Clostridium tetani* exotoxin epitopes. Plant Cell Rep 30:417–424. doi:10.1007/s00299-010-0973-yPubMed

148. Spolaroe S, Trainotti L, Casadoro G (2001) A simple protocol for transient gene expression in ripe fleshy fruit mediated by *Agrobacterium*. J Exp Bot 357:845–850. doi:10.1093/jexbot/52.357.845

149. The Tomato Genome Consortium (2012) The tomato genome sequence provides insights into fleshy fruit evolution. Nature 485:635–641. doi:10.1038/nature11119

150. Tieman DM, Zeigler M, Schmelz EA, Taylor MG, Bliss P, Kirst M, Klee HJ (2006) Identification of loci affecting flavour volatile emissions in tomato fruits. J Exp Bot 4:887–896. doi:10.1093/jxb/erj074

151. Tyburski J, Tretyn A (1999) Organogenetic response of photomorphogenic mutants of tomato. J Plant Physiol 155:568–575. doi:10.1016/S0176-1617(99)80056-X

152. Velcheva M, Faltin Z, Flaishman M, Eshdat Y, Perl A (2005) A liquid culture system for Agrobacterium-mediated transformation of tomato (*Lycopersicon esculentum L.* Mill.). Plant Sci 168:121–130. doi:10.1016/j.plantsci.2004.07.037

153. Vu T, Choudhury NR, Mukherjee SK (2013) Transgenic tomato plants expressing artificial microRNAs for silencing the pre-coat and coat proteins of a begomovirus, Tomato leaf curl New Delhi virus, show tolerance to virus infection. Virus Res 172:35–45. doi:10.1016/j.virusres.2012.12.008PubMed

154. Waller JC, Akhtar TA, Lara-Nunez A, Gregory JF, McQuinn RP, Giovannoni JJ, Hanson AD (2010) Developmental and feedforward control of the expression of folate biosynthesis genes in tomato fruit. Mol Plant 1:66–77. doi:10.1093/mp/ssp057

155. Wang H, Jones B, Li Z, Frasse P, Delalande C, Regad F, Chaabouni S, Latche A, Pech JC, Bouzayen M (2005) The tomato Aux/IAA transcription factor IAA9 is involved in fruit development and leaf morphogenesis. Plant Cell 10:2676–2692. doi:10.1105/tpc.105.033415

156. Wang Y, Wisniewski M, Meilan R, Cui M, Fuchigami L (2006) Transgenic tomato (*Lycopersicon esculentum*) overexpressing *cAPX* exhibits enhanced tolerance to UV-B and heat stress. J Appl Hortic 2:87–90

157. Wang S, Liu J, Feng Y, Niu X, Giovannoni J, Liu Y (2008) Altered plastid levels and potential for improved fruit nutrient content by downregulation of the tomato

DDB1-interacting protein CUL4. Plant J 55:89–103. doi:10.1111/j.1365-313X.2008.03489.xPubMed

158. Wang BQ, Zhang QF, Liu JH, Li GH (2011) Overexpression of *PtADC* confers enhanced dehydratation and drought tolerance in transgenic tobacco and tomato: effect on ROS elimination. Biochem Biophys Res Commun 413:10–16. doi:10.1016/j.bbrc.2011.08.015PubMed

159. Wayase UR, Shitole MG (2014) Effect of plant growth regulators on organogenesis in tomato (*Lycopersicon esculentum* Mill.) cv. Dhanashri. Int J Pure Appl Sci Technol 2:65–71

160. Wiktorek-Smagur A, Hnatuszko-Konka K, Gerszberg A, Kowalczyk T, Łuchniak P, Kononowicz AK, (2012) Green way of biomedicine—how to force plants to produce new important proteins. In: Yelda Ozden Çiftçi (ed) Transgenic Plants - Advances and Limitations, PhD. ISBN: 978-953-51-0181-9, InTech, doi: 10.5772/31145

161. Wróblewski T, Tomczak A, Michelmore R (2005) Optimization of *Agrobacterium-mediated* transient assays of gene expression in lettuce, tomato, *Arabidopsis*. Plant Biotechnol J 2:259–273. doi:10.1111/j.1467-7652.2005.00123.x

162. Wu Z, Sun S, Wang F, Guo D (2011) Establishment of regeneration and transformation system of *Lycopresicon esculentum* Micro tom. Br Biotechnol J 3:53–60.www.sciencedomain.org/download.php?f=1311487212-Guo.pdf

163. Wurbs D, Ruf S, Bock R (2007) Contained metabolic engineering in tomatoes by expression of carotenoid biosynthesis genes from the plastid genome. Plant J 49:276–288. doi:10.1111/j.1365-313X.2006.02960.xPubMed

164. Yanez M, Caceres S, Orellana S, Bastias A, Verdugo I, Luiz-Lara S, Casaretto JA (2009) An abiotic stress-responsive bZIP transcription factor from wild and cultivated tomatoes regulates stress-related genes. Plant Cell Rep 10:1497 507. doi:10.1007/s00299-009-0749-4

165. Yang L, Shen H, Pan A, Chen J, Huang C, Zhang D (2005) Screening and construct-specific detection methods of transgenic Hufan No 1 tomato by conventional and real-time PCR. J Sci Food Agric 85:2159–2166. doi:10.1002/jsfa.2193

166. Yang S, Vanderbeld Wan J, Huang Y (2010) Narrowing down the targets: towards successful genetic engineering of drought-tolerant crop. Mol Plant 3:469–490. doi:10.1093/mp/ssq016PubMed

167. Yarra R, He SJ, Abbagani S, Ma B, Bulle M, Zhang WK (2012) Overexpression of wheat Na^+/H^+ antiporter gene (*TaNHX2*) enhances tolerance to salt stress in transgenic tomato plants (*Solanum lycopersicum* L.). Plant Cell Tissue Organ Cult 111:49–57. doi:10.1007/s11240-012-0169-y

168. Yasmeen A (2009) An improved protocol for the regeneration and transformation of tomato (cv. Rio Grande). Acta Physiol Plant 31:1271–1277

169. Yasmeen A, Mirza B, Inayatullah S, Safdar N, Jamil M, Ali S, Choudry MF (2009) In planta transformation of tomato. Plant Mol Biol Rep 27:20–28. doi:10.1007/s11105-008-0044-5

170. Youm JW, Jeon JH, Kim H, Kim YHK, Ko K, Joung H, Kim HS (2008) Transgenic tomatoes expressing human beta-amyloid for use as a vaccine against Alzheimer's disease. Biotechnol Lett 30:1839–1845. doi:10.1007/s10529-008-9759-5PubMedCentralPubMed

171. Zanor MI, Osorio S, Nunes-Nesi A, Carrari F, Lohse M, Usadel B, Kuhn C, Bleiss W, Giavalisco P, Willmitzer L, Sulpice R, Zhou YH, Fernie AR (2009) RNA interference of LIN5 in tomato confirms its role in controlling Brix content, uncovers the influence of sugars on the levels of fruit hormones, and demonstrates the importance of sucrose cleavage for normal fruit development and fertility. Plant Physiol 3:1204–1218. doi:10.1104/pp.109.136598

172. Zhang H, Zhao L, Chen Y, Cui L, Ren W, Tang K (2007) Expression of human coagulation Factor IX in transgenic tomato (*Lycopersicon esculentum*). Biotechnol Appl Biochem 48:101–107. doi:10.1042/BA20060224PubMed

173. Zhang C, Liu J, Zhang Y, Cai X, Gong P, Zhang J, Wang T, Li H, Ye Z (2011) Overexpression of SlGMEs leads to ascorbate accumulation with enhanced oxidative stress, cold, and salt tolerance in tomato. Plant Cell Rep 30:389–398. doi:10.1007/s00299-010-0939-0PubMed

174. Zhang W, Hou L, Zhao H, Li M (2012) Factors affecting regeneration of tomato cotyledons. Biosci Methods 4:27–33

175. Zhou F, Badillo-Corona JA, Karcher D, Gonzalez-Rabade N, Piepenburg K, Borchers AM, Maloney AP, Kavanagh TA, Gray JC, Bock R (2008) High-level expression of human immunodeficiency virus antigens from the tobacco and tomato plastid genomes. Plant Biotechnol J 9:897–913. doi:10.1111/j.1467-7652.2008.00356.x

176. Zhou T, Zhang H, Lai T, Qin C, Shi N, Wang H, Jin M, Zhong S, Fan Z, Liu Y, Wu Z, Jackson S, Giovannoni JJ, Rolin D, Gallusci P, Hong Y (2012) Virus-induced gene complementation reveals a transcription factor network in modulation of tomato fruit ripening. Sci Rep 2:836. doi:10.1038/srep00836

CHAPTER 6

DNA METHYLATION CHANGES IN TRITICALE DUE TO IN VITRO CULTURE PLANT REGENERATION AND CONSECUTIVE REPRODUCTION

J. Machczyn´ska R. Orłowska P. T. Bednarek[1], D. R. Man´kowski[2], J. Zimny[3] Andrzej K. Kononowicz

[1]Department of Plant Physiology and Biochemistry, Plant Breeding and Acclimatization Institute, National Research Institute, 05-870 Błonie, Radziko´w, Poland

[2]Department of Seed Science and Technology, Plant Breeding and Acclimatization Institute, National Research Institute, 05-870 Błonie, Radziko´w, Poland

[3]Department of Plant Biotechnology and Cytogenetics, Plant Breeding and Acclimatization Institute, National Research Institute, 05-870 Błonie, Radziko´w, Poland

INTRODUCTION

Triticale is an artificial species that originated ca. 130 years ago from a cross between wheat and rye, with the first cultivars useful for breeders available in the 1960s (Mergoum et al. 2009). It exhibits high yield potential, grain quality, resistance to pathogens, favourable amino acid composition and adaptation to adverse conditions (Heger and Eggum 1991). Hybrid breeding based on cytoplasmic male sterility is a promising direction in its cultivation (Oettler et al. 2005). However, homozygous, genetically uniform and stable doubled haploids (DHs) are necessary to maximize heterosis. Such forms could be derived via androgenesis employing anther cultures or isolated microspores (Eudes and Amundsen 2005; Żur et al. 2008; Lantos et al. 2013; Würschum et al. 2013).

Tissue culture induced variation (TCIV) in regenerants (Li et al. 2007; González et al.2013) and somaclonal variation (SV) in progeny of such regenerants (Linacero and Vazquez 1992; Ivanov et al. 1998) have been observed in crops. TCIV may be manifested at the morphological level of regenerants (Jaligot et al. 2000). Usually, off-type plants are eliminated from the breeding population. However, epigenetic changes without obvious morphological effect, are mostly not easily recognized (Bednarek et al. 2007) and may spontaneously come into being even after several generations (Brettell and Dennis 1991). Evidently, tissue culture may induce alterations in DNA methylation pattern (González et al. 2011; Shan et al. 2013) since it may impose stressful conditions (González et al. 2013).

Additionally, cell reprogramming may require resetting of the DNA methylation pattern (Baroux et al. 2011) causing epimutations. Relatively low levels of epigenetic change related to tissue culture were described in barley (Bednarek et al. 2007). Those changes occurred whether regeneration was through either androgenesis or somatic embryogenesis. However, it is not evident whether genome methylation of triticale regenerants derived via androgenesis and somatic embryogenesis and their progeny behaves similarly. Possibly, the complex genome of triticale may react differently as suggested by genomic studies (Bento et al. 2011) as well as molecular and cytological analyses (Brettell et al. 1986).

In triticale, off-type forms derived from tissue culture regeneration are not rare (Pauk et al. 2000). Cytogenetic analysis showed that triticale is genetically unstable during tissue culture manipulation (Lapitan et al. 1984). Chromosome analysis of the root tip cells of regenerated plants revealed that the rye genome was more frequently involved in chromosomal rearrangements than wheat genome (Charmet et al. 1986). Not only tissue culture regeneration, but also generative reproduction may cause genome instability (Suenaga and Nakajima 1993) of the progeny derived from regenerants.

Citation: Joanna Machczyńska, Renata Orłowska, Dariusz Rafał Mańkowski, Janusz Zimny, Piotr Tomasz Bednarek, DNA methylation changes in triticale due to in vitro culture plant regeneration and consecutive reproduction, DOI: 10.1007/s11240-014-0533-1.

To study the extent of changes induced by tissue culture and inherited by the progeny different methods have been used (Hossain et al. 2003; Peredo et al. 2009; Linacero et al.2011; Díaz-Martínez et al. 2012). Most of them involved isoschizomers differing in their sensitivity towards DNA methylation sites (Ochogavía et al. 2009; Wang et al. 2013; Machczyńska et al. 2014). The limitation of the system to restriction sites recognized by the endonucleases is apparent (Schrey et al. 2013). Contrary to marker-based approaches, high performance liquid chromatography (HPLC) has been applied in studies of DNA methylation of plants regenerated via tissue culture (Renau-Morata et al. 2005; Rival et al.2013). This technique supports general information on global DNA methylation and could be applied for the evaluation of differences among plant materials that could be related to spontaneously arising off-type plants (Johannes et al. 2009; Yi et al. 2010).

Tissue culture is a stressful environment that may lead to numerous epigenetic changes. These changes may be transmitted to the next generations. However, it is not clear how long such changes may persist in the progeny. The purpose of the study was to evaluate the level of global epigenetic changes related to genomic DNA methylation among donor triticale plants, their regenerants derived via different in vitro culture methods and consecutive generative progenies of the regenerants using RP-HPLC approach.

MATERIALS AND METHODS

Plant Material

Four lines of DH donor plants (each line was represented by vegetatively multiplicated clones, Table 1) of winter triticale cv. Bogo (Oleszczuk et al. 2004) were used to generate regenerants via shed microspore culture (M), anther culture (A), and immature zygotic embryo culture (E). The regenerants were used to obtain the first, the second and the third generative progenies. Donor plants, their homozygous regenerants and their consecutive progenies formed a set. Each plant within the set had its complete pedigree to a specific donor from which it was derived. There were four of such sets encoded S^1, S^2, S^3 and S^4. The arrangement of the plant material is given in Table 1.

Table 1: Number of plants encompassing four sets used for the RP-HPLC analyses

Plant material	Four sets derived from four lines of donor plants			
	S^1	S^2	S^3	S^4
	No. of regenerants and progenies comprising the sets			
D	9	8	12	10
R_M	13	8	6	12
P_M^1	23	14	14	18
P_M^2	20	19	16	16
P_M^3	18	12	18	15
R_A	25	20	22	14
P_A^1	20	13	14	22
P_A^2	15	17	18	24
P_A^3	15	15	18	15
R_E	26	15	15	17
P_E^1	22	14	14	15
P_E^2	15	12	18	18
P_E^3	17	9	18	9
Total	238	176	203	205

S^1, S^2, S^3, S^4 stands for the first, the second, the third and the fourth set, respectively. Each set consists of donors, regenerants and progeny. D, vagetatively multiplicated doubled haploid donor plants; R_M, regenerants derived from shed microspore; R_A, regenerants derived from anthers; R_E, regenerants derived from immature zygotic embryos; P_M^1, P_A^1 and P_E^1, the first progeny derived from R_M, R_A and R_E, respectively. P_M^2, P_A^2, P_E^2 and P_M^3, P_A^3, P_E^3 stands for the second and third generative progeny, respectively

Androgenesis

Anther Culture

Plants of winter triticale cv. Bogo have grown under controlled conditions in a chamber room at a photoperiod of 16 h day and 8 h night at 16 and 12 °C, respectively. The tillers were collected, when the microspores were at the mid to the late uninucleate stage and pretreated in a cool room for 4 weeks at 4 °C. Next, the spikes were sterilized (70 % ethanol then 10 % sodium hypochlorite) and rinsed three times for 1 min with sterile water. Subsequently, anthers from one half of the spikes were excised and cultured on solid induction medium 190-2 (Zhuang and Xu 1983) with Phytagel and supplemented with 90 g/l maltose, 400 mg/l glutamine, 2 mg/l 2,4-D and 0.5 mg/l kinetin at 26 °C in the dark, for a maximum of 8 weeks. Calli and embryo-like structures were transferred onto regeneration medium (190-2 with 0.5 mg/l NAA and 0.5 mg/l kinetin), when they were 2–5 cm in size. Afterwards, green plantlets were kept in an Erlenmeyer flasks, in rooting medium (190-2 with 2 mg/l IAA). After roots appeared, plants were vernalized for 6 weeks in 4 °C and transferred to soil in a greenhouse, then grew to maturity.

Shed Microspore Culture

The same protocol as for anther culture was used for shed microspore culture from anthers taken from the second half of the same spikes except with a liquid induction medium. Then, appearing callus pieces, as well as proembryos were transferred to solid regeneration medium.

Somatic Embryogenesis

Immature Zygotic Embryo Culture

Caryopses obtained after self-pollination of the DH mother plants were cut to remove the embryos at the coleoptile stage under aseptic conditions. Immature embryos were isolated and placed with the scutellum exposed onto MS medium (Murashige and Skoog 1962) with 30 g saccharose and 30 µM dicamba. Developing embryos were transferred onto medium supplemented with 0.9 µM 2,4-D for conversion of somatic embryos to plantlets. Cultures were maintained in controlled conditions at 26 °C under 16/8 h (day/night) photoperiod. Afterwards, the plantlets were transferred to the regeneration medium. The 5 cm long plantlets were transferred on a root formation medium (regeneration and rooting media are the same as for androgenesis). Next, small plants were kept in 4 °C for 6 weeks, then they were placed into soil in a greenhouse and grown to maturity.

Progeny

Individual DH regenerants derived via anther cultures and shed microspore cultures as well as regenerants derived from immature zygotic embryo cultures were selfed by bag isolation. Seeds of those regenerants were advanced to obtain progeny. Some plants were again bag isolated to obtain the second and third generations of progeny.

Flow Cytometry

Ploidy status of regenerants after in vitro androgenesis was determined by flow cytometry (Partec; PAII; HBO lamp; Germany). Young regenerated plants were prepared according to Galbraith et al. (1983) with minor modification. Nuclei were isolated from leaves which were chopped with a razor blade in 2 ml of a lysis buffer containing DAPI, afterwards the suspension was filtered through a 30 µm pore diameter filter prior to analysis. Subsequently, histograms of the cell nuclei were generated. The ploidy level of regenerants was determined by comparing the G1 peaks of each sample to a reference (diploid plant).

DNA isolation

Total genomic DNA isolation was performed from young leaves (with flag leaf emerging) of donors, regenerants and their progeny following Plant DNeases MiniKit 250 (Qiagen) manufacturer protocol. DNA quantity was evaluated spectrophotometrically (NanoDrop 1000 Overview). The samples were tested for their integrity and purity on agarose gel with ethidium bromide (1.2 % gel, 1× TBE buffer, 160 V).

RP-HPLC ANALYSIS

The Plant Material used for the RP-HPLC is Arranged in Table 1.

Sample Preparation

DNA samples (6 µg) were dissolved in 100 µl deionised water and denatured by heating to 100 °C for 2 min following immediate placing on ice for 5 min. DNA hydrolysis was performed in a total volume of 135 µl consisting of 5 µl of 10 mM $ZnSO_4$ and 10 µl of 1.0 U ml^{-1} of P1 nuclease (in 30 mM NaOAc pH 5.4), the samples were gently stirred and incubated for 17 h at 37 °C. After nuclease digestion 10 µl 0.5 M Tris pH 8.3 and 10 µl of 10 U ml^{-1} of alkaline phosphatase (in 2.5 M $(NH_4)_2SO_4$) were added, followed by gentle mixing and incubating at 37 °C for 2 h. After digestion, the samples were centrifuged at 12,000 rpm for 5 min.

Chromatographic Assessment of Global DNA Methylation

RP-HPLC quantification of overall DNA methylation was performed according to the procedure of Johnston et al. (2005) using Waters 625 LC System. Briefly, the 4u Max-RP C12 (250 × 4.6 mm, Phenomenex) column combined with 4u Max-RP C12 pre-column were applied using linear gradient of the eluent A comprising of 0.5 % methanol plus 10 mM KH_2PO_4 at a pH of 3.7 (v/v) and eluent B comprising of 10 % methanol in 10 mM KH_2PO_4, pH 3.7 (v/v) that changed from 100 % of A to 100 % of B within 10 min following 15 min of 100 % B eluent and ending with 100 % of the A eluent for 5 min at a flow rate equal to 1 ml min^{-1}. The absorbance was measured spectrophotometrically (UV detector, λ = 280 nm). Chromatograph was calibrated with external standards consisting of each DNA nucleosides (5–50 µM), 15–150 µM for RNA nucleosides and 1.5–15 µM of 5-methyldeoxycytidine dissolved in deionised water. On the basis of that data the calibration curves were evaluated and used for the quantification of the nucleosides in the samples with Millennium 32v 4.0 software. The percentage of deoxycytidine methylation in relation to the total content of cytidine was calculated according to the following equation: 5mdC % = [5mdC/(5mdC + dC)] × 100, where 5mdC and dC states for 5-methyldeoxycytidine and deoxycytidine, respectively. Each RP-HPLC analysis was conducted in three repeats using DNA samples from plant material arranged in Table 1.

Statistics

One-way analysis of variance (ANOVA) with Tukey's contrasts analysis were performed in R CRAN software (*stats* and *multcomp* packages, respectively). The regression analysis was performed in R CRAN software using *stats* package and polynomial regression was evaluated. For all statistical analyses significance level $\alpha = 0.01$ was applied.

All RP-HPLC results (including three repeats for each sample) based on the whole plant material arranged in Table 1 were used for Tukey's contrasts and regression analysis.

RESULTS

Donor Plants

Each donor's genotype was represented by several vegetatively multiplicated individuals (Table 1). No morphological variation among multiplicated plants (representing the given genotype as well as all genotypes) was observed.

Plant Regeneration

Embryos and embryo-like structures were observed after 4–8 weeks of culture in darkness in shed microspore culture (Fig. 1a). Microspore development resulted in complete embryo formation with visible cotyledons and embryo axis (Fig. 1b). Androgenesis in anther cultures resulted in lumpy embryo-like structures visible after ca. 6 weeks of culture (Fig. 1c). After transmission of such structures into the light some albino plants were detected (Fig. 1d). Somatic embryogenesis from cultured immature zygotic embryos is shown in Fig. 1e, f. Embryos appeared after 4 weeks of culture on MS medium with dicamba (Fig. 1e) and converted to plants on regeneration medium (Fig. 1f). Self-pollination of regenerants (Fig. 1g) resulted in formation of the first generative progeny (Fig. 1h).

Figure. 1: *Triticosecale* Wittmack cv. Bogo regenerants derived via in vitro culture and their progeny; a androgenesis on liquid induction medium, b embryo formation from shed microspore culture, c androgenesis on solid induction medium, d albino plant derived from anther culture, e somatic embryogenesis from immature zygotic embryo, fgerminating embryo, g fertile anther culture-derived plants, h progenies of regenerants

With the exception of albinos, no obvious morphological changes were observed among fertile regenerants as well as among the first, second and third generations of the progeny. However, morphological differences were visible between haploids and diploids. The haploid plants were smaller, had thinner stems and narrower leaves compared to DH plants.

Analysis of ploidy level

There were 63.3 and 71.7 % haploids among regenerants derived via shed microspore and anther culture, respectively (Table 2) with spontaneous diploidization amounting to 27.1 and 27.7 % for the two approaches, severally. Aneuploids were observed for 0.6 and 0.62 % of cases, respectively (Table 2).

Table 2: Flow cytometry data. R_M and R_A states for the regenerants derived from shed microspore and anthe culture, respectively

Plant material	No. of plants analyzed	No. of plants with various ploidy levels		
		Haploid	Doubled haploid	Aneuploid
R_M	166	105	45	1
R_A	318	228	88	2

Changes in Global DNA Methylation

The RP-HPLC approach resulted in clearly distinguishable, well-separated peaks corresponding to dC and 5mdC, with retention time equal to 5.44 and 8.43 min, respectively.

The greatest degree (25.4 %) of average global DNA methylation was observed for donors. In case of all regenerants (R) the level of DNA methylation was equal to 24.1 % and was the lowest for the first generative progeny (P^1) (23.6 %). Starting from the second generative progeny, global DNA methylation was slightly greater than for P^1 (Table 3).

Table 3: Arrangement of the mean values of the genomic cytosine methylation evaluated based on RP-HPLC analyses of donor plants, their regenerants divided according to the tissue culture approach and consecutive progeny of those regenerants

	Plant materials				
	D	R	P^1	P^2	P^3
Regeneration method	Mean value of total genomic cytidine methylation (±standard error)				
M	25.39 ± 0.14	24.68 ± 0.41	23.81 ± 0.41	23.96 ± 0.64	24.02 ± 0.22
A		24.43 ± 0.47	23.53 ± 0.47	23.88 ± 0.89	23.90 ± 0.39
E		23.19 ± 0.44	23.40 ± 0.44	23.52 ± 0.86	23.84 ± 0.35

D, Donor plants; R, regenerants; P^1, P^2, P^3, the first, the second, the third generations of regenerants; M, shed microspore culture; A, anther culture; E, immature zygotic embryo culture

Comparison of global DNA methylation for D, R, P^1, P^2 and P^3without considering sets and regeneration method

According to Tukey's grouping (Fig. 2) donor plants differed from regenerants due to genomic DNA methylation. No difference between R and P^1 was observed, however, R and P^1 differed from P^2 and P^3 that formed separate group (Table 4). Polynomial regression (Table 5) calculated for the abovementioned data demonstrated that global DNA methylation decreased from donors to the first generative progeny and then reversed (Fig. 3).

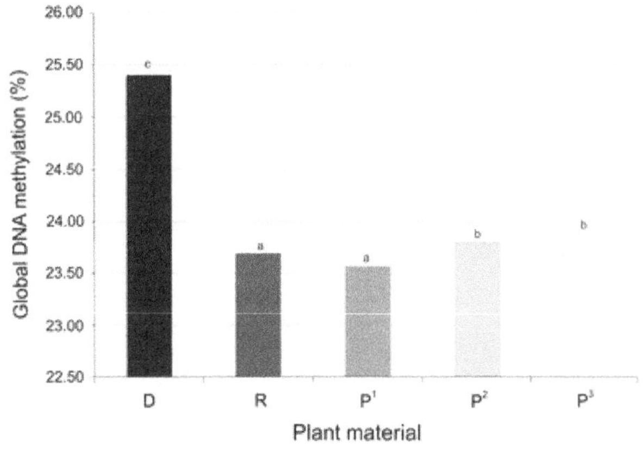

Figure. 2: Tukey's grouping of donors (*D*), regenerants (*R*), the first (*P* [1]), the second (*P* [2]) and the third (*P* [3]) generative progeny of the regenerants without considering sets or tissue culture regeneration method, based on global DNA methylation evaluated via RP-HPLC. Global DNA methylation is expressed as the amount of methylated cytosine to the total amount of cytosine converted to percentages. Materials forming the same group are indicated by either '*a*', '*b*' or '*c*' letters

Table 4: Arrangement of the ANOVA for means comparisons

Plant material	ANOVA for means comparison
$D–R–P^1–P^2–P^3$	$F = 37.94; p < 0.01$
$R_M–R_A–R_E$	$F = 99.67; p < 0.01$
$P_M^1–P_A^1–P_E^1$	$F = 55.11; p < 0.01$
$P_M^2–P_A^2–P_E^2$	$F = 16.67; p < 0.01$
$P_M^3–P_A^3–P_E^3$	$F = 3.57; p < 0.01$
S^1	$F = 61.14; p < 0.01$
S^2	$F = 51.91; p < 0.01$
S^3	$F = 70.5; p < 0.01$
S^4	$F = 53.09; p < 0.01$

For the explanation of the abbreviations see Tables 1 and 3

Table 5: Arrangement of the polynomial regression models

Plant material	Polynomial regression model
$D–R–P^1–P^2–P^3$	$F = 31.56; p < 0.01; Y = 24.81031 - 20.74299x + 0.11423x^2$
$D–R_A–P_A^1–P_A^2–P_A^3$	$F = 44.03; p < 0.01; Y = 26.45975 - 21.522122x + 0.20587x^2$
$D–R_M–P_M^1–P_M^2–P_M^3$	$F = 85.61; p < 0.01; Y = 26.40839 - 21.22876x + 0.14997x^2$
$D–R_E–P_E^1–P_E^2–P_E^3$	$F = 44.51; p < 0.01; Y = 25.04682 - 21.13266x + 0.18040x^2$
S^1	$F = 5.698; p < 0.01; Y = 24.65436 - 20.52956x + 0.07611x^2$
S^2	$F = 79.79; p < 0.01; Y = 26.91769 - 21.94924x + 0.27377x^2$
S^3	$F = 17.65; p < 0.01; Y = 25.4608 - 20.9356x - 20.1274x^2$
S^4	$F = 73.56; p < 0.01; Y = 26.72317 - 2.28005x + 0.35598x^2$

For the explanation of the abbreviations see Tables 1 and 3

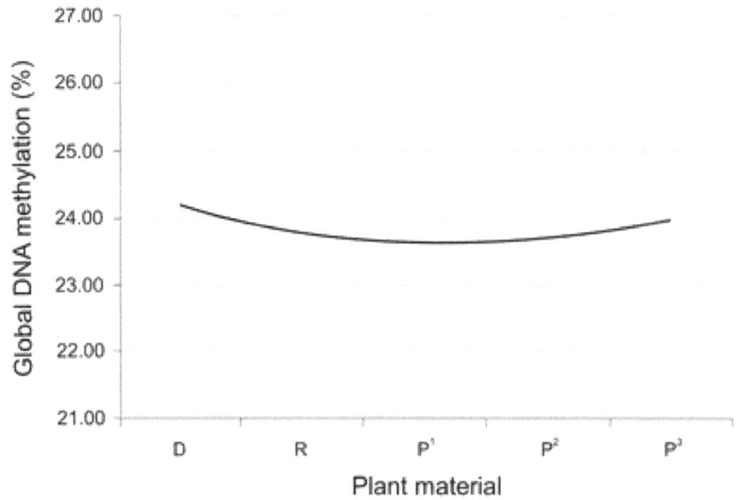

Figure. 3: Polynomial regression of donors (*D*), regenerants (*R*), the first (P^1), the second (P^2) and the third (P^3) progenies of the regenerants (without considering sets and regeneration method). Global DNA methylation is expressed as the amount of methylated cytosine to the total amount of cytosine converted to percentages

Comparison of global DNA methylation for regenerants and their three generations of progeny by regeneration method

Comparison (ANOVA) of the global DNA methylation of the regenerants derived via shed microspore (M), anther (A) and immature zygotic embryos (E) tissue culture approaches revealed (Fig. 4) that plants obtained via androgenesis differed from those regenerated via somatic embryogenesis (Table 4). The first generative progeny of shed microspore derived regenerants differed from those obtained via anthers and the first progeny from somatic embryogenesis derived regenerants (Table 4). $P_M^{\ 2}$ and $P_A^{\ 2}$ did not differ from each other according to 5mdC content, although they were distinct from $P_E^{\ 2}$ (Table 4). In the case of the third generative offspring, overall DNA methylation in $P_M^{\ 3}$ differed from $P_E^{\ 3}$(Table 4), whereas DNA methylation level in $P_A^{\ 3}$ was similar both to $P_M^{\ 3}$ and $P_E^{\ 3}$.

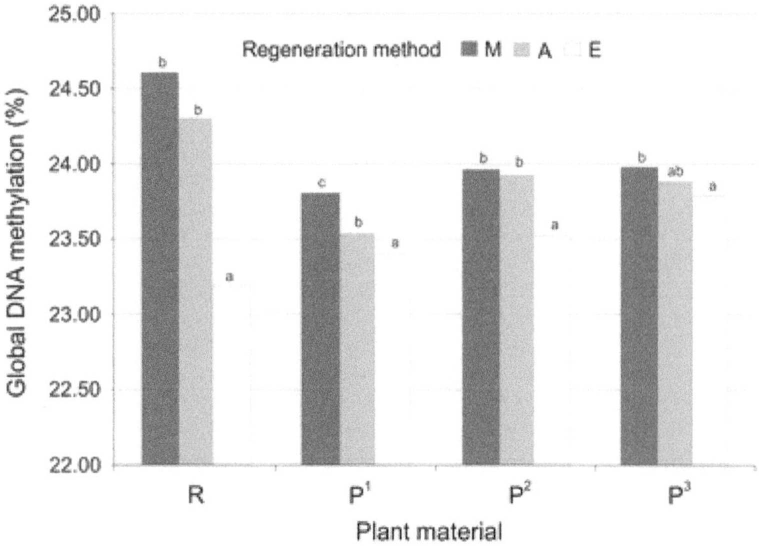

Figure. 4: Tukey's grouping of the regenerants (*R*), the first (*P* [1]), the second (*P* [2]) and the third (*P* [3]) generative progeny of the regenerants based on tissue culture method used. Grouping is indicated by *small letters*. The materials with the *same letter* (e.g. '*a*') form the same group and are distinct from those classified as '*b*' within the group of *R*, *P* [1], *P* [2] and *P* [3], respectively. Global DNA methylation is expressed as the amount of methylated cytosine to the total amount of cytosine converted to percentages. The regenerants derived either via M (*dark grey*), A (*grey*) or E (*white*) approach and their progeny are indicated by*colours*

Regression analysis (Fig. 5) performed for donors, R_A, R_M, R_E and their three generations demonstrated that global genomic DNA methylation decreased in regenerants compared to the donors. The decrease in DNA methylation was observed up to the first/second generative progeny following by reversion of the process. Independently of the tissue culture approach used for plant regeneration the observed trends were significant and followed the polynomial function (Table 5). A little bit higher level of DNA demethylation for R_E than for R_A and R_M was observed.

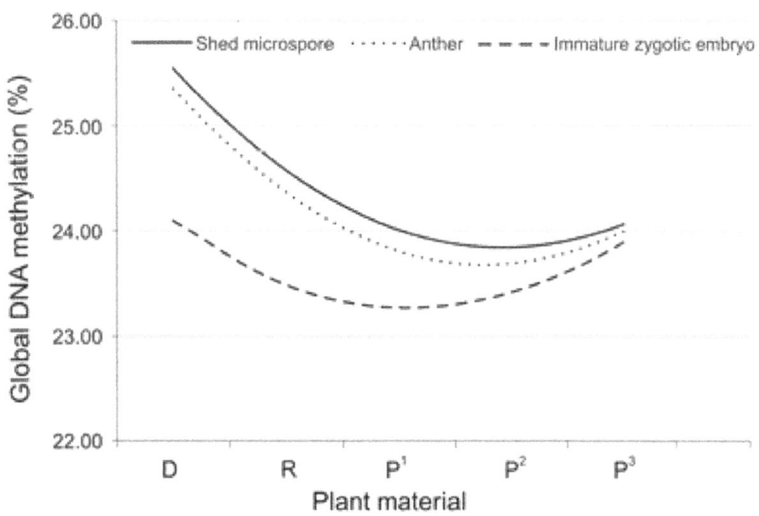

Figure. 5: Polynomial regression analysis of global DNA methylation changes exhibited between *D*,*R*, *P* [1], *P* [2] and *P* [3] materials according to the tissue culture regeneration method. The regression line for $D–R_M–P_M^{\ 1}–P_M^{\ 2}–P_M^{\ 3}$, $D–R_A–P_A^{\ 1}–P_A^{\ 2}–P_A^{\ 3}$ and $D–R_E–P_E^{\ 1}–P_E^{\ 2}–P_E^{\ 3}$ are given in *black, dotted* and *dashed lines*, respectively

Comparison of Global DNA Methylation for Sets

Tukey's test revealed significant differences in global DNA methylation of plant material encompassing each of the S^1, S^2, S^3 and S^4 sets (Fig. 6). In detail, in the first set D, R, P^2differed from each other, whereas differences between P^1 and P^3 were insignificant (Table 4). In the S^2–D, P^1, P^2, P^3 differed from each other, whereas R were similar both to P^1 and P^3 (Table 4). In the third set D, R, P^1, P^2, P^3 differed from each other (Table 4) while in the fourth one D, R, P^2, P^3 differed from each other, whereas P^1 were similar both to R and P^2 (Table 4).

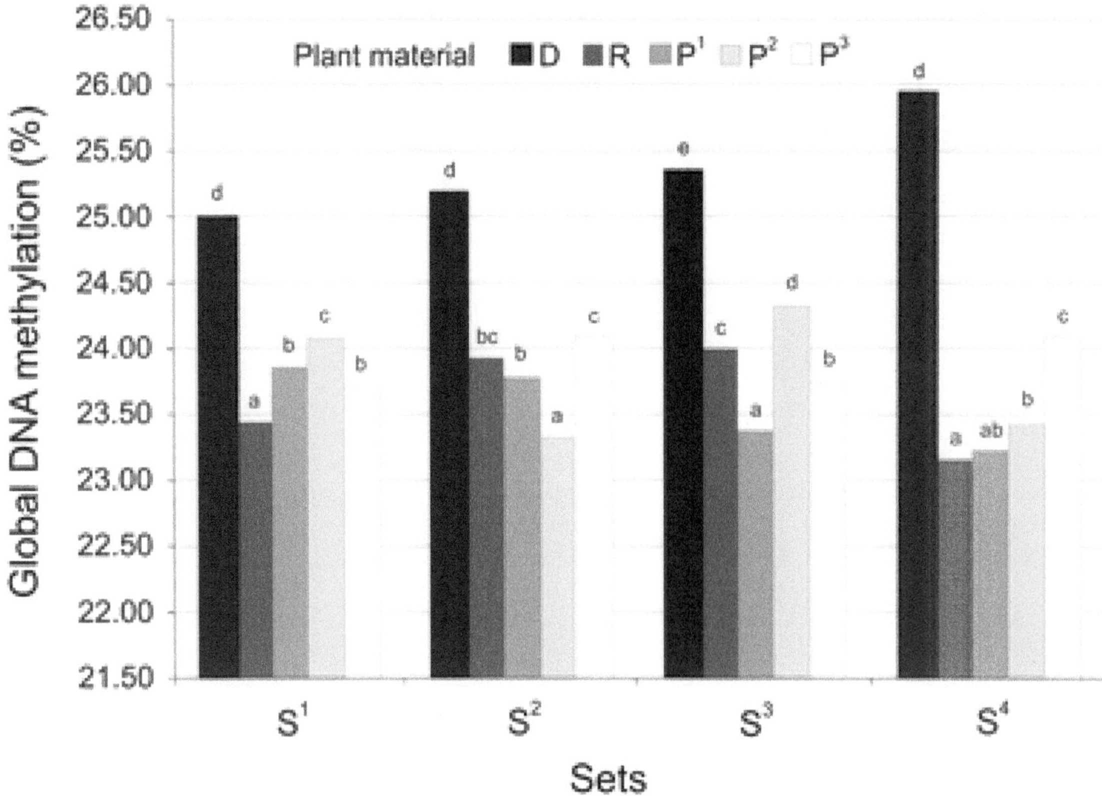

Figure. 6: Tukey's grouping of the donor plants, regenerants and consecutive generative progeny within sets without considering tissue culture approach used for plant regeneration. *S* [1], *S*[2], *S* [3] and *S* [4] state for sets comprising *D, R, P* [1], *P* [2] and *P* [3]. For detailed description of the abbreviation see Fig 2. Materials with the same *small letter* (indicated on graph) of the given set belong to the same group

Global genomic DNA methylation changes followed polynomial regression in the case of all sets with local minimum at the level of the first/second generative progeny (Table 5; Fig. 7). All trends were significant. In the case of S^2 and S^4 local minimum was deeper compared with S^1 and S^3.

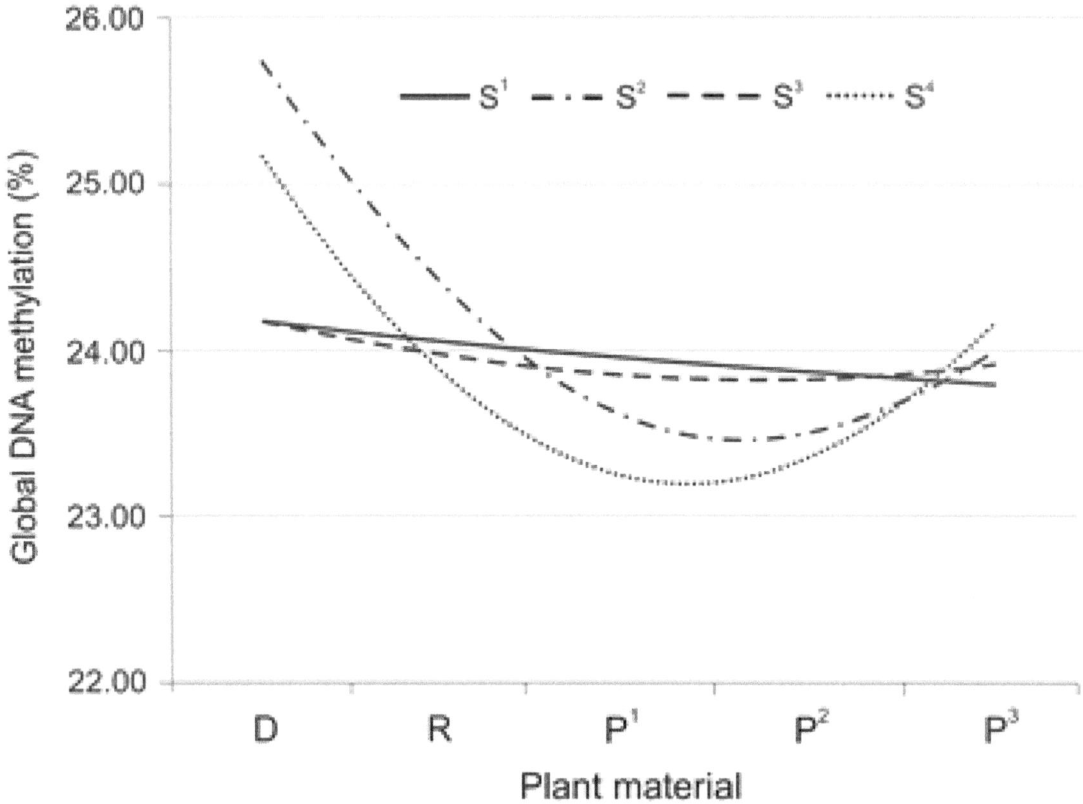

Figure. 7: Polynomial regression reflecting global DNA methylation changes among donors (*D*), regenerants (*R*), their successive generative progeny (*P* [1], *P* [2] and *P* [3]) within each of the *S*[1], *S* [2], *S* [3] and *S* [4] sets

DISCUSSION

There is ample evidence suggesting that tissue culture is a stressful environment that may be responsible for the induction of abiotic stress during plant regeneration (Miguel and Marum 2011). Such stress may cause the regeneration of the off-type plants that could be detected at the morphological (Li et al. 2010), cytological (Linacero and Vazquez 1992), genetic or epigenetic level (Bednarek et al. 2007; Vining et al. 2013). Some of the changes may subsequently arise among progeny originated from tissue culture-derived regenerants even after many generative cycles (Bregitzer et al. 1998; Kirikovich et al. 2003).

Thus, it is essential to have several sets of plant material encompassing homozygous donor plants, the regenerants derived from them and the progenies obtained from the regenerants to be able to identify epigenetic differences among such materials. The plant material used in our study originated from DH plants that became the source of tissues for the regeneration. The donor plants represented distinct genotypes but did not exhibit any morphological variation. Moreover, generative progeny was obtained by self-pollination under controlled conditions. Thus, any variation among our materials within sets (donor plants, regenerants and their progenies) should be attributed to the changes induced in tissue cultures and transmitted to the next generations. However, possibly due to highly homozygous plant material, as well as the limited number of plants analyzed in our study, we failed to observe at the morphological level any off-type plants among regenerants and their progeny within as well as and among sets. Nevertheless, the differences were revealed for haploids and a few aneuploids detected via flow cytometry among regenerants. The level of spontaneously derived DH plants and aneuploids was similar to prior report (Oleszczuk et al. 2004).

The lack of morphological variation or even cytological uniformity does not, however, exclude changes at the DNA level and especially those related to DNA methylation, which may take place under tissue culture conditions, as showed in barley (Bednarek et al. 2007). Our RP-HPLC study demonstrated that the triticale genome was affected by global DNA demethylation which occurred during tissue culture regeneration. The same direction of DNA methylation changes was observed in our preliminary study on triticale cv. Bogo with metAFLP approach (Machczyńska et al. 2014), where demethylation prevailed over de novo methylation. Our data are congruent with the MSAP studies of genomic DNA methylation changes performed between rye donor plants and their regenerants where the level of demethylation was greater for the regenerants (González et al. 2013). Similar results were obtained also in rye using *Hpa*II and *Msp*I digested DNA amplified with ISSR markers (Linacero et al. 2011). It was demonstrated that the decrease in DNA methylation at the level of regenerants may be caused by the necessity of genes activation during cell reprogramming to facilitate plant regeneration (Kaeppler and Phillips 1993).

Our RP-HPLC results showed no difference in the genomic DNA methylation level between shed microspore and anther culture approaches while androgenic regenerants exhibited greater methylation than somatic embryogenesis derived ones. It is well documented that a callus phase induces variation during plant regeneration (Bouman and Klerk 2001). In our experiment immature zygotic embryo plant regeneration encompassed an obvious callus stage whereas androgenesis, especially in the case of shed microspore cultures, largely circumvents callus. Thus, the longer callus duration of somatic embryogenesis compared to androgenesis may have been responsible for the differences in the level of DNA methylation evaluated between regenerants derived via two paths (Xu et al. 2004). Alternatively, the tissue culture medium composition could be also responsible for the difference in global DNA methylation. Plant growth regulators may enhance in vitro induced variation resulting from epigenetic modifications (Nehra et al. 1992; Bairu et al. 2006). Study on *Malus xiaojinensis* revealed increased DNA methylation with increasing 2,4-D (Huang et al. 2012). To induce the androgenic response in our experiment 2,4-D was supplemented to the induction medium, which may have been responsible for the greater genome-wide methylation in androgenic than in somatic embryogenesis derived regenerants. Immature zygotic embryos were cultured on a medium containing dicamba reported to cause more negative changes than 2,4-D (Rakoczy-Trojanowska 2002). It might be suggested that higher demethylation in somatic embryogenesis derived regenerants in triticale was caused by the presence of dicamba. Alternatively, the differences in global DNA methylation observed between androgenesis and somatic embryogenesis derived regenerants could have been due to cold stress used to convert the gametophytic pathway to a sporophytic one. Studies on stress-induced epigenetic changes confirm that cold treatment induces DNA demethylation (Fan et al.2013; Shan et al. 2013). Moreover, the differences in DNA methylation observed between androgenesis and somatic embryogenesis derived regenerants could be also attributed to the ploidy level. The relationship between DNA methylation and haploids versus diploids was studied in rice (Zhang et al. 2006). The authors concluded that the higher methylation level in haploids might be a readjusting reaction to the decrease in ploidy because of survival. If DH plants inherit the increased DNA methylation e.g. via replication mechanisms then our data is congruent with the abovementioned results.

In our study significant methylation change induced in vitro and revealed by ANOVA affected not only regenerants but also their consecutive generative progeny. Reversion of the DNA demethylation in the first/second generation demonstrated by polynomial regression analysis may have reflected the very beginning of the genomic DNA methylation re-establishment. Apparently, plant regeneration via tissue cultures resulted in epigenetic changes that were not fully compensated even after several generative cycles. Based on our experiment, however, it was not possible to predict whether further generative cycles would result in DNA methylation comparable to that of donor plants or whether the process would behave differently. Although, in general, the trend of global DNA demethylation and its reversion was observed for all sets, the differences among them were obvious. Most probably this may reflect genotypic differences existing between donor plants. Obviously, revealed DNA methylation changes observed among sets suggest that triticale plant material derived via in vitro cultures and its progeny may be prone to epimutations dependent on DNA methylation that maybe exhibited among e.g. regenerants' progeny.

Based on our HPLC analysis, we demonstrated for the first time that the level of genomic DNA methylation alterations between donor triticale plants, their regenerants derived via different in vitro culture methods and consecutive generative progenies of the regenerants was affected by in vitro culture. Moreover, genomic DNA demethylation of the regenerants depended on both regeneration method and genotype of donor plants. Evidently, re-establishment of the triticale genomic DNA methylation level after tissue culture treatment may take time and special care needs to be taken to control epimutations in regenerants.

ACKNOWLEDGMENTS

The authors wish to thank Dr. Sylwia Oleszczuk for providing seeds of triticale cv. Bogo and for her substantial help in plant regeneration.

REFERENCES

1. Bairu MW, Fennell CW, van Staden J (2006) The effect of plant growth regulators on somaclonal variation in Cavendish banana (Musa AAA cv. 'Zelig'). Sci Hortic 108(4):347–351. doi:10.1016/j.scienta.2006.01.039
2. Baroux C, Raissig MT, Grossniklaus U (2011) Epigenetic regulation and reprogramming during gamete formation in plants. Curr Opin Genet Dev 21(2):124–133. doi:10.1016/j.gde.2011.01.017
3. Bednarek PT, Orłowska R, Koebner RMD, Zimny J (2007) Quantification of the tissue-culture induced variation in barley (*Hordeum vulgare* L.). BMC Plant Biol 7:10. doi:10.1186/1471-2229-7-10
4. Bento M, Gustafson JP, Viegas W, Silva M (2011) Size matters in Triticeae polyploids: larger genomes have higher remodeling. Genome 54(3):175–183. doi:10.1139/G10-107
5. Bouman H, Klerk G-J (2001) Measurement of the extent of somaclonal variation in begonia plants regenerated under various conditions. Comparison of three assays. Theor Appl Genet 102:111–117
6. Bregitzer P, Halbert SE, Lemaux PG (1998) Somaclonal variation in the progeny of transgenic barley. Theor Appl Genet 96(3–4):421–425. doi:10.1007/s001220050758
7. Brettell RS, Dennis E (1991) Reactivation of a silent Ac following tissue culture is associated with heritable alterations in its methylation pattern. Mol Gen Genet 229(3):365–372. doi:10.1007/BF00267457
8. Brettell RIS, Pallotta MA, Gustafson JP, Appels R (1986) Variation at the Nor loci in triticale derived from tissue culture. Theor Appl Genet 71(4):637–643. doi:10.1007/BF00264268
9. Charmet G, Bernard S, Bernard M (1986) Origin of aneuploid plants obtained by anther culture in triticale. Can J Genet Cytol 28(3):444–452. doi:10.1139/g86-067
10. Díaz-Martínez M, Nava-Cedillo A, Guzmán-López JA, Escobar-Guzmán R, Simpson J (2012) Polymorphism and methylation patterns in *Agave tequilana* Weber var. 'Azul' plants propagated asexually by three different methods. Plant Sci 185–186:321–330. doi:10.1016/j.plantsci.2012.01.004
11. Eudes F, Amundsen E (2005) Isolated microspore culture of Canadian 6× triticale cultivars. Plant Cell Tissue Organ Cult 82(3):233–241. doi:10.1007/s11240-005-0867-9
12. Fan H, Wei J, Li T, Li Z, Guo N, Cai Y, Lin Y (2013) DNA methylation alterations of upland cotton (*Gossypium hirsutum*) in response to cold stress. Acta Physiol Plant 35(8):2445–2453. doi:10.1007/s11738-013-1278-x
13. Galbraith DW, Harkins KR, Maddox JM, Ayres NM, Sharma DP, Firoozabady E (1983) Rapid flow cytometric analysis of the cell cycle in intact plant tissues. Science 220(4601):1049–1051. doi:10.1126/science.220.4601.1049
14. González R, Ricardi M, Iusem N (2011) Atypical epigenetic mark in an atypical location: cytosine methylation at asymmetric (CNN) sites within the body of a non-repetitive tomato gene. BMC Plant Biol 11(1):1–11. doi:10.1186/1471-2229-11-94
15. González AI, Sáiz A, Acedo A, Ruiz ML, Polanco C (2013) Analysis of genomic DNA methylation patterns in regenerated and control plants of rye (*Secale cereale* L.). Plant Growth Regul 70(3):227–236. doi:10.1007/s10725-013-9794-7
16. Heger J, Eggum BO (1991) The nutritional values of some high-yielding genotypes of triticale. J Cereal Sci 14:63–71
17. Hossain AM, Konisho K, Mineo M, Kazuhiro N (2003) Somaclonal variation of regenerated plants in chili pepper (*Capsicum annuum* L.). Euphytica 130(2):233–239. doi:10.1023/A:1022856725794
18. Huang H, Han S, Wang Y, Zhang X, Han Z (2012) Variations in leaf morphology and DNA methylation following in vitro culture of *Malus xiaojinensis*. Plant Cell Tissue Organ Cult 111(2):153–161. doi:10.1007/s11240-012-0179-9
19. Ivanov P, Atanassov Z, Milkova V, Nikolova L (1998) Culture selected somaclonal variation in five *Triticum aestivum* L. genotypes. Euphytica 104(3):167–172. doi:10.1023/A:1018656903559
20. Jaligot E, Rival A, Beulé T, Dussert S, Verdeil JL (2000) Somaclonal variation in oil palm (*Elaeis guineensis* Jacq.): the DNA methylation hypothesis. Plant Cell Rep 19(7):684–690. doi:10.1007/s002999900177
21. Johannes F, Porcher E, Teixeira FK, Saliba-Colombani V, Simon M, Agier N, Bulski A, Albuisson J, Heredia F, Audigier P, Bouchez D, Dillmann C, Guerche P, Hospital F, Colot V (2009) Assessing the impact of transgenerational epigenetic variation on complex traits. PLoS Genet 5(6):e1000530. doi:10.1371/journal.pgen.1000530
22. Johnston JW, Harding K, Bremner DH, Souch G, Green J, Lynch PT, Grout B, Benson EE (2005) HPLC analysis of plant DNA methylation: a study of critical methodological factors. Plant Physiol Biochem 43(9):844–853. doi:10.1016/j.plaphy.2005.07.015
23. Kaeppler SM, Phillips RL (1993) DNA methylation and tissue culture-induced variation in plants. In Vitro Cell Dev Biol Plant 29(3):125–130. doi:10.1007/BF02632283

24. Kirikovich SS, Svirshchevskaya AM, Levites EV (2003) Variation at isozyme loci in seed offspring of sugar beet gynogenetic lines. Sugar Tech 5(4):289–292. doi:10.1007/BF02942487

25. Lantos C, Bóna L, Boda K, Pauk J (2013) Comparative analysis of in vitro anther—and isolated microspore culture in hexaploid Triticale (×*Triticosecale* Wittmack) for androgenic parameters. Euphytica 1–11. doi:10.1007/s10681-013-1031-y

26. Lapitan NL, Sears RG, Gill BS (1984) Translocations and other karyotypic structural changes in wheat x rye hybrids regenerated from tissue culture. Theor Appl Genet 68(6):547–554. doi:10.1007/bf00285012

27. Li X, Yu X, Wang N, Feng Q, Dong Z, Liu L, Shen J, Liu B (2007) Genetic and epigenetic instabilities induced by tissue culture in wild barley (*Hordeum brevisubulatum* (Trin.) Link). Plant Cell Tissue Organ Cult 90(2):153–168. doi:10.1007/s11240-007-9224-5

28. Li R, Bruneau AH, Qu R (2010) Tissue culture-induced morphological somaclonal variation in St. Augustinegrass [*Stenotaphrum secundatum* (Walt.) Kuntze]. Plant Breed 129:96–99. doi:10.1111/j.1439-0523.2009.01647x

29. Linacero R, Vazquez AM (1992) Cytogenetic variation in rye regenerated plants and their progeny. Genome 35(3):428–430. doi:10.1139/g92-063

30. Linacero R, Rueda J, Esquivel E, Bellido A, Domingo A, Vázquez A (2011) Genetic and epigenetic relationship in rye, *Secale cereale* L., somaclonal variation within somatic embryo-derived plants. In Vitro Cell Dev Biol Plant 47(5):618 628. doi:10.1007/s11627-011-9407-y

31. Machczyńska J, Orłowska R, Zimny J, Bednarek P (2014) Extended metAFLP approach in studies of tissue culture induced variation (TCIV) in triticale. Mol Breed 1–10. doi:10.1007/s11032-014-0079-2

32. Mergoum M, Singh PK, Peña RJ, Lozano-del Río AJ, Cooper KV, Salmon DF, Gómez Macpherson H (2009) Triticale: a "new" crop with old challenges. In: Carena MJ (ed) Cereals. Handbook of plant breeding, vol 3. Springer, Berlin, pp 267–287. doi:10.1007/978-0-387-72297-9_9

33. Miguel C, Marum L (2011) An epigenetic view of plant cells cultured in vitro: somaclonal variation and beyond. J Exp Bot 62:3713–3725. doi:10.1093/jxb/err155

34. Murashige T, Skoog F (1962) A revised medium for rapid growth and bio assays with tobacco tissue cultures. Physiol Plant 15(3):473–497. doi:10.1111/j.1399-3054.1962.tb08052.x

35. Nehra N, Kartha K, Stushnott C, Giles K (1992) The influence of plant growth regulator concentrations and callus age on somaclonal variation in callus culture regenerants of strawberry. Plant Cell Tissue Organ Cult 29(3):257–268. doi:10.1007/BF00034361

36. Ochogavía A, Cervigni G, Selva J, Echenique V, Pessino S (2009) Variation in cytosine methylation patterns during ploidy level conversions in *Eragrostis curvula*. Plant Mol Biol 70(1–2):17–29. doi:10.1007/s11103-009-9454-5

37. Oettler G, Tams SH, Utz HF, Bauer E, Melchinger AE (2005) Prospects for hybrid breeding in winter triticale. Crop Sci 45(4):1476–1482. doi:10.2135/cropsci2004.0462

38. Oleszczuk S, Sowa S, Zimny J (2004) Direct embryogenesis and green plant regeneration from isolated microspores of hexaploid triticale (×*Triticosecale* Wittmack) cv. Bogo. Plant Cell Rep 22(12):885–893. doi:10.1007/s00299-004-0796-9

39. Pauk J, Puolimatka M, Lökös TK, Monostori T (2000) In vitro androgenesis of triticale in isolated microspore culture. Plant Cell Tissue Organ Cult 61(3):221–229. doi:10.1023/A:1006416116366

40. Peredo EL, Arroyo-García R, Revilla MÁ (2009) Epigenetic changes detected in micropropagated hop plants. J Plant Physiol 166(10):1101–1111. doi:10.1016/j.jplph.2008.12.015

41. Rakoczy-Trojanowska M (2002) The effect of growth regulators on somaclonal variation in rye (*Secale cereale* L.) and selection of somaclonal variants with increased agronomic traits. Cell Mol Biol Lett 7(4):1111–1120

42. Renau-Morata B, Nebauer SG, Arrillaga I, Segura J (2005) Assessments of somaclonal variation in micropropagated shoots of Cedrus: consequences of axillary bud breaking. Tree Genet Genomes 1(1):3–10. doi:10.1007/s11295-004-0001-x

43. Rival A, Ilbert P, Labeyrie A, Torres E, Doulbeau S, Personne A, Dussert S, Beulé T, Durand-Gasselin T, Tregear J, Jaligot E (2013) Variations in genomic DNA methylation during the long-term in vitro proliferation of oil palm embryogenic suspension cultures. Plant Cell Rep 32(3):359–368. doi:10.1007/s00299-012-1369-y

44. Schrey AW, Alvarez M, Foust CM, Kilvitis HJ, Lee JD, Liebl AL, Martin LB, Richards CL, Robertson M (2013) Ecological epigenetics: beyond MS-AFLP. Integr Comp Biol 53(2):340–350. doi:10.1093/icb/ict012

45. Shan X, Wang X, Yang G, Wu Y, Su S, Li S, Liu H, Yuan Y (2013) Analysis of the DNA methylation of maize (*Zea mays* L.) in response to cold stress based on methylation-sensitive amplified polymorphisms. J Plant Biol 56(1):32–38. doi:10.1007/s12374-012-0251-3

46. Suenaga K, Nakajima K (1993) Variation in doubled haploid plants of wheat obtained through wheat (*Triticum aestivum*) x maize (*Zea mays*) crosses. Plant Breed 111(2):120–124

47. Vining K, Pomraning K, Wilhelm L, Ma C, Pellegrini M, Di Y, Mockler T, Freitag M, Strauss S (2013) Methylome reorganization during in vitro dedifferentiation and regeneration of *Populus trichocarpa*. BMC Plant Biol 13(1):92. doi:10.1186/1471-2229-13-92

48. Wang X, Wu R, Lin X, Bai Y, Song C, Yu X, Xu C, Zhao N, Dong Y, Liu B (2013) Tissue culture-induced genetic and epigenetic alterations in rice pure-lines, F1 hybrids and polyploids. BMC Plant Biol 13(1):77

49. Würschum T, Tucker M, Maurer H (2013) Stress treatments influence efficiency of microspore embryogenesis and green plant regeneration in hexaploid triticale (×*Triticosecale* Wittmack L.). In Vitro Cell Dev Biol Plant 1–6. doi:10.1007/s11627-013-9539-3

50. Xu M, Li X, Korban S (2004) DNA-methylation alterations and exchanges during in vitro cellular differentiation in rose (*Rosa hybrida* L.). Theor Appl Genet 109(5):899–910. doi:10.1007/s00122-004-1717-6

51. Yi C, Zhang S, Liu X, Bui HT, Hong Y (2010) Does epigenetic polymorphism contribute to phenotypic variances in *Jatropha curcas* L.? BMC Plant Biol 10:259. doi:10.1186/1471-2229-10-259

52. Zhang H, Peng H, Li Y, Xu P, Wang X, Wu X (2006) Patterns of DNA cytosine methylation between haploids and corresponding diploids in rice. Chin Sci Bull 51(14):1721–1728. doi:10.1007/s11434-006-2048-6

53. Zhuang JJ, Xu J (1983) Increasing differentiation frequencies in wheat pollen callus. In: Hu H, Vega MR (eds) Cell and tissue culture techniques for cereal crop improvement, vol 431. Science Press, Beijing

54. Żur I, Dubas E, Golemiec E, Szechyńska-Hebda M, Janowiak F, Wędzony M (2008) Stress-induced changes important for effective androgenic induction in isolated microspore culture of triticale (×*Triticosecale* Wittm.). Plant Cell Tissue Organ Cult 94(3):319–328. doi:10.1007/s11240-008-9360-6

CHAPTER 7

ASSESSMENT OF GENETIC AND EPIGENETIC CHANGES DURING CELL CULTURE AGEING AND RELATIONS WITH SOMACLONAL VARIATION IN COFFEA ARABICA

Roberto Bobadilla Landey[1] Alberto Cenci[2] Romain Guyot[3] Benoı̂t Bertrand[1] Fre´de´ric Georget[1] Eveline Dechamp[1] Juan-Carlos Herrera[4] Jamel Aribi[3] Philippe Lashermes[5] Herve´ Etienne[1]

[1]UMR IPME, CIRAD, 911 Avenue Agropolis, BP 64501, 34394 Montpellier, France

[2]Bioversity, Montpellier, France

[3]UMR IPME, IRD, 911 Avenue Agropolis, BP 64501, 34394 Montpellier, France

[4]Centro Nacional de Investigaciones de Cafe´, CENICAFE, Manizales, Caldas, Colombia

[5]UMR DIADE, IRD, 911 Avenue Agropolis, BP 64501, 34394 Montpellier, France

INTRODUCTION

In vitro culture has long been associated with the occurrence of somaclonal variation (SV), ranging from moderate to strikingly high rates (Karp 1991; Skirvin et al. 1994). The term SV defines the phenotypic variation observed in clonally propagated plants derived from tissue cultures. Although some SV have been positively evaluated at agronomic level (Skirvin et al. 1994; Biswas et al. 2009), the loss of genetic fidelity is detrimental for commercial purposes when the main objective is strictly the clonal propagation of elite plant material to ensure the maintenance of selected traits.

Little information is yet available about SV-related mechanisms. Growth regulators such as auxin and/or cytokinin are generally indispensable to sustain cell and embryo multiplication. Unfortunately their involvement in SV occurrence has been reported. Studies have been undertaken either directly on cell cultures (Tanurdzic et al. 2008) or on plants derived from cell cultures (Hao and Deng 2002; Smýkal et al. 2007; Rodriguez-Lopez et al. 2010). Early attempts to explain SV revealed the implication of chromosomal abnormalities (Karp 1991). Mutations affect the primary sequence of DNA and include numerical and structural chromosome changes (chromosome deletions, translocations and changes in ploidy), somatic recombination, point mutations, along with deletions and transpositions also occurring in mitochondrial and chloroplast genomes (Orton 1983; Karp 1991; Kaeppler et al. 2000; Neelakandan and Wang 2012). All of these classes of mutations would be expected to give rise to stable sexually heritable variation (Orton1983). Chromosomal instabilities and ploidy alterations have been observed in cell cultures (Hao and Deng 2002; Giorgetti et al. 2011).

The involvement of epigenetic events in SV has also been proposed because some variants reverted back to the normal phenotype after some cycles of vegetative propagation or sometime after field planting (Biswas et al. 2009). A wide range of mechanisms that are genetic or supposedly epigenetic in origin have been recently described as possible causes of SV including TE (Transposable Element) transposition and small RNA directed DNA methylation (Smulders and de Klerk 2011; Neelakandan and Wang 2012). DNA methylation in the form of 5-Methylcytosine (5mC) is a major epigenetic mark involved in gene expression and plays an important role in plant regulation and development (Vanyushin and Ashapkin 2011). DNA methylation in plants commonly occurs at cytosine bases in all sequence contexts, the symmetric CG and CHG contexts (where H indicates A, T or C) and the asymmetric CHH contexts. DNA hypomethylation can produce severe phenotypic assignment as noted in the progeny of *ddm1* Arabidopsis mutants (Kakutani et al. 1996). Tissue culture induces sequence-specific methylation variation that

Citation: Roberto Bobadilla Landey, Alberto Cenci, Romain Guyot, Benoît Bertrand, Frédéric Georget, Eveline Dechamp, Juan-Carlos Herrera, Jamel Aribi, Philippe Lashermes, Hervé Etienne, Assessment of genetic and epigenetic changes during cell culture ageing and relations with somaclonal variation in Coffea Arabica, DOI: 10.1007/s11240-015-0772-9.

can be detected by DNA molecular markers such as Met-AFLP (Bednarek et al. 2007; Fiuk et al.2010) or MSAP (Peraza-Echeverria et al. 2001; Diaz-Martínez et al. 2012).

The loss of DNA methylation in cell cultures can be maintained during plant regeneration, and be stably inherited over subsequent generations (Stroud et al. 2013). However, the effect of DNA methylation deregulation associated with SV remains unclear. An elevated transposition of TEs (class I retrotransposons and class II transposons) has been positively correlated to demethylating events induced by tissue culture in Arabidopsis and maize (Barret et al. 2006; Smith et al. 2012). TEs could potentially alter gene expression when insertions occur inside or near encoding protein genes, finally affecting phenotype (Sato et al. 2011).

Previous studies on *Coffea arabica* (2n = 4x = 44) have shown that somatic embryogenesis can generate variable and relatively high SV levels (5–90 %) depending on in vitro culture conditions (Söndahl and Bragin 1991; Etienne and Bertrand 2003). Nevertheless, when applied over a short 6-month period in the presence of a low growth regulator supply, embryogenic cell cultures gave rise to very low SV rates, as observed through massive phenotypic observations in field plots (0.74 % from 200,000 plants) (Bobadilla Landey et al. 2013). Molecular analyses performed on the regenerated plants revealed that genetic and epigenetic alterations were particularly limited. The involvement of culture ageing in the induction of SV has been proposed in coffee (Etienne and Bertrand 2003). Whatever the genotype, the degree of SV was low (1.3 %) in plants directly produced from embryogenic callus and increased progressively in line with the cell culture age, reaching 25 % on average in plants produced from 12-month-old suspensions in the presence of 4.52 μM 2,4-D (2,4-dichlorophenoxyacetic acid). However, the mechanisms occurring during cell culture ageing and underlying SV have yet to be elucidated in coffee.

The objectives of this work were to identify the phenotypic, chromosomal, genetic and epigenetic changes associated with SV occurrence during cell culture ageing, as well as the stochastic character of the phenomenon. To that end, by establishing three independent embryogenic cell lines, we investigated the effects of culture ageing on the derived plants after more or less extended cell proliferation periods (up to 27 months). Molecular analyses were performed with AFLP (amplified fragment length polymorphism) and two related techniques, SSAP (sequence-specific amplification polymorphism retrotransposon-based molecular marker) and MSAP (methylation-sensitive amplified polymorphism), in order to investigate changes in the activity of different TEs and DNA methylation patterns.

MATERIALS AND METHODS

Plant Material and Somatic Embryogenesis

The plant material corresponded to somatic embryo-derived plants (somatic seedlings) of*C. arabica* var. Caturra regenerated from three independent embryogenic cell lines (A, B and C) established with distinct embryogenic calli. Somatic seedlings were regenerated after three different culture periods, i.e. 4, 11 and 27 months from the establishment of the embryogenic cell line (Fig. 1).

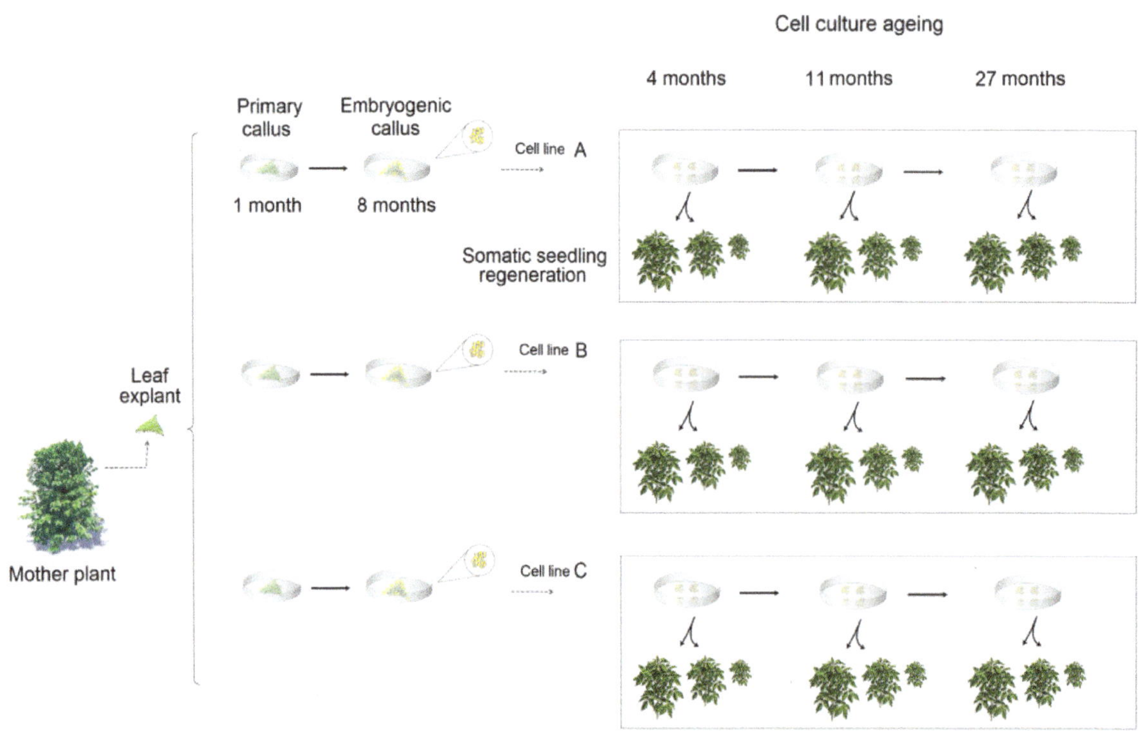

Figure. 1: Schematic representation of the experimental design based on the establishment and long-term maintenance of independent embryogenic cell lines and further regeneration of somatic embryos-derived plantlets at different culture ages (4, 11 and 27 months)

Establishment of Embryogenic Cell Cultures and Plant Regeneration

Embryogenic calli were induced from leaf explants as previously described by Etienne (2005). Three long-term cell cultures (Fig. 2a) were successfully established in baby food jars by initially isolating three distinct embryogenic calli that proliferated independently on half-strength MS (Murashige and Skoog 1962) embryogenic callus production medium (Etienne 2005) enriched with 4.5 μM 2,4-D and 12 μM 6-benzylaminopurine (6-BA) and solidified with 2.8 g/L phytagel. Long-term maintenance of the cell cultures was achieved over 3 years by 1-month proliferation cycles in the dark at 27 °C. Plants could be regenerated during the first 27 proliferation months but, after 36 months, all cell cultures lost the ability to regenerate somatic embryos.

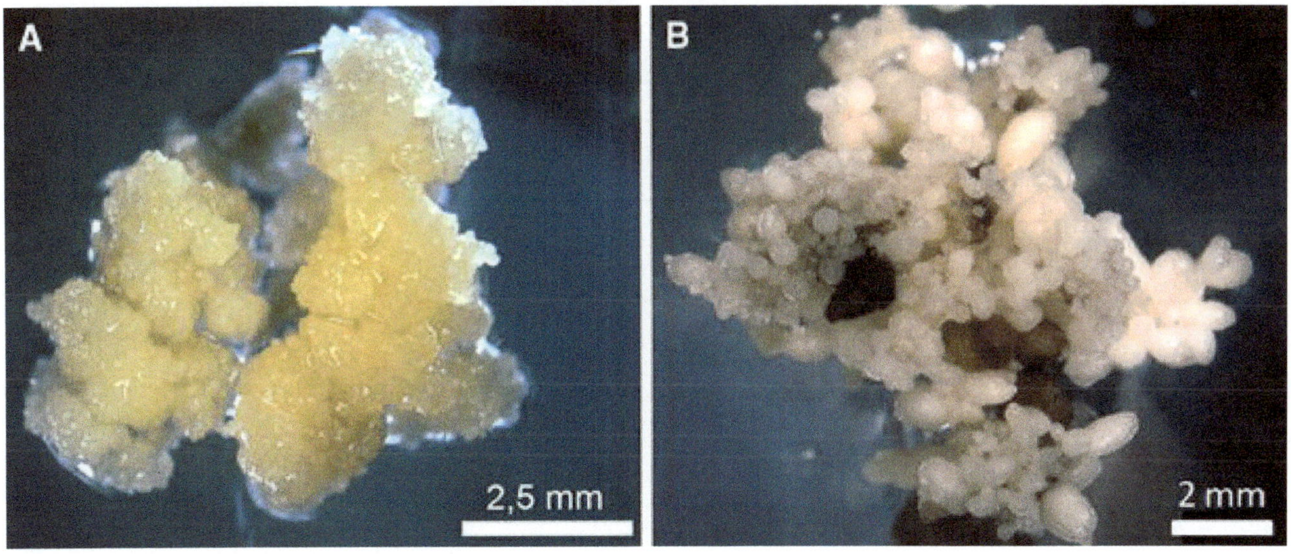

Figure. 2: Long-term cell cultures of *Coffea arabica*. **a** An embryogenic callus belonging to the cellular line A after 11 month proliferation. **b** Regeneration of globular and heart shaped-somatic embryos from an 11 months old cell line

Cell cultures were subjected to embryo regeneration conditions (Fig. 2b) by subculturing approximately 200 mg callus every 4 weeks twice on 'R' regeneration medium (Etienne2005) containing 17.76 μM 6-BA and twice on 'M' maturation medium containing 1.35 μM 6-BA until plantlets developed. Plantlets exhibiting one leaf pair were transferred to the same maturation medium devoid of plant growth regulators and enriched with 1 g/L active charcoal until acclimatization. During 1 month, the plantlets were maintained on a horticultural substrate under a 14 h photoperiod (20 μmol m^{-2} s^{-1} light intensity) at 26 °C and 80–90 % relative humidity. Afterwards, they were moved to nursery (2000–4000 lux).

Phenotype Characterization

The detection of variants was performed on 199 somatic seedlings from the three cell lines (50–65 plants per cell line) after 12 and 24 months in the nursery. For each culture age, 40–75 somatic seedling plants were observed. Phenotypic characterization consisted in assessing plant height, internode number and length, leaf length, leaf width and collar diameter. Other characteristics such as presence or absence of domatia, leaf shape, canopy shape and orthotropic or plagiotropic development were also observed and compared to those of normal plants.

Slide Preparation and Chromosome Counting

Root tips were collected and placed in an aqueous solution of 8-hydroxyquinoline (2.5 mM) used as the pre-treatment, for 4 h in darkness (2 h at 4 °C plus 2 h at 21 °C). A solution of Carnoy (absolute ethanol and glacial acetic acid, 31 v/v) was used to fix the tissues for at least 24 h at −20 °C. Fixed material was then stored in 70 % ethanol at 4 °C pending slide preparation. The stored root tips were used for slide preparations by employing the technique for cell dissociation of enzymatically macerated roots, as described previously by Herrera et al. (2007). Preparations were frozen in liquid nitrogen to remove the coverslips, stained with 1 μg/mL 4', 6-Diamidino-2-Phenylindole, Dihydrochloride (DAPI), and mounted in Vectashield (Vector Laboratories, Peterborough, UK). During slide examination, mitotic cells at metaphase or prometaphase stages were used for chromosome counting. Between 4 and 8 mitotic cells from each individual were analysed to determine the chromosome number. The best examples were photomicrographed at metaphase to document the chromosome number and morphology. A Nikon Eclipse 90i epi-fluorescence microscope equipped with a digital, cooled B/W CCD camera (VDS 1300B Vosskühler®) was used with the appropriate filter (UV-2E/C excitation wavelength 340–380).

Molecular Analyses

One hundred and twenty four plants derived from the three cell cultures and from all culture ages were subjected to molecular markers to estimate their genetic and epigenetic conformity in comparison with mother plants (Table 1). To that end, AFLP, MSAP and SSAP analyses were performed on both the mother plants and on the somatic embryogenesis progeny to reveal polymorphisms in the molecular marker patterns. Genomic DNA was isolated from 100 mg of lyophilized fully developed young leaves using Dellaporta buffer (Dellaporta et al. 1983) containing sodium dodecyl sulphate (SDS) detergent and sodium bisulphite 1 % w/v to avoid leaf oxidation. DNA was purified on spin-column plates as described in the DNeasy plant kit protocol from QIAGEN.

Table 1: Plant material used in molecular marker analyses

Cell culture age (months)	Cellular lines	No. of plants analyzed
4	A	10
	B	20
	C	15
11	A	ND
	B	16
	C	17
27	A	15
	B	20
	C	11
Total		124

Somatic embryogenesis derived plants from *C. arabica* var. caturra were regenerated from cell cultures after variable proliferation periods. Plants were grouped and analyzed based on the embryogenic cell line (A, B, C) and cell culture age (4, 11, 27)

AFLP Markers

AFLP analysis was carried out as described by Vos et al. (1995). Selective amplification included a total of five *Eco*R1/*Msp*I primer combinations using 5-FAM or 5-HEX fluorescently labelled *Eco*R1 (+3) and unlabelled *Mse*I (+3) primers (S1). The PCR program for selective amplification was performed under the following conditions: 3 min at 94 °C, followed by 12 cycles of 45 s at 94 °C, 45 s at 65 °C and 1 min at 72 °C; the annealing temperature was decreased by 0.7 °C per cycle from a starting point of 65 °C during this stage (Touchdown), with a final round of 25 cycles of 94 °C for 45 s, 56 °C for 45 s, 72 °C for 1 min and a final extension of 72 °C for 1 min. MSAP and S-SAP molecular markers were used under the same conditions. Two replications using different DNA extractions were performed for each primer combination and for all molecular markers to reduce the possibility of technical artefacts.

MSAP Molecular Markers

The MSAP protocol (Reyna-López et al. 1997) is an adaptation of the AFLP method for the evaluation of different states of methylation in the symmetric sequence CCGG. In this protocol, the frequent cutting endonuclease (*Mse*I) is replaced by the two restriction enzymes isoschizomers *Hpa*II and *Msp*I with different sensitivities to the methylation state of the CCGG sequence (Table 2). Specifically, *Hpa*II is able to recognize and cut when the CCGG sequence is unmethylated or hemi-methylated on the external cytosine. *Msp*I is able to cut when the CCGG sequence is unmethylated or if the internal cytosine is fully or hemi-methylated. Both *Hpa*II and *Msp*I are unable to cut if the external cytosines are fully methylated. Selective amplification included a total of eight primer combinations per isoschizomer (Supplementary Table S1). *Hpa*II (+2) primers were fluorescently labelled with 5-FAM or 5-HEX while *Eco*R1 (+3) remained unlabelled. The set of fragments from both digestions was grouped by primer combination, fragment size and sample to determine all individual profiles. After fragment classification three possible *Eco*RI/*Hpa*II and *Eco*RI/*Msp* MSAP patterns were found by presence (1) or absence (0).

Table 2: MSAP DNA electrophoretic pattern as revealed by the specificity of the restriction enzymes *Hpa*II and *Msp*I to different methylation states of the symmetric sequence CCGG

Restriction enzymes	MSAP DNA electrophoretic patterns after enzymatic digestion			
	Pattern 1	Pattern 2	Pattern 3	Pattern 4
*Hpa*II	1	1	0	0
*Msp*I	1	0	1	0
Methylation state	Unmethylated	Hemi-methylated	Fully-methylated	Fully-methylated
Methylation position	None	External cytosine	Internal cytosines	External cytosines
Schematic representation	CCGG GGCC CH_3	CH_3 CCGG GGCC	CH_3 CCGG GGCC CH_3	CH_3 CCGG GGCC CH_3

SSAP Markers

The SSAP protocol (Waugh et al. 1997) is an adaptation of AFLP markers to discover polymorphisms associated with transposable element mobility. The TE sequences used in this study were identified in the *C. canephora* genome sequence (Denoeud et al. 2014). TE were identified based on sequence similarities (BLAST) against Repbase (Jurka et al.2005) (http://www.girinst.org/), fine structure analysis using dotter alignments (Sonnhammer and Durbin 1995) or identified through assembly using the AAARF algorithm (DeBarry et al. 2008). Lastly, a sequence database of transposable elements comprising Class I retrotransposons and Class II transposons was constructed and used for SSAP-based primer design. Different elements were used for primer design including 13 different *Copia* and *Gypsy* LTR retrotransposon families, one SINE (Short INterspersed Elements and 2 different transposon families (MuLEs and Mariner families; Supplemenatry Table S2). Primers were designed within a maximum distance of 200 bp at both extremities of each type of elements to find nearby polymorphic TE insertions associated with *Eco*RI restriction sites. Specific TE primers were fluorescently labelled either with 5-FAM, 5-HEX or 5-Tamra while *Eco*R1 (+1) remained unlabelled (Supplementary Table S2). In all, 35 primers were designed and used in screening prior to final analysis; only clear and repetitive profiles were used in the final analysis.

Capillary Electrophoresis and Data Analysis

PCR products were separated by capillary electrophoresis with Pop 7™ polymer in a 16 capillary 3130 XL Genetic Analyzer from Applied Biosystems. Fragment size was estimated using an internally manufactured 524 ROX fluorophore as the sizing standard. The fragments used for fingerprinting were visualized as electropherograms using applied Biosystems software GeneMapper® version 3.7. Informative fragments were mostly found in the 100–450 bp range. All amplified fragments were classified based on the primer combination and their size. The sample fingerprint data were converted to binary code, with (1) denoting the presence of the fragment and (0) its absence. Different binary matrices were constructed for comparative analysis depending on the kind of molecular marker. Global genetic and epigenetic levels were estimated as total polymorphisms. Total polymorphism = [no. of polymorphic fragments/(no. of fragments x no. emblings)] × 100. A 3σ confidence limit for binomial distribution was calculated to analyse the statistical difference between the global genetic and epigenetic levels derived from different molecular markers using the formula $p \pm 3\sqrt{p(1-p)/n}$ with a confidence level of 99 %. The variable analysed was the proportion (p) of polymorphism ($p = x/n$), where "x" was the number of polymorphic fragments and "n" the total number of fragments observed.

RESULTS

Effect of Cell Culture Ageing on Somaclonal Variation

Batches of somatic seedlings from three different culture ages (4, 11 and 27-month-old cell cultures) and regenerated from three different cell lines (A, B and C) were phenotyped (Fig. 1). All the three cell lines were able to regenerate plants after 4, 11 and 27 months of proliferation. Based on plant architecture and leaf morphology we identified a clear pattern of phenotypic abnormalities among somatic seedlings after 12 months of development in the nursery (Fig. 3a). We found a predominant aberrant phenotype characterized by smaller size, oval-elliptic leaves and closed canopy (Fig. 3a, b, d, e), corresponding to "Bullata" coffee mutants described by Krug and Carvalho (1951).

Figure. 3: Morphological differences between normal and *C. arabica* variant phenotypes observed in somatic seedlings regenerated after different culture times (4 and 27 months). **a**Differences in plant size between somatic seedlings derived from 4 (*on the right*) and 27 months (*on the left*) old cell cultures. **b** Size differences after 8 months of parallel development in nursery; Bullata variant from 27 months old cultures (*left*) and phenotypically normal plant from 4 months old culture (*right*). **c** Phenotypically normal somatic seedling from 4 months old cell culture. **d** Bullata variant phenotype from 27 months old culture. **e** Leaves from normal and Bullata phenotype. Bullata leaves (*left*) are rounder and smaller than those from true-to-type coffee somatic seedlings (*right*)

For the three independent cell lines, somaclonal variants were observed in somatic seedlings from long-term cell cultures i.e. 11 and 27-month-old calli affecting in the latter almost all the regenerated plants (Fig. 3a), although all plants derived from younger cell cultures (4 months) exhibited normal phenotypes (Fig. 3c; Table 3). Thirty percent of somatic seedlings from 11-month-old cultures corresponded to abnormal phenotypes (20 and 40 % for cell lines B and C, respectively). From a total of nine variant plants out of 30 plants observed, eight were Dwarf and one Bullata. Out of 75 somatic seedlings from 27-month-old cultures, ninety-four percent displayed an abnormal phenotype (100, 92 and 92 % for cell lines A, B and C, respectively). Only four plants displayed normal phenotypes. The mutants corresponded to four Dwarf phenotypes, one Angustifolia and the rest of the plants were Bullata phenotypes.

Table 3: Frequency of abnormal plants found in somatic seedlings derived from cell cultures of 4, 11 and 27 months

Culture age (months)	Proportion of abnormal plants for each cell line (%)			Somaclonal variation frequency (%)	Confidence interval[a]
	Cell line A	Cell line B	Cell line C		
4	0 *(0/25)* [b]	0 *(0/25)*	0 *(0/25)*	0	0
11	nd	21 *(6/28)*	42 *(9/21)*	30	12.3
27	100*(25/25)*	92*(23/25)*	92*(23/25)*	94.7	5.2

Data were obtained from 199 observed plants regenerated from three independent cellular lines (A, B, C)

nd no data available

[a]We checked whether average frequencies calculated were different. The variable analyzed was the proportion (p) (p = X/n), where X was the number of abnormal plants and n the number of plants observed by culture age. A 3σ confidence limit for binomial distribution was calculated using the formula $p \pm 3\sqrt{pq/n}$

[b]Data in italics represent the No.of abnormal plants/No. of plants observed

Abnormal Chromosome Numbers Limited to Phenotypic Variants

Chromosome counts were performed on 15 somatic seedlings obtained from all cell culture ages (Fig. 4; Table 4). All plants showing a normal phenotype systematically presented normal chromosome numbers. This outcome was even found in the rare case of one phenotypically normal plant regenerated from a 27 month-old cell culture. Conversely, the chromosome counts performed on eight abnormal phenotype plants derived from the two longer cell culture periods revealed aneuploidy. All phenotypic variants presented monosomies i.e. the lack of one or more chromosomes. The Dwarf variants presented 2n-1, 2n-2 and 2n-3 monosomies while the three Bullata phenotypic variants showed 2n-1 monosomy (43 chromosomes).

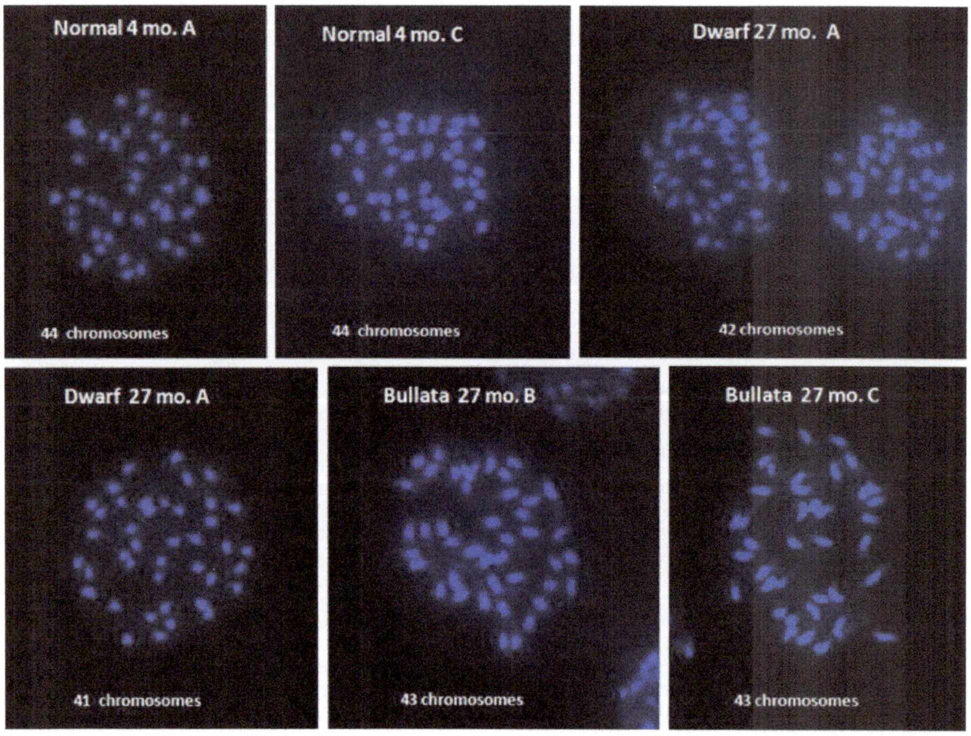

Figure. 4: Mitotic cells at metaphase or prometaphase stages and observed ploidy levels of some normal and variant somatic seedlings from the allotetraploid *C. arabica* species (2n = 4x = 44). Plants were regenerated from 4 and 27 month old cell cultures. Karyotype analyses were performed by counting chromosomes on four to eight clear metaphase spreads obtained from root tips of 12 month-old nursery plants

Table 4: Summary of chromosome countings in some normal *versus* variant *C. arabica* somatic seedlings derived from cell cultures

Cell culture age (months)	Cell line	Somatic seedling phenotype	Chromosome number
4	C	Normal	44
4	A	Normal	44
11	B	Normal	44
11	C	Normal	44
11	B	Normal	44
11	B	Normal	44
11	B	Dwarf variant	43
27	A	Dwarf variant	42
27	A	Dwarf variant	41
27	A	Dwarf variant	43
27	A	Bullata variant	43
27	C	Bullata variant	43
27	B	Bullata variant	43
27	C	Dwarf variant	42
27	B	Normal	44

The chromosome numbers obtained from root tips are indicated for the specific culture age of cell cultures (4, 11 and 27 months) from which somatic seedlings derived

No Genetic Changes Revealed by AFLP and SSAP in the Somatic Seedlings

AFLP analyses were conducted to reveal genetic polymorphisms associated with somatic embryogenesis and more particularly with the age of the embryogenic cell culture. Five primer combinations produced 182 total fragments with an average of 36 fragments per primer combination (Supplementary Table S3). No differences were found between the electrophoretic profiles of mother plants and the 124 somatic seedlings studied. SSAP molecular markers were performed to uncover polymorphisms associated with TE activity (i.e. produced by insertions), possibly activated by tissue culture. In all, 1003 SSAP fragments were generated after 45 selective SSAP amplifications resulting in 22 fragments per primer combination (Supplementary Table S4). No polymorphic fragments were observed between the electrophoretic profiles of mother plants and those of the in vitro progeny (Fig. 5). The number of fragments produced after selective amplification was in a range of 25–133 depending on the element tested. Both SSAP and AFLP analyses of somatic seedlings indicated that the level of genetic variations was not significantly enhanced by either short or long-term cell multiplication periods.

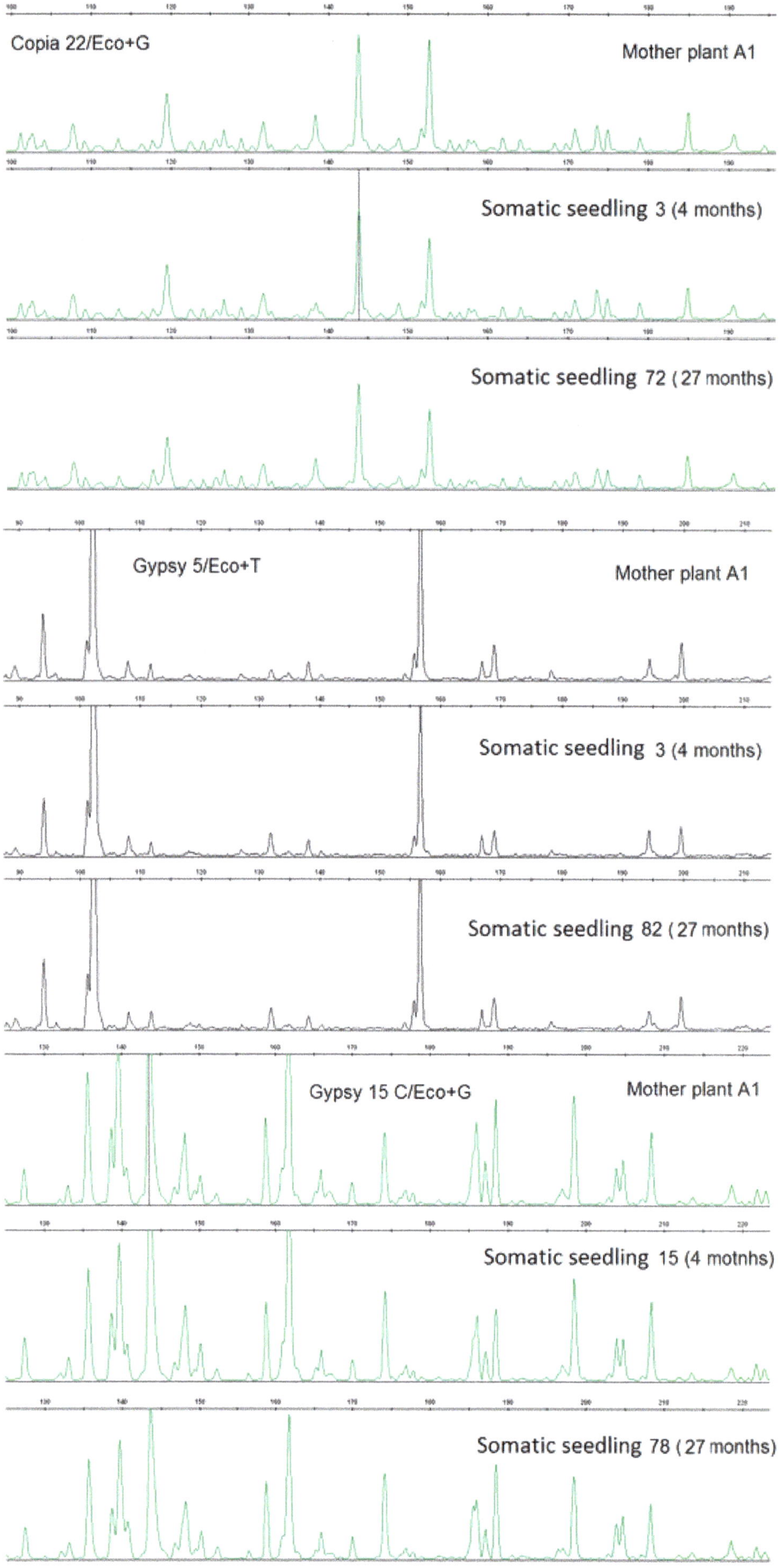

Figure. 5: Examples of SSAP electropherograms observed for coffee mother plants and somatic seedlings derived from 4 and 27 months cell cultures. Illustration of monomorphic electrophoretic profiles

Low Induction of Methylation Changes in short and Long-Term Cell Cultures

MSAP markers were used on coffee somatic seedlings to investigate the methylation changes induced by tissue culture and more particularly by the cell culture age. Eight primer combinations from the separate *Eco*RI/*Hpa*II and *Eco*RI/*Msp*I digestions produced a total of 402 MSAP fragments (Table 5). The pattern distribution resulted in 348 non-methylated sites with pattern 1 (1/1), 52 fragments with internally methylated cytosines with pattern 3 (0/1) and only 2 externally hemi-methylated cytosines with pattern 2 (1/0). The interpretation of the possible methylation state of these patterns indicated that 86.6 % of the fragments corresponded to non-methylated sites (pattern 1) whereas 13.4 % were methylated (patterns 2 and 3).

Table 5: Overall MSAP data and methylation polymorphisms among *C. arabica* mother plants cv. Caturra and somatic seedlings derived from three different cell lines of different ages (4, 11 and 27 months)

Culture age (months)	No. of analyzed somatic seedlings	No. of fragments	Polymorphic MSAP fragments		Total no. of changes	**Total polymorphism (%)[a]**	**3σ Confidence intervals[b]**
			No.	**(%)**			
4	45	402	9	2.2	27	0.149	(0.063–0.235)
11	33	402	6	1.5	15	0.113	(0.026–0.201)
27	46	402	9	2.2	16	0.087	(0.022–0.151)

[a]Total polymorphism = [no. of polymorphic fragments/ (no. of fragments × no. somatic seedlings)] × 100

[b]A 3σ confidence limit for binomial distribution was calculated using the formula $P \pm 3\sqrt{P(1-P)/n}$ with a confidence level of 99 %. The variable analyzed was the proportion (P) of methylation polymorphisms (P = X/n), where X was the number of methylation polymorphisms and n the total number of fragments

The mother plants' profiles and those of somatic seedlings were then compared to find polymorphic methylation fragments (Fig. 6). A total of 58 methylation changes associated with 14 polymorphic fragments were detected in the whole plant population analysed (Table 6). Somatic seedlings derived from 4-month-old cell cultures presented 27 methylation changes accounting for 46.5 % of the overall changes (Table 5). Plant batches from 11 and 27-month-old cell cultures displayed a similar number of changes, i.e. 15 and 16 changes representing 26 and 28 % respectively. The rate of epigenetic variation in somatic seedlings from 4, 11 and 27-month-old cell cultures was calculated as total methylation polymorphism for global estimation. Total polymorphism ranged between 0.087 and 0.149 % whatever the culture age (Table 5). These rates were not statistically different for somatic seedlings derived from short and long-term cell cultures indicating no significant effect of cell culture age. We also noted that methylation changes arose from both methylated and unmethylated sites. The methylation patterns of the 14 polymorphic fragments are indicated in Table 6. Eight fragments showing pattern 3 in mother plants changed into pattern 1 in somatic seedlings indicating demethylation events (the internal methylation of $CC^{Met}GG$ changing to the unmethylated state of the CCGG sequence). The remaining six fragments passed from pattern 1 in mother plants to pattern 3 in somatic seedlings revealing methylation events (the unmethylated state of CCGG changing to the methylated state of the $CC^{Met}GG$ sequence).

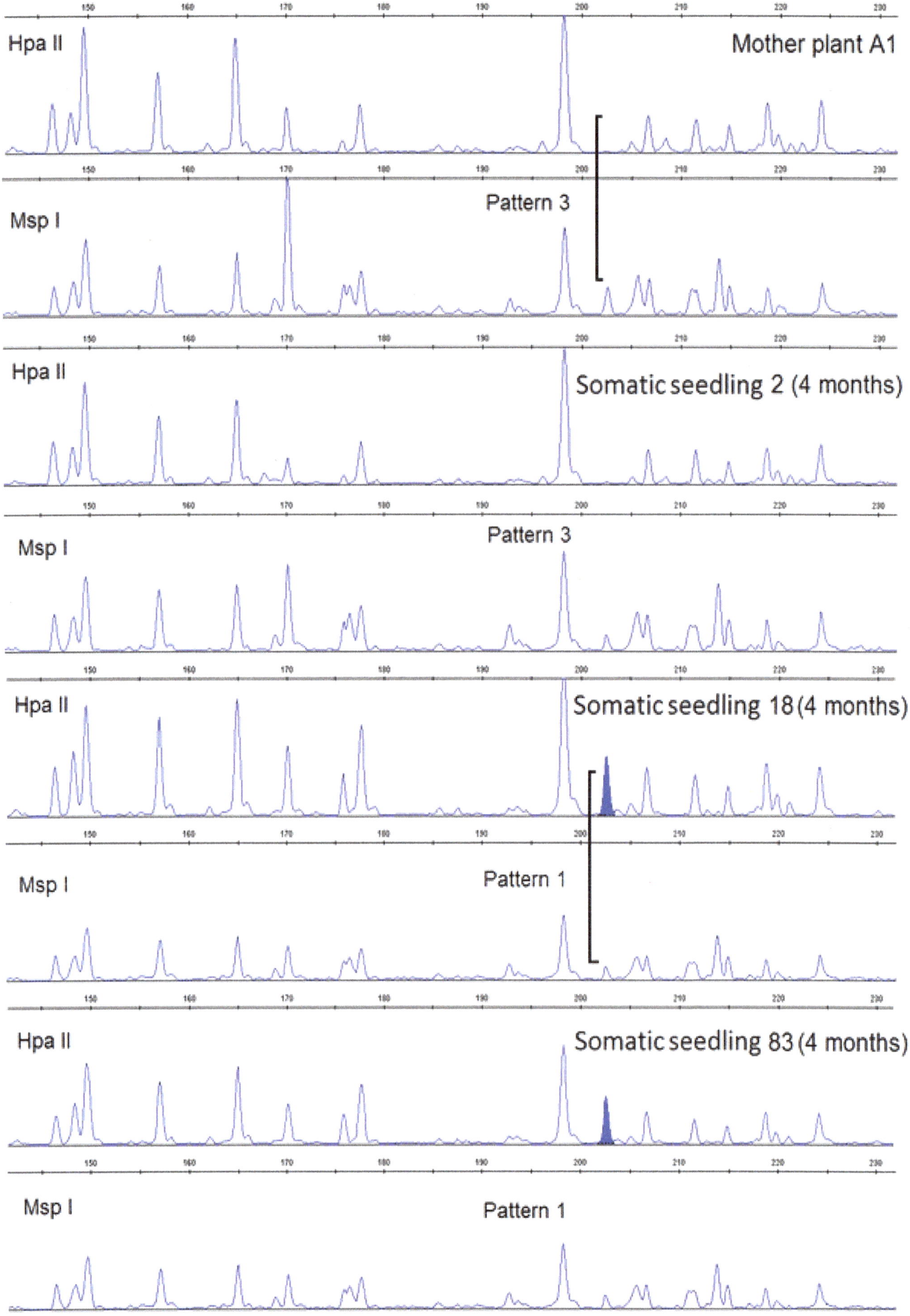

Figure. 6: Examples of MSAP electropherograms observed for coffee mother plants and somatic seedlings progeny derived from 4 month cell cultures using the isoschizomers *Hpa*II and*Msp*I. Illustration of the variation patterns obtained for the phenotypically normal plants

Table 6: MSAP methylation patterns in *C. arabica* mother plants cv. Caturra and modified patterns in regenerated somatic seedlings

Polymorphic MSAP fragments (size in bp)	MSAP methylation patterns[a]		Methylation changes relation with a particular culture age (month) and cell line (A, B, C)			Presence of the methylation change depending on the plant phenotype[b]		Number of methylation changes per fragment	
	Mother plants	**Somatic seedlings**	**4**	**11**	**27**	**Normal**	**Variant**	**No.**	**(%)**
C3 107	Pattern 3	Pattern 1	B	C		+	−	10	17.2
C8 115	Pattern 3	Pattern 1			C	−	+	1	1.7
C3 122	Pattern 3	Pattern 1	B			+	−	1	1.7
C2 127	Pattern 1	Pattern 3	B		A	+	+	4	6.9
C1 133	Pattern 1	Pattern 3	B		A	+	+	5	8.6
C7 182	Pattern 1	Pattern 3	B		A	+	+	2	3.4
C3 202	Pattern 1	Pattern 3	C	B, C	C	+	+	9	15.5
C3 229	Pattern 3	Pattern 1		B, C		+	−	4	6.9
C1 243	Pattern 3	Pattern 1		B		+	−	2	3.4
C4 264	Pattern 3	Pattern 1	B		C	+	+	6	10.3
C7 305	Pattern 3	Pattern 1			C	−	+	3	5.2
C6 323	Pattern 1	Pattern 3	B		A	+	+	2	3.4
C7 386	Pattern 3	Pattern 1		C		+	−	1	1.7
C5 454	Pattern 1	Pattern 3	C	A, B, C	C	+	+	8	13.8
No. changes								58	100

Relation with a specific culture age (4, 11 and 27 months) and phenotype. Presence of a methylation change in a specific cell line was indicated by the cellular line code (A, B or C)

[a]Pattern 1 DNA site restricted by both *Hpa*II and *Msp*I (1, 1); Pattern 3 DNA site not restricted by *Hpa*II and restricted by *Msp*I (0, 1)

[b]Relation with a particular plant phenotype is indicated with (+) for presence and (−) for absence

The number of methylation changes accumulated by somatic seedlings was between 1 and 4 (Table 7). We did not find any somatic seedlings accumulating a large quantity of methylation changes or any effect of cell culture age on their accumulation. The numbers of accumulated methylation changes were similar for somatic seedlings regenerated from all proliferation durations (4, 11, 27 months) (Table 7). Most of the methylation changes appeared and disappeared with no clear relation to a particular cell line or cell culture age (Table 6). Two polymorphic molecular markers (C3 202 bp and C5 454 bp) were observed in some somatic seedlings derived from cell line C and for the three culture ages analysed. However, changes from these two fragments C3 were also shared with other plants derived from different cell lines and without any relation to cell culture age.

Table 7: Accumulation of methylation changes expressed in plant frequencies (%) in 124 coffee somatic seedlings derived from three different cell lines depending on the time of culture (4, 11 and 27 months)

Culture age (months)	No. of plant analyzed	No. of methylation changes/plant [plant frequency (%)]				
		0	1	2	3	4
4	45	56 ± 6	31 ± 5	11 ± 6	0	2 ± 2
11	33	63 ± 9	27 ± 7	10 ± 5	0	0
27	46	78 ± 8.9	13 ± 6	7 ± 6	0	2 ± 2

4Values are means of 3–4 independent measurements ± SD

DISCUSSION

Elevated levels of SV affecting phenotypes have been frequently reported with long-term cell cultures (Tremblay et al. 1999; Jambhale et al. 2001; Biswas et al. 2009). We clearly showed that the occurrence of SV is closely related to cell culture age. The phenotypic evaluation of regenerated plants derived from three independent cell lines showed that no SV or intermediate levels (20–40 %) were found with short and medium-term cultures, respectively. On the contrary, most of the plants (94 %) were variants when regeneration was induced from 2-year-old cultures. Observation of the same behaviour for the three independent cell lines strongly supported the existence of a non-random mechanism for SV. Low levels of SV (around 1 %) were previously found in *C. arabica* plants regenerated from recently obtained callus or from 6-month-old cell cultures (Etienne and Bertrand2001, 2003; Bobadilla Landey et al. 2013).

The first studies on plants derived from tissue culture identified many chromosomal abnormalities as one of the main factors involved in SV (D'Amato 1985; Kaeppler et al.2000). Many altered mechanisms observed during meiosis and mitosis can produce chromosomal alterations, such as non-disjunction (i.e. the failure of sister chromatids to separate during meiosis or mitosis), multipolar mitotic spindles, lagging anaphase chromosomes and chromosome missegregation into micronuclei (D'Amato 1985; Kaeppler et al. 2000; Holland and Cleveland 2009). Such mechanisms can produce a different extent and severity of chromosomal abnormalities in in vitro plants, including segmental or complete aneuploidy (Kaeppler and Phillips 1993; Tremblay et al. 1999). Aneuploidy has frequently been reported in tissue cultures during the callus induction and cell suspension phases (Hao and Deng 2002; Kumar and Mathur 2004; Giorgetti et al. 2011).

The karyotype analysis of coffee somatic seedlings regenerated from old cell cultures revealed different types of aneuploidy exclusively in the monosomic form. Such abnormalities were systematically found in the analysed somaclonal variants derived from 11 and 27-month-old cultures. Normal ploidy levels were always observed in plants showing a normal phenotype. In coffee, many cytological abnormalities were described in leaf explants and cell cultures by Menéndez-Yuffá et al. (2000), such as non-disjunction, double prophase cells, lagging chromosomes, micronuclei and binucleated cells, also including polyploid and aneuploid cells. Our results showed that cytological abnormalities are enhanced by long-term cultures and that aneuploidy plays a major role in *C. arabica*SV.

The mechanisms of non-disjunction of chromosomes during anaphase could be the best explanation for the production of aneuploid cells and plants during tissue culture (Holland and Cleveland 2009). The gene imbalance associated with specific altered chromosomes results in well defined and sometimes predictable variant phenotypes (Henry et al. 2010; Makarevitch and Harris 2010). In this study different chromosome numbers can be found for the same Dwarf variant phenotype. We recently reported the same tendency for other variant phenotypes such as Angustifolia, Giant and Variegata observed in commercial batches of coffee somatic seedlings (Bobadilla Landey et al. 2013). Consequently, it is not possible to predict a coffee variant phenotype on the sole basis of karyotype analyses limited to chromosome counting. Moreover, such analyses did not lead to any observation of other chromosomal alterations associated with undetected deletions, duplications, inversions or translocations of specific chromosomal segments.

Different types of molecular markers were used to explore a genetic or epigenetic origin of SV and reveal any accumulation of polymorphisms over time. AFLP markers did not show any polymorphism in the whole plant population analysed, indicating low levels of genetic variation during ageing of coffee cell cultures. Similar approaches with long-term cultures using different kinds of molecular markers found contrasting results. Polymorphisms were not found using RAPD markers in 2-year-old cultures of hop (Peredo et al. 2006). Using SSR markers Smýkal et al. (2007) reported no genetic variation in 24-year-old multiple shoot cultures of pea. However, in the same work, AFLP molecular markers successfully revealed 22 polymorphic and singleton fragments out of a total of 436 fragments.

Extended coffee cell proliferation periods did not reveal TE activity. These findings are in agreement with AFLP results indicating no or low genetic variation. On the contrary, clear evidence of *Tos 17* and *Tos 19* retrotransposon activity was provided in rice cell cultures (Hirochika et al. 1996). The first hypothesis is that these elements were not reactivated under stressful somatic embryogenesis conditions or are inactive. Although TEs constitute a large percentage of the plant genome, only a few elements were reported active in the literature. A second hypothesis is that the inactivity of the chosen elements resulted from their transcriptional inactivation through methylation or epigenetic post-transcriptional mechanisms preventing their insertion (Mirouze et al. 2009).

Many studies on vitroplants have reported high and unexpected levels of epigenetic variation related to cytosine methylation (Miguel and Marum 2011). A first approach was aimed at exploring specific sequence contexts with methylation-sensitive molecular markers (e.g. CCGG for MSAP) and the other at measuring global methylation levels in all three cytosine contexts CG, CHG and CHH with HPLC methods. In our conditions, the number of methylation changes per plant was small and the resulting total polymorphism was <1 % independently of the age of the cell culture, indicating little disruption to the epigenetic status. Similarly, low levels of methylation variation were found in freesia somatic seedlings derived from short-term cultures (Gao et al. 2010) and pea long-term multiple shoot cultures (Smýkal et al. 2007).

Many studies with methylation-sensitive molecular markers that reported a high level of epigenetic variation reported no effect on phenotype or morpho-agronomic traits (Bednarek et al. 2007; Li et al. 2007; Schellenbaum et al. 2008; Fiuk et al. 2010). Methylation changes contribute poorly to global estimations when they do not occur at high frequencies (Smulders and Klerk 2011). The individual accumulation of methylation changes could be a more reliable indicator of tissue culture stress or plant epigenetic instability. In the epigenetic evaluation made by molecular markers (e.g. MSAP) most in vitro plants do not generally tend to accumulate large numbers of methylation changes (Li et al. 2007; Bednarek et al. 2007; Schellenbaum et al. 2008; Fiuk et al. 2010). However, few plants could individually accumulate many methylation changes indicating higher epigenetic instability. Coffee plants from short and long-term cell cultures accumulated comparable small numbers of methylation changes (1-4 per individual) independently of their normal *vs* variant phenotypes. Similar observation was done in both normal and variant coffee somatic seedlings propagated industrially (Bobadilla Landey et al. 2013).

The elevated SV rates frequently reported with long-term cell cultures offer a unique opportunity to understand the underlying molecular and physiological mechanisms. Long-term cultures were used to establish the full range of cytological, genetic and epigenetic mechanisms leading to SV in coffee. Cell culture ageing was proved to be highly mutagenic. Although the molecular marker approach used cannot generate exhaustive information, it can be concluded that genetic polymorphisms and epigenetic changes are particularly limited during cell culture ageing, while aneuploidy plays a major role in SV.*C. arabica* is a young allopolyploid that still has most of its genes in duplicated copies (Cenci et al. 2012). It could be hypothesized that the impact of genetic or epigenetic polymorphisms has been restricted because of the buffer effect due to polyploidy. However the allopolyploid structure also allowed aneuploid cells to survive and regenerate viable plants, which would certainly not be the case with non-polyploid species. The absence of one or more chromosomes was probably buffered by the other homologous chromosomes. Hence, *C. arabica* is an interesting species for studying the occurrence and consequences of chromosomal abnormalities in relation to SV.

ACKNOWLEDGMENTS

Financial support for this study was provided by the Mexican Government through a grant to Roberto Bobadilla Landey by the CONACyT (CVU:1623391) program and by the CIRAD funds for doctoral support. The authors thank Peter Biggins (CIRAD) for critically reviewing the manuscript.

REFERENCES

1. Barret P, Brinkman M, Beckert M (2006) A sequence related to rice Pong transposable element displays transcriptional activation by in vitro culture and reveals somaclonal variations in maize. Genome 49:1399–1407
2. Bednarek PT, Orłowska R, Koebner RMD, Zimny J (2007) Quantification of tissue-culture induced variation in barley (*Hordeum vulgare* L.). BMC Plant Biol 7:10
3. Biswas MK, Dutt M, Roy UK, Islam R, Hossain M (2009) Development and evaluation of in vitro somaclonal variation in strawberry for improved horticultural traits. Sci Hort 22:409–416
4. Bobadilla Landey R, Cenci A, Georget F, Bertrand B, Camayo G, Dechamp E, Simpson J, Herrera JC, Santoni S, Lashermes P, Etienne H (2013) High genetic and epigenetic stability in *Coffea arabica* plants derived from embryogenic suspensions and secondary embryogenesis as revealed by AFLP, MSAP and the phenotypic variation rate. PLoS One 8(2):e56372. doi:10.1371/journal.pone.0056372
5. Cenci A, Combes M-C, Lashermes P (2012) Genome evolution in diploid and tetraploid*Coffea* species as revealed by comparative analysis of orthologous genome segments. Plant Mol Biol 78:135–145
6. D'Amato F (1985) Cytogenetics of plant cells and tissue cultures and their regenerates. Plant Sci 3:73–112
7. DeBarry JD, Liu R, Bennetzen JL (2008) Discovery and assembly of repeat family pseudomolecules from sparse genomics sequence data using the assisted automated assembler of repeat families (AAARF) algorithm. BMC Bioinform 13(9):235. doi:10.1186/1471-2105-9-235
8. Dellaporta SL, Wood J, Hicks JB (1983) A plant DNA minipreption: version II. Plant Mol Biol Rep 1:19–21
9. Denoeud F, Carretero-Paulet L, Dereeper A, Droc G, Guyot R, Pietrella M, Zheng C, Alberti A, Anthony F et al (2014) The coffee genome provides insight into the convergent evolution of caffeine biosynthesis. Science 345:1181–1184. doi:10.1126/science.1255274
10. Díaz-Martínez M, Nava-Cedillo A, Guzmán-López JA, Escobar-Guzmán R, Simpson J (2012) Polymorphism and methylation patterns in *Agave tequilana* Weber var. 'Azul' plants propagated asexually by three different methods. Plant Sci 185–186:321–330
11. Etienne H (2005) Protocol of somatic embryogenesis: coffee (*Coffea arabica* L. and *C. canephora* P.). In: Jain SM, Gupta PK (eds) Protocols for somatic embryogenesis in woody plants. Series: Forestry Sciences, vol 77. 590 p, Springer, Dordrecht, pp 167–179, ISBN: 1-4020-2984-5
12. Etienne H, Bertrand B (2001) The effect of the embryogenic cell suspension micropropagation technique on the trueness to type, field performance,

bean biochemical content and cup quality of *Coffea arabica* trees. Tree Physiol 21:1031–1038

13. Etienne H, Bertrand B (2003) Somaclonal variation in *Coffea arabica*: effects of genotype and embryogenic cell suspension age on frequency and phenotype of variants. Tree Physiol 23:419–426
14. Fiuk A, Bednarek PT, Rybczýnski JJ (2010) Flow cytometry, HPLC-RP and met AFLP analyses to assess genetic variability in somatic embryo-derived plantlets of *Gentiana pannonica* scop. Plant Mol Biol Rep 28:413–420
15. Gao X, Yang D, Cao D, Ao M, Sui X, Wang Q, Kimatu JN, Wang L (2010) In vitro micropropagation of *Freesia hybrida* and the assessment of genetic and epigenetic stability in regenerated plantlets. J Plant Growth Reg 29:257–267
16. Giorgetti L, Castiglione M, Turrini A, Ronchi V, Geri C (2011) Cytogenetic and histological approach for early detection of "mantled" somaclonal variants of oil palm regenerated by somatic embryogenesis: first results on the characterization of regeneration system. Caryologia 64:223–234
17. Hao YI, Deng XX (2002) Occurrence of chromosomal variations and plant regeneration from long-term-cultured *citrus* callus. In vitro Cell Dev Biol Plant 38:472–476
18. Henry IM, Dilkes BP, Miller ES, Burkart-Waco D, Comai L (2010) Phenotypic consequences of aneuploidy in *Arabidopsis thaliana*. Genetics 186:1231–1245
19. Herrera JC, D'hont A, Lashermes P (2007) Use of fluorescence *in situ* hybridization as a tool for introgression analysis and chromosome identification in coffee (*Coffea arabica*L.). Genome 50:619–626
20. Hirochika H, Sugimoto K, Otsuki Y, Tsugawa H, Kanda M (1996) Retrotransposons of rice involved in mutations induced by tissue culture. Proc Natl Acad Sci 93:7783–7788
21. Holland AJ, Cleveland DW (2009) Boveri revisited: chromosomal instability, aneuploidy and tumorigenesis. Nat Rev Mol Cell Biol 10:478–487
22. Jambhale ND, Patil SC, Jadhav AS, Pawar SV, Waghmode BD (2001) Effect of numer of subcultures on *in vitro* multiplication of four banana clones. Infomusa 10:38–39
23. Jurka J, Kapitonov VV, Pavlicek A, Klonowsky P, Kohany O, Walichiewicz J (2005) Repbase update, a database of eukaryotic repetitive elements. Cytogenet Genome Res 110:462–467
24. Kaeppler SM, Phillips RL (1993) DNA methylation and tissue culture-induced DNA methylation variation in plants. In Vitro Cell Dev Biol Plant 29:125–130
25. Kaeppler SM, Kaeppler HF, Rhee Y (2000) Epigenetic aspects of somaclonal variation in plants. Plant Mol Biol 43:179–188
26. Kakutani T, Jeddeloh JA, Flowers SK, Munakata K, Richards EJ (1996) Developmental abnormalities and epimutations associated with DNA hypomethylation mutations. Proc Natl Acad Sci USA 93:12406–12411
27. Karp A (1991) On the current understanding of somaclonal variation. In: Miflin BJ (ed) Oxford surveys of plant molecular and cell biology, vol 17. Oxford University Press, Oxford, pp 1–58
28. Krug CA, Carvalho A (1951) The genetics of *Coffea*. Adv Genet 4:127–158
29. Kumar PS, Mathur VL (2004) Chromosomal instability in callus culture of *Pisum sativum*. Plant Cell Tissue Organ Cult 78:267–271
30. Li X, Yu X, Wang N, Feng Q, Dong Z, Liu L, Shen J, Liu B (2007) Genetic and epigenetic instabilities induced by tissue culture in wild barley (*Hordeum brevisubulatum* (Trin.) Link). Plant Cell Tissue Organ Cult 90:153–168
31. Makarevitch I, Harris C (2010) Aneuploidy causes tissue specific qualitative changes in global gene expression patterns in maize. Plant Physiol 152:927–938
32. Menéndez-Yuffá A, Fernandez R, Rios L, Xena N (2000) Mitotic aberrations in coffee (*Coffeea arabica* cv. 'Catimor') leaf explants and their derived embryogenic calli. Electron J Biotech 3:1–6
33. Miguel C, Marum L (2011) An epigenetic view of plant cells cultured in vitro: somaclonal variation and beyond. J Exp Bot 62:3713–3725
34. Mirouze M, Reinders J, Bucher E, Nishimura T, Schneeberger K, Ossowski S, Cao J, Weigel D, Paskowski J, Mathieu O (2009) Selective epigenetic control of retro transposition in Arabidopsis. Nature 461:427–430
35. Murashige T, Skoog FA (1962) Revised medium for rapid growth and bioassays with tobacco tissue cultures. Physiol Plant 15:473–497
36. Neelakandan AK, Wang K (2012) Recent progress in the understanding of tissue culture-induced genome level changes in plants and potential applications. Plant Cell Rep 31:597–620
37. Orton TJ (1983) Experimental approaches to the study of somaclonal variation. Plant Mol Biol Rep 1:67–76
38. Peraza-Echeverria S, Herrera-Valencia VA, Kay A-J (2001) Detection of DNA methylation changes in micropropagated banana plants using methylation-sensitive amplification polymorphism (MSAP). Plant Sci 161:359–367
39. Peredo EL, Angeles Revilla M, Arroyo-Garcia R (2006) Assessment of genetic and epigenetic variation in hop plants regenerated from sequential subcultures of organogenic calli. J Plant Physiol 163:1071–1079
40. Reyna-Lopez GE, Simpson J, Ruiz-Herrera J (1997) Differences in DNA methylation pattern are detectable during the dimorphic transition of fungi by amplification of restriction polymorphisms. Mol Gen Genet 253:703–710
41. Rodríguez López CM, Wetten AC, Wilkinson MJ (2010) Progressive erosion of genetic and epigenetic variation in callus-derived. New Phytol 186:856–868
42. Sato M, Hosokawa M, Doi M (2011) Somaclonal variation is induced de novo via the tissue culture process: a study quantifying mutated cells in Saintpaulia. PLoS One 6(8):e23541. doi:10.1371/journal.Pone.0023541

43. Schellenbaum P, Mohler V, Wenzel G, Walter B (2008) Variation in DNA methylation patterns of grapevine somaclones (*Vitis vinifera* L.). BMC Plant Biol 8:78

44. Skirvin RM, McPheeters KD, Norton M (1994) Sources and frequency of somaclonal variation. HortScience 29:1232–1237

45. Smith AM, Hansey CN, Kaeppler SM (2012) TCUP a novel *h*AT transposon active in maize tissue culture. Front Plant Sci 3:6

46. Smulders MJM, de Klerk GJ (2011) Epigenetics in plant tissue culture. Plant Growth Reg 63:137–146

47. Smýkal P, Valledor L, Rodriguez R, Griga M (2007) Assessment of genetic and epigenetic stability in long-term in vitro shoot culture of pea (*Pisum sativum* L.). Plant Cell Rep 26:1985–1998

48. Söndahl MR, Bragin A (1991) Somaclonal variation as a breeding tool for coffee improvement. In: Proceedings of 14th Colloquium International Coffee Science Association, San Francisco, USA. ASIC Publishers, Vevey, pp 701–710

49. Sonnhammer EL, Durbin R (1995) A dot-matrix program with dynamic threshold control suited for genomic DNA and protein sequence analysis. Gene 167:1–10

50. Stroud H, Ding B, Simon SA, Feng S, Bellizzi M, Pellegrini M, Wang GL, Meyers BC, Jacobsen SE (2013) Plants regenerated from tissue culture contain stable epigenome changes in rice. Elife 2:e00354. doi:10.7554/eLife.00354

51. Tanurdzic M, Vaughn MW, Jiang H, Lee T-J, Slotkin RK, Sosinski B, Thompson WF, Doerge RW, Martienssen RA (2008) Epigenomic consequences of immortalized plant cell suspension culture. PLoS Biol 6:2880–2895

52. Tremblay L, Levasseur C, Tremblay FM (1999) Frequency of somaclonal variation in black spruce (*Picea mariana,* Pinaceae) and white spruce (*P. glauca,* Pinaceae) derived from somatic embryogenesis and identification of some factors involved in genetic instability. Am J Bot 86:1373–1381

53. Vanyushin BF, Ashapkin VV (2011) DNA methylation in higher plants: past, present and future. Biochim Biophys Acta 1809:360–368

54. Vos P, Hogers R, Bleeker M, Reijans M, Van de Lee T, Homes M, Frijters A, Pot J, Peleman J, Kuiper M, Zabeau M (1995) AFLP: a new technique for DNA fingerprinting. Nucleic Acids Res 23:4407–4414

55. Waugh R, McLean K, Flavell AJ, Pearce SR, Kumar A, Thomas BB, Powell W (1997) Genetic distribution of Bare-1-like retrotransposable elements in the barley genome revealed by sequence-specific amplification polymorphisms (S-SAP). Mol Gen Genet 253:687–694

CHAPTER 8

RHAPONTICUM CARTHAMOIDES REGENERATION THROUGH DIRECT AND INDIRECT ORGANOGENESIS, MOLECULAR PROFILES AND SECONDARY METABOLITE PRODUCTION

Ewa Skała[1] Renata Gra¸bkowska[1] Przemysław Sitarek[1] Łukasz Kuz´ma[1] Andrzej Błauz_ [2] Halina Wysokin´ska[1]

[1].Department of Biology and Pharmaceutical Botany, Medical University of Ło´dz´, Muszyn´skiego 1, 90-151 Ło´dz´, Poland

[2] Cytometry Laboratory, Department of Molecular Biophysics, University of Lo´dz´, ul. Pomorska 141/143, 90-236 Ło´dz´, Poland

INTRODUCTION

Rhaponticum carthamoides (Willd.) Iljin (known as "maral root" or Russian leuzea) is an endemic, perennial, herbaceous plant belonging to the family Asteraceae. The plants originate from the mountains of South Siberia, Central Asia and China (Timofeev et al.2006; Kokoska and Janovska 2009). The roots and rhizomes of *R. carthamoides* have been used for a long time in Siberian folk medicine in the treatment of overstrain and weakness after illness. They also have adaptogenic, antioxidant, immunomodulatory, anticancerogenic and antimicrobial properties (Kokoska et al. 2002; Biskup and Lojkowska 2009; Kokoska and Janovska 2009). The raw material (roots and rhizomes) of*R. carthamoides* is used to eliminate physical and mental weariness, and to promote muscle growth and sexual function (Kokoska and Janovska 2009). The major chemical constituents in this plant are ecdysteroids, flavonoids, phenolic acids, triterpenoid glycosides, polyacetylenes and sesquiterpene lactones (Kokoska and Janovska 2009). Harvesting the roots and underground parts requires the plant to be cut down, which has led to the depletion of *R. carthamoides* from its natural habitat. The development of biotechnological methods such as micropropagation is one solution which may overcome this problem. There are few reports on the in vitro propagation of *R. carthamoides*. Duškova and Dušek (1995) regenerated the plant species from cotyledon-derived callus, while Zand et al. (2014) obtained shoots of *R. carthamoides* from both leaf-derived callus and directly from root explants. However, the genetic stability of in vitro regenerated plants has not been studied so far.

Clonal fidelity is an imperative among in vitro regenerated plants. A major problem associated with in vitro cultures might be the occurrence of somaclonal variation, mainly in callus-derived plants. Of several molecular markers used for assessment of genetic fidelity of in vitro regenerated plants, random amplified polymorphic DNA (RAPD) and inter-simple sequence repeats (ISSR) marker analyses are simple, quick and cost-effective methods. The use of techniques to successfully assess the genetic fidelity of several plant species like *Saussurea involucrate* (Yuan et al. 2009), *Hydrangea macrophylla* (Liu et al.2011), *Garcinia* sp. (Mohat et al. 2012) or *Withania somnifera* (Nayak et al. 2013) has been recently reported. Flow cytometry for genome size estimation can be used to complement molecular marker analyses (Slazak et al. 2015).

The present study describes the regeneration of *R. carhamoides* plants through direct (from leaves of in vitro culture) and indirect organogenesis (from hypocotyl, cotyledon and root of 3-week-old seedling). The assessment of genetic variation of the regenerated plants by RAPD and ISSR markers, and flow cytometry was performed. To evaluate whether or not the regeneration pathway altered the biosynthesis of secondary

Citation: Ewa Skała, Renata Grąbkowska, Przemysław Sitarek, Łukasz Kuźma, Andrzej Błauż, Halina Wysokińska, Rhaponticum carthamoides regeneration through direct and indirect organogenesis, molecular profiles and secondary metabolite production, DOI 10.1007/s11240-015-0816-1.

metabolites in *R. carthamoides*, the contents of chlorogenic acid and 20-hydroxyecdysone (20-HE) in leaves and roots of *R. carthamoides* plants derived through direct and indirect organogenesis were determined using UHPLC method, the results being compared to the respective parts of seed-derived plants. Chlorogenic acid and 20-hydroxyecdysone are known to be the principle compounds responsible for therapeutic effects of extracts of *R. carthamoides*. Chlorogenic acid is a free radical and metal scavenger; it may interfere with glucose absorption and has been shown to modulate gene expression of antioxidant enzymes (Clifford 1999; Olthof et al. 2001). 20-HE is known to have antioxidant and anti-inflammatory properties (Zhou et al. 2012), and to exert a stimulatory effect on protein synthesis and increase metabolic activity (Sláma and Lafont 1995).

MATERIALS AND METHODS

Plant Material

Rhaponticum carthamoides seeds were obtained from the Medicinal Plant Garden of the Department of Pharmacognosy, Medical University of Łódź (Poland). The botanical identity of plants was confirmed by E. Skała according to Flora of China (www.eFloras.org). The seeds were sterilized with 2 % sodium hypochlorite solution for 10 min and then rinsed three times (15 min, each) in sterile distilled water. For germination, seeds were placed on Murashige and Skoog (MS; 1962) agar (0.7 %) medium supplemented with 3 % sucrose, gibberellic acid (0.1 mg L^{-1}) and kinetin (0.02 mg L^{-1}).

Indirect Shoot Regeneration

The cotyledons, hypocotyl and roots of the 3-week-old single seedling were used as explants for callus induction. The cotyledons were divided into quarters, the hypocotyl and roots were cut into sections of 5 mm in length. The explants were placed horizontally on MS agar (0.7 %) medium supplemented with auxin: 2,4-D, NAA or IBA in combination with cytokinin, BA. The growth regulators were used at concentrations of 0.2 or 0.5 mg L^{-1}. The ratio between cytokinin and auxin in the media was 2.5:1 or 1:2.5. Organogenic calluses generated from different seedling explants were subcultured every 5 weeks on fresh MS medium of the same composition. After 5 weeks, the percentage of calluses inducing buds/shoots (organogenesis frequency) and the average number of adventitious buds/shoots per callus were determined for three successive passages (3–5). 5–15 calluses were used for each explant type and medium combination. The results are presented in Table 1.

Table 1: Indirect shoot organogenesis of *R. carthamoides* from callus culture initiated from different seedling explants. Calluses were cultured on MS agar medium supplemented with different plant growth regulators for 5 weeks

Growth regulators (mg L^{-1})	Type of explants	Organogenesis frequency (%)	Mean number of buds/shoots per callus ± SE
2,4-D 0.5	Root	0	0^{a}
BA 0.2	Hypocotyl	0	0^{a}
	Cotyledon	0	0^{a}
2,4-D 0.2	Root	0*	0^{a}
BA 0.5	Hypocotyl	0	0^{a}
	Cotyledon	0	0^{a}
NAA 0.5	Root	82.6	3.47 ± 0.41^{de}
BA 0.2	Hypocotyl	88.9	2.25 ± 0.45^{b}
	Cotyledon	92.3	3.83 ± 0.36^{e}
NAA 0.2	Root	70.0	2.79 ± 0.26^{bc}
BA 0.5	Hypocotyl	76.9	2.70 ± 0.52^{bc}
	Cotyledon	80.0	3.44 ± 0.52^{de}
IBA 0.5	Root	66.7	2.25 ± 0.41^{b}
BA 0.2	Hypocotyl	60.0	2.22 ± 0.32^{b}
	Cotyledon	69.2	3.22 ± 0.46^{cde}
IBA 0.2	Root	64.3	2.44 ± 0.41^{bc}
BA 0.5	Hypocotyl	42.9	2.33 ± 0.33^{b}
	Cotyledon	73.3	2.91 ± 0.49^{bcd}

The values represent the mean ± standard error (SE). The experiments were conducted with three replicates. 5–15 calluses were used for each explant type and treatment for each experiment. Values within a column followed by the same letter are not significantly different from each other at a 5 % level according to the Kruskal–Wallis test and multiple comparisons of average ranks

* Callus regenerated the numerous roots (about 1 cm of length)

Direct Shoot Organogenesis

The shoot tip (about 5 mm long) excised from the same 3-week-old seedling of *R. carthamoides*, which was used to initiate callus cultures was transferred into MS agar medium containing 0.1 mg L^{-1} IAA and 0.2 mg L^{-1} BA. Following the establishment of the shoot culture, the shoots were subcultured regularly at 5-week-intervals on MS medium, as described above. After 5 passages, leaves of the shoots were taken as starting material. The leaves were cut into lamina (about 0.5 × 0.5 cm; length × width) and petiole (about 1 cm in length). The explants were placed horizontally with their abaxial side down on MS agar (0.7 %) medium supplemented with BA (0.2 or 0.5 mg L^{-1}) alone or in combination with auxin IBA or NAA (0.2 or 0.5 mg L^{-1}). The ratio of cytokinin to auxin in the media was 2.5:1 or 1:2.5. After 5 weeks, the percentage of explants forming adventitious shoots (organogenesis frequency) and the average number of buds/shoots per explant were recorded (Table 2). The measurements were repeated three times, using 20 of each type explants for each replicate.

Table 2: Direct shoot organogenesis from leaf explants of *R. carthamoides*

Growth regulators (mg L^{-1})					Mean number of buds/shoots per explant ±SE
NAA	IBA	BA	Type of explant	Organogenesis frequency (%)	
–	–	0.2	Lamina	96.9	3.48 ± 0.39^{ab}
–	–	0.2	Petiole	96.8	3.97 ± 0.30^{cd}
–	–	0.5	Lamina	90.6	3.24 ± 0.32^{ab}
–	–	0.5	Petiole	96.8	3.47 ± 0.27^{ab}
0.2	–	0.5	Lamina	96.8	3.83 ± 0.34^{bcd}
0.2	–	0.5	Petiole	100	3.61 ± 0.21^{abc}
0.5	–	0.2	Lamina	100	4.48 ± 0.38^{d}
0.5	–	0.2	Petiole	93.5	4.38 ± 0.38^{d}
–	0.2	0.5	Lamina	85.7	3.08 ± 0.31^{a}
–	0.2	0.5	Petiole	82.8	3.33 ± 0.27^{ab}
–	0.5	0.2	Lamina	92.3	4.25 ± 0.34^{d}
–	0.5	0.2	Petiole	100	4.08 ± 0.33^{cd}

The explants were cultured on MS medium for 5 weeks

The values represent the mean ± SE of three independent experiments. Each experiment comprised 20 each type explants. Values within a column followed by the same letter are not significantly different from each other at a 5 % level according to the Kruskal–Wallis test and multiple comparisons of average ranks

Culture Conditions

Incubation was performed in a growth chamber at 26 ± 2 °C and 16 h photoperiod (light intensity of 40 μmol m^{-2} s^{-1} provided by fluorescent tubes).

Shoot Rooting and Plantlet Acclimatization

Excised shoots regenerated from root-derived callus and leaf explants were transferred into MS agar (0.7 %) medium with half-strength concentration of macro- and micro-nutrients (½ MS) without growth regulators or supplemented with IBA at concentrations of 0.2, 0.5, 1.0, 1.5 or 2.0 mg L^{-1}. The cultures were incubated under the same conditions as above. After 5 weeks, the percentage of shoots forming roots, the average number roots per rooted shoot and the average length of roots (cm) were recorded. The results presented in Fig. 1, are means of three replications: each replicate comprising 15 shoots. The rooted plantlets were washed in tap water to remove agar from the roots and transplanted into 10-cm-diameter-pots containing a sterilized mixture of soil, sand, peat and perlite (4:3:3:3 v/v/v/v). To maintain high humidity, the potted plantlets were covered with glass, which were gradually opened after 3 days and completely removed after 2 weeks. The potted plantlets were kept in the greenhouse at 26 ± 2 °C and natural light. The percentage survival rate of the plants and their morphology were recorded after 3 months. The following characteristics were measured and are presented in Table 3: the mean plant length (determined as greatest leaf length), the number of leaves/plant, the width and length of leaf lamina, length of leaf petiole, maximum root length and average plant fresh weight. The twenty-five plants regenerated from leaf explants (DSO-plants) and 25 plants regenerated from callus (IDSO-plants) were used in this experiment.

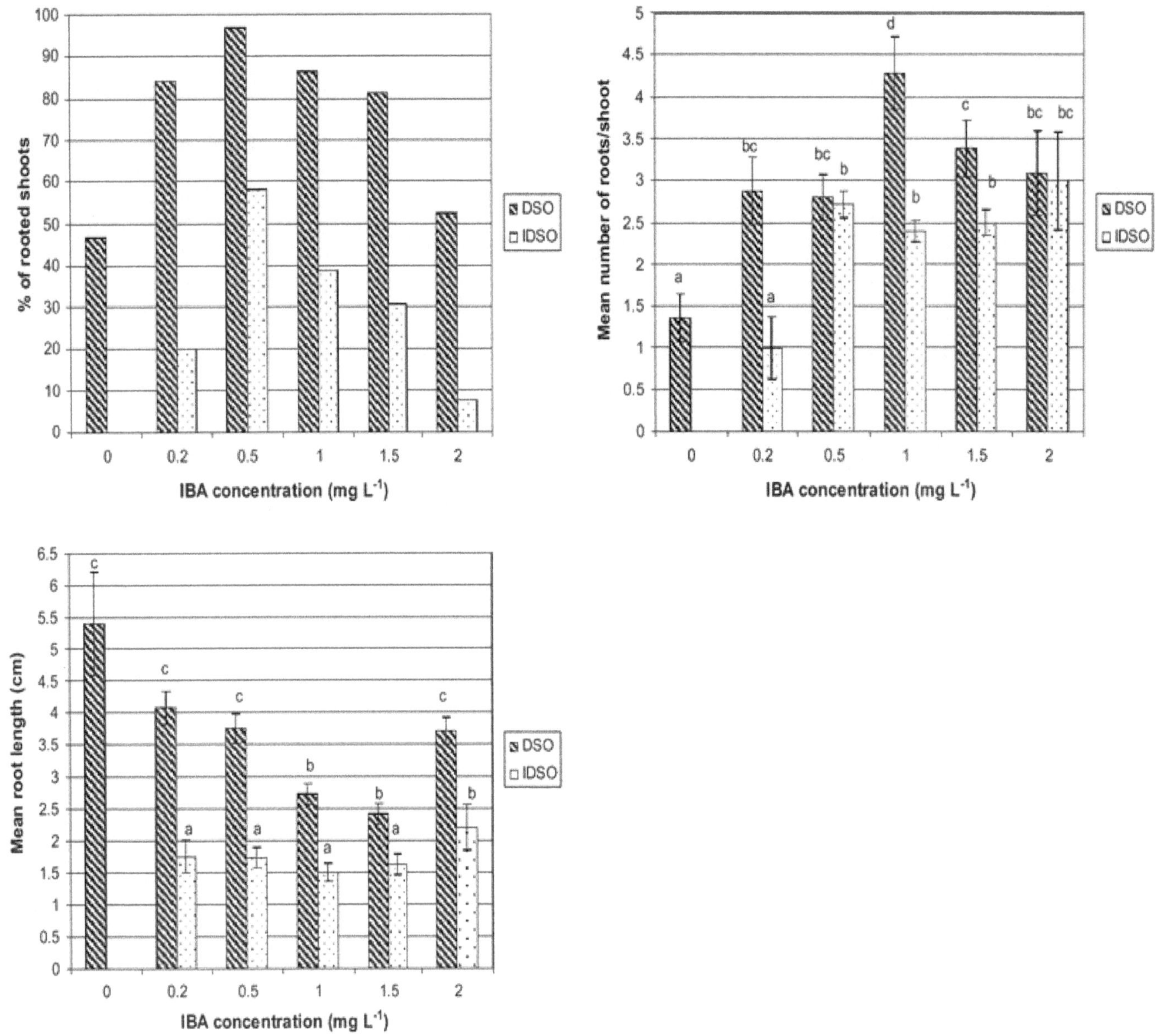

Figure. 1: Rooting of *R. carthamoides* shoots on ½ MS medium without or supplemented with IBA. The shoots were obtained through direct organogenesis (DSO) and through indirect organogenesis (IDSO). For rooting the shoots were cultured for 5 weeks. Results are mean ± standard error from three independent experiments; means marked by the *same letter* are not significantly different from each other at a 5 % level according to the Kruskal–Wallis test and multiple comparisons of average ranks

Table 3: Comparison of morphological features of *R. carthamoides* plants derived in vitro through direct organogenesis (DSO-plants) and indirect organogenesis (IDSO-plants) and in vivo from seeds (SP-plants)

Measuring parameters	DSO-plants	IDSO-plants	SP-plants
Plant fresh weight (g)	19.7 ± 0.48^{b}	13.7 ± 0.22^{a}	21.1 ± 0.47^{b}
Mean plant length (cm)	9.95 ± 0.22^{b}	5.14 ± 0.19^{a}	10.76 ± 0.19^{b}
Maximum root length (cm)	24.4 ± 0.51^{b}	12.3 ± 0.11^{a}	25.1 ± 0.53^{b}
Number of leaves/plant	11.2 ± 0.59^{b}	6.70 ± 0.40^{a}	13.70 ± 0.49^{b}
Length of leaf lamina (cm)	6.98 ± 0.26^{b}	3.44 ± 0.18^{a}	7.6 ± 0.15^{b}
Width of leaf lamina (cm)	2.31 ± 0.06^{b}	1.55 ± 0.07^{a}	2.51 ± 0.04^{b}
Length of leaf petiole (cm)	2.96 ± 0.09^{b}	1.7 ± 0.06^{a}	3.15 ± 0.06^{b}

Plants grown in soil, in greenhouse for 3 months

Values represent mean ± SE. Values within a rows followed by the same letter are not significantly different from each other at a 5 % level according to the Kruskal–Wallis test and multiple comparisons of average ranks

For comparison, plants of *R. carthamoides* derived directly from seeds (SP-plants) were used. They were grown under identical conditions as the in vitro regenerated plants i.e. in the greenhouse for 3 months.

Statistical Analysis

The results presented in Tables 1, 2, 3 and 6 and Fig. 1 were subjected to analysis of variance. Differences in mean values were calculated using the Kruskal–Wallis test, with$p$ = 0.05 being the level of significance. The significance of the differences between means was assessed using multiple comparisons of average ranks. All statistical analyses were performed by Statistica 10.0PL for Windows (StatSoft Inc., Poland).

MOLECULAR ANALYSIS

Plant Materials used for the Genetic Analysis

Genomic DNA was isolated from fresh roots of ten randomly-selected 3-month-old plants of *R. carthamoides* regenerated through direct organogenesis from lamina explants (DSO-L-plants), petiole explants (DSO-P-plants), and plants regenerated through indirect organogenesis from root-derived callus (IDSO-plants). Additionally, we included in the analysis genomic DNA isolated from fresh roots of the mother plant (MP; a field grown plant which was the source of seeds used for in vitro culture initiation), roots of ten randomly selected plants obtained through clonal propagation (CP; plants obtained from axillary buds) and root-callus, which was used for induction of shoot organogenesis (RC-line maintained on MS medium supplemented with 0.5 mg L^{-1} NAA and 0.2 mg L^{-1} BA for one year). The genomic DNA of MP and CP plants was used as the control.

DNA Isolation and PCR Amplification

Genomic DNA was extracted using a NucleoSpin Plant II Kit (Macherey–Nagel, Germany). The plant materials (200 mg) were powdered in liquid nitrogen and stored at −80 °C until extraction. The quantity and quality of DNA samples were estimated by comparing band intensities on 1.5 % agarose gel. Three replicate DNA extractions from the tissue plant were used to assess the consistency of the band profiles. RAPD and ISSR amplifications were repeated at last three times and only the reproducible PCR products were scored.

RAPD Analysis

RAPD amplification was carried out in a 25 µL reaction volume containing 2.5 µg DNA (2 µL), 2.5 µL of 10× PCR buffer (100 mM Tris–HCl, pH 8.3, 500 mM KCl, 11 mM $MgCl_2$, 0.1 % gelatin), 1.5 µL $MgCl_2$, 0.75 µL of 100 µM dNTP (Blirt, Poland), 1.5 µL of RAPD primer, 0.75 µL of *Taq*Nova DNA polymerase (Blirt, Poland) and MilliQ water to make up the volume. RAPD amplification was performed with random decamer primers obtained from Blirt (Poland). Nine RAPD primers were tested for PCR amplification. The PCR was performed at an initial denaturation at 95 °C for 5 min followed by 40 cycles of 1 min denaturation at 95 °C, 45 s annealing at 37 °C (and next depending on the primer, Table 4) and 2 min extension at 72 °C with a final extension of 72 °C for 7 min using a thermal cycler (BioRad, UK).

Table 4: Total number and size range of amplified fragments and number of polymorphic fragments generated by nine random primers in in vitro regenerated plants, callus and mother plant of *R. carthamoides* by RAPD analysis

Serial no	Primer	Sequence (5′–3′)	Annealing temperature (°C)	Number of amplified fragments	Number of polymorphic fragments	Approximate fragment size range (bp)
1	OPA 1	CAGGCCCTTC	32	11	3	200–1500
2	OPA 2	AGCCAGGCAA	35	8	2	300–1000
3	OPA 3	GGGTAACGCC	34	7	2	200–1200
4	OPA 4	ATTCGGGTCG	34	9	0	100–1500
5	OPA 8	GTGACGTAGG	32	6	2	300–1200
6	OPA 9	GGGTAACGCC	34	14	2	100–2000
7	OPA 10	GTGATCGCAG	34	10	3	300–2500
8	OPA 13	CAGCACCCAC	35	5	0	100–1000
9	OPA 17	CAGCAACGAA	33	7	3	300–2000
Total				77	19	

ISSR Analysis

In case of ISSR primers, the optimal annealing temperature was found to vary according to the base composition of the primers. Eight ISSR primers (UBC primer) (Blitr, Poland) were used in the study. Amplification was carried out in 25 µL reaction volume containing 2.5 µg genomic DNA (2 µL) as template, 1.5 µL $MgCl_2$, 0.75 µL of 100 µM dNTP (Blirt, Poland), 3.5 µL of 10× PCR buffer (Taq buffer B devoid of $MgCl_2$), 2 µL of ISSR primers (10 pM), 0.75 µL of *Taq*Nova DNA polymerase (Blirt, Poland) and MilliQ water to make up the volume. PCR amplifications were performed with initial denaturation at 95 °C for 5 min followed by 35 cycles of 30 s denaturation at 94 °C, 1 min at the annealing temperature (and next depending on the primer, Table 5), 2 min extension at 72 °C with a final extension of 72 °C for 10 min using a thermal cycler (BioRad, UK).

Table 5: Total number and size range of amplified fragments and number of polymorphic fragments generated by eight random primers in in vitro regenerated plants, callus and mother plant of *R. carthamoides* by ISSR analysis

Serial no.	Primer	Sequence (5′–3′)	Annealing temperature (°C)	Number of amplified fragments	Number of polymorphic fragments	Approximate fragment size range (bp)
1	UBC 807	AGAGAGAGA GAGAGAGT	53	12	3	100–1000
2	UBC 811	GAGAGAGAG AGAGAGAA	54	10	3	200–1500
3	UBC 812	GAGAGAGAG AGAGAGAC	55	8	3	200–1200
4	UBC 813	CTCTCTCTC TCTCTCTT	54	8	1	200–1500
5	UBC 830	TGTGTGTGT GTGTGTGG	52	6	2	100–1500
6	UBC 834	AGAGAGAGA GAGAGAGCT	50	7	4	100–2000
7	UBC 835	AGAGAGAGA GAGAGAGTC	54	5	0	200–1200
8	UBC 836	AGAGAGAGA GAGAGAGCA	55	9	3	300–2500
Total				65	23	

The amplification RAPD–PCR and ISSR–PCR products were separated on 1.5 % agarose (Bioline, UK) gel through electrophoresis using 100 bp and 3 kb ladder (Fermentas, Lithuania) as the band size standard and stained with ethidium bromide and photographed using a DNR Bio-Imaging System MiniBIS Pro (Israel). The results are presented in Tables 4, 5 and Fig. 2.

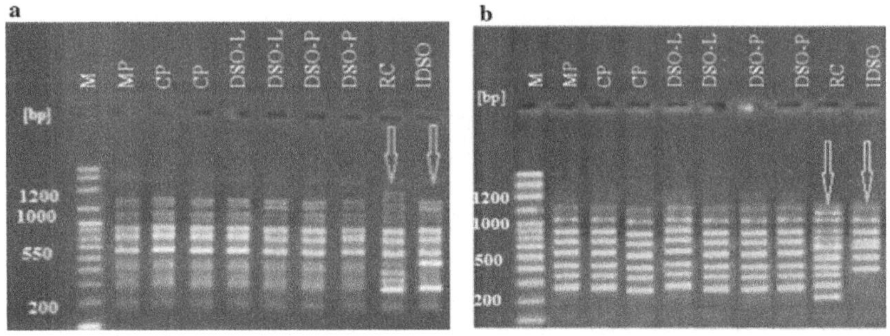

Figure. 2: Gel electrophoresis of **a** RAPD fragments obtained with primer OPA1 and **b** ISSR fragments obtained with primer UBC812. *Lane M* DNA marker 100–3000 bp DNA ladder; *lane MP* DNA of *R. carthamoides* mother plant (control); *lane CP* plants obtained from clonal propagation (control); *lane DSO-L* and *lane DSO-P* plants regenerated by direct shoot organogenesis from leaf lamina and petiole, respectively; *lane RC* root-derived callus; *lane IDSO* plants regenerated by indirect shoot organogenesis

Data Analysis

Two independent amplifications were performed with all the selected ISSR and RAPD primers (Tables 4, 5), and only clear and reproducible bands in the size range of 100 bp–2.5 kb were scored. The bands were transformed into a binary character matrix for each plant material: "1" for the presence and "0" for the absence of a DNA band. A similarity matrix and dendrogram were constructed using the NTSYSpc 2.11X (Exeter software, Setauket, N.Y.) software package (Rohlf 2000). Genetic similarities between samples were measured by the Jaccard's similarity coefficient (Jaccard 1908). Similarity coefficients were used to construct the dendrogram using the UPGMA and SHAN routine in the NTSYS program. The results are presented in Figs. 3, 4.

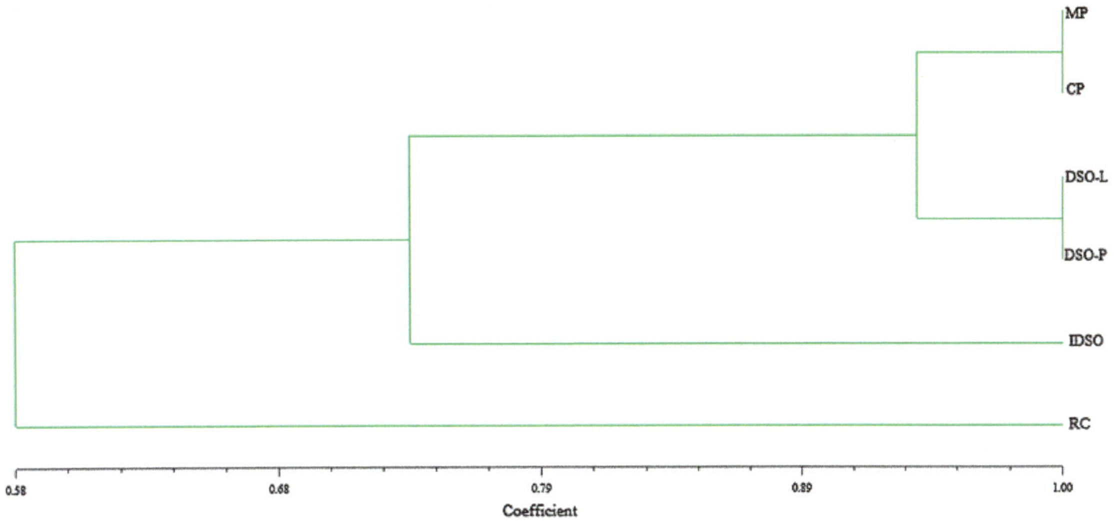

Figure. 3: Dendrogram illustrating coefficient similarities among in in vitro regenerated plants, callus line and mother plant of *R. carthamoides* by UPGMA cluster analysis (NTSYS) of the RAPD profiles derived from nine selected primers using Jaccard's similarity coefficient. *MP* mother plant (field-grown plant), *CP* plants obtained from axillary buds,*DSO-L* and *DSO-P* plants regenerated by direct shoot organogenesis from leaf lamina and petiole, respectively, *IDSO* plants regenerated by indirect shoot organogenesis, *RC* root-derived callus

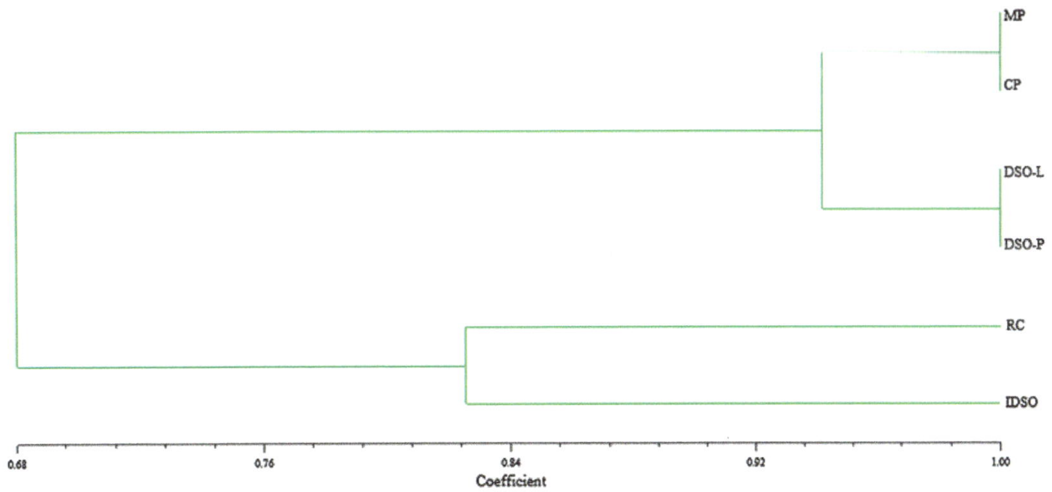

Figure. 4: Dendrogram illustrating coefficient similarities among in in vitro regenerated plants, callus line and mother plant of *R. carthamoides* by UPGMA cluster analysis (NTSYS) of the ISSR profiles derived from eight selected primers using Jaccard's similarity coefficient. *Symbols* are given in Fig. 3

Flow Cytometry for Genome size

Plant materials were prepared as previously described by Thiem and Śliwińska (2003). As materials were used organogenic root-callus (which was used for induction of shoot organogenesis; RC-line callus) and young, fresh leaves of *R. carthamoides* plants regenerated through direct organogenesis (DSO-plants), plants regenerated through indirect organogenesis from root-derived callus (IDSO-plants), plants obtained through clonal propagation (CP-plants), and mother plant (MP). Leaves of the mother plant were used as a control. Chicken (*Gallus domesticus* L.) erythrocytes (2C DNA content = 2.33 pg) (Hiddemann et al. 1984; Larsen et al. 1984; Ronot et al. 1986) were used as an internal standard for genome size for cell cycle estimation BD DNA QC Particles. Samples were prepared using nuclei isolation buffer (0.1 M Tris, 2.5 mM $MgCl_2{\cdot}6H_2O$, 85 mM NaCl, 0.1 % Triton X-100; pH 7.0) supplemented with 1 % PVP-10 (Thiem and Śliwińska 2003). The plant tissues blade in plastic Petri dish with 1 mL nucleus-isolation buffer. The suspension was filtered through nylon filter with 50 μm mesh diameter. Samples were incubated in PBS with 75 μM propidium iodide and ribonuclease A (50 Kunitz unit/mL) for 30 min (37 °C in the dark) (Larsen et al. 1984; Ronot et al. 1986). Stainewere chopped with a sharp razord specimens were analyzed using LSR II flow cytometer

(Becton–Dickinson Biologicals, USA) in channel for PE (excitation blue laser 488 nm, emission PE channel 575/26). For each sample 5000 events were obtained. Total nuclear DNA content was calculated using linear relationship between the ratio of the 2C peak positions of*Rhaponticum*/chicken erythrocytes on a histogram of fluorescence intensities. Initial histogram analysis were performed using BD FACS DIVA software ver. 6.2, further inquiries were done using FlowJo software.

QUANTIFICATION OF CHLOROGENIC ACID AND 20-HYDROXYECDYSONE

Extraction Procedure

The leaves and roots of DSO- and IDSO-plants, as well as seed-derived plants (SP-plants) of *R. carthamoides* were lyophilized and pulverized. All plants were grown for 3 months in pots, in the greenhouse. First, 200 mg of plant material was extracted with *n*-hexane. After filtration, the *n*-hexane extract was discarded. The defatted sample was sonicated for 15 min with 80 % (v/v) aqueous methanol (15 mL) at 35 °C using an ultrasonic bath, and then twice with 5 mL of the same solvent for 15 min. The combined extracts were diluted with methanol to 25 mL, filtered through a PTFE syringe filter (25 mm, 0.2 µm, AlChem, Czech Republic) and the filtrate was directly injected into the UHPLC–DAD system for quantitative determination of chlorogenic acid and 20-hydroxyecdysone.

UHPLC–DAD Analysis

Chromatographic analysis was performed using an Agilent Technologies 1290 Infinity UHPLC apparatus equipped with the diode array detector (DAD), the binary solvent delivery pump with included a vacuum degasser, autosampler and thermostated column compartment. During the UHPLC analyses, the samples of extracts were maintained in the autosampler at 15 °C in the dark. The separations of the samples were performed on the Zorbax Eclipse Plus C18 column (100 × 3 mm id; 1.8 µm Agilent Technologies, Cat. no 928975-302) at 30 °C. The mobile phase consisted of 0.5 % phosphoric acid in acetonitrile (v/v) (solvent A) and 0.5 % phosphoric acid (v/v) in water (solvent B). A gradient program was applied as follows: 0–1 min 1 % A; 1–1.4 min 1–5 % A; 1.4–3.5 min 5 % A; 3.5–4 min 5–19 % A; 4–6.5 min 19 % A; 6.5–6.9 min 19–30 % A; 6.9–7.5 min 30–35 % A; 7.5–9.5 min 35–95 % A; 9.5–11 min 95 % A. The column was equilibrated with 1 % of the solvent A for 4 min between injections. The flow rate was 0.6 mL min^{-1}. The injection volume was 5 µL. The detection wavelength was set at 245 nm with a 4 nm bandwidth. Chlorogenic acid and 20-hydroxyecdysone (20-HE) were identified by comparing their UV spectra and retention times of their peaks (Rt 5.55 and 7.94 min, respectively) (Fig. 5) with those of the reference compounds. Authentic standards of chlorogenic acid and 20-HE were purchased from Sigma-Aldrich (Germany/USA). Chlorogenic acid and 20-HE contents were estimated by the interpolation of the peaks area with calibration curves constructed for the standard compounds. The content of the chlorogenic acid and 20-HE were estimated as mg per gram of dry weight (mg g^{-1} DW) of leaves or roots. The results are presented in Table 7.

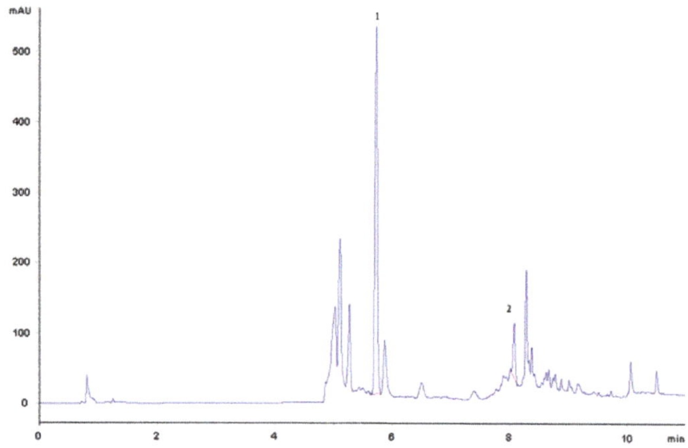

Figure. 5: UHPLC chromatogram at 245 nm of methanolic extract from leaves of *R. carthamoides*seed-derived plants. *1* chlorogenic acid; *2* 20-hydroxyecdysone

RESULTS AND DISSCUSION

Indirect Shoot Regeneration

Organogenesis from *R. carthamoides* seedling explants (hypocotyl, cotyledons, roots) was induced indirectly via an intervening callus phase. The shoot regeneration ability of *R. carthamoides* in terms of the percentage of explants producing organogenic calluses and number of buds/shoots per responding callus was tested by culturing segments of hypocotyl, cotyledons and roots on MS basal medium containing a combination of BA (0.2 or 0.5 mg L^{-1}) with one of three auxins: 2,4-D, NAA or IBA (0.2 or 0.5 mg L^{-1}; Table 1). The choice of media for induction of shoot-regenerating callus was based on our own preliminary experiments and a review of earlier studies (Li et al. 2013; Zand et al. 2014). Zand et al. (2014) report the highest shoot regeneration (89 %) from *R. carthamoides* leaf explants cultured on MS medium supplemented with 2,4-D (0.25 mg L^{-1}) and BA (1.5 mg L^{-1}). However, in the current study, only induction of non-organogenic callus or callus forming roots was observed in the presence of 2,4-D (0.2 or 0.5 mg L^{-1}) with BA (0.2 or 0.5 mg L^{-1}). *R carthamoides* shoot regeneration was possible when calluses were cultured in the presence of NAA or IBA as auxin and BA as cytokinin (Table 1). Following 5 weeks of culture on the media, most seedling explants formed compact, green or yellow-green regenerative callus with buds and shoots. The calluses were transferred into fresh medium of the same composition as that used for callus induction to allow the growth of callus biomass and further development of shoots. Of the two auxins (NAA and IBA), NAA was found to be more effective for shoot regeneration than IBA. The maximum number of buds/shoots (almost 4 per callus) and the highest percentage cultures producing buds/shoots (92 %) were achieved when cotyledon-derived calluses were cultured on MS medium containing NAA (0.5 mg L^{-1}) and BA (0.2 mg L^{-1}). Root-derived calluses cultured on the same medium composition showed a similar ability to bud/shoot regeneration and more than 3 buds/shoots per callus were formed at the frequency of organogenesis of 83 %. Significant reductions in the number of buds/shoots (about 2 per callus) was obtained when hypocotyl explants were used for callus initiation (Table 1). It was also observed that with each subculture, the ability of *R. carthamoides* seedling-derived calluses to shoot differentiation declined. Only calluses initiated from roots and cultured for 1 year on MS medium containing NAA and BA maintained competence to form shoots (Fig. 6a) and the shoots excised from the calluses were used for the rooting stage.

Figure. 6: Regeneration of *R. carthamoides* through indirect organogenesis (**a–d**) and direct organogenesis (**e–h**). Adventitious shoot formation from root-derived callus (**a**) and from leaf lamina (**e**). Shoot rooted on ½ MS medium supplemented with 0.5 mg L^{-1} IBA after 5 weeks of culture (**b**, **f**). Plants after 3 months of growth in pots (**c**, **g**) and after removing from pots (**d**, **h**). *Bars* 1 cm

Direct Shoot Organogenesis

Direct shoot organogenesis was achieved by using leaves from in vitro shoot culture of *R. carthamoides*, as initial explants. Lamina and petiole explants were incubated on six variations of MS media. The media contained BA alone and in combination with NAA or IBA as shown in Table 2. Combination of 2,4-D (0.2 or 0.5 mg L^{-1}) and BA (0.2 or 0.5 mg L^{-1}) was also tested but responses induced under these conditions were inefficient, therefore the variants were not shown. Shoot organogenesis did not occur from leaf explants cultured in the absence of plant growth regulators. Formation of adventitious buds and shoots from the cut ends or along the whole surface of petiole explants, and the mid-vein or whole surface of lamina explants (Fig. 6e), was observed after 5 weeks of culture with high frequency (83–100 %) and a mean number of 3–4.5 buds/shoots per explant, both in response to BA alone and in combination with NAA or IBA. The maximum number of buds/shoots (approximately 4.5 per responsive explant) was achieved on MS medium supplemented with 0.2 mg L^{-1} BA and 0.5 mg L^{-1} NAA. On this medium high bud/shoot regeneration frequency (93.5–100 %) was also achieved. The lowest organogenesis frequency (83–86 %) and number of buds/shoots (3.1–3.3 per explant responding) were obtained on lamina and petiole explants cultured on MS medium with BA (0.5 mg L^{-1}) and IBA (0.2 mg L^{-1}) as well as on explants grown on MS medium supplemented with 0.5 mg L^{-1} BA alone (Table 2). There was no significant difference between bud/shoot number in petiole and lamina explants (Table 2).

Differences in morphogenetic responses of seedling (hypocotyl, cotyledon and root) and leaf (lamina and petiole) explants indicate that source of explants was crucial for regenerative response of *R. carthamoides* in terms of induction of shoot regenerative callus or direct shoot differentiation. Plant tissues are known to contain different levels of endogenous growth hormones. As a result they may respond differently to culture medium (Palla and Pijut 2011).

Shoot Rooting and Plant Acclimatization

The adventitious shoots regenerated from root-derived callus and directly developed from leaf explants were rooted on ½ MS agar medium without auxin or supplemented with different concentrations of IBA (0.2–2.0 mg L^{-1}; Fig. 1). More shoots formed roots on ½ MS supplemented with IBA and the plantlets showed more roots per shoot (up to 4.3) than those rooted in auxin-free ½ MS medium (1.4/shoot) but the roots were shorter (1.5–4.1 vs. 5.4 cm; Fig. 1). The maximum root induction (97 % after 5 weeks of culture) was shown in the presence of 0.5 mg L^{-1} IBA. It was significantly higher (twice) than that achieved for shoots rooted on the auxin–free medium (max. 47 %). An increase in concentration of IBA from 0.5 to 1.0 mg L^{-1} decreased the rooting percentage to 86 % but increased the mean number of roots per shoot (4.3 vs. 2.8; Fig. 1). The positive effect of IBA on root induction has also been reported for other species in Asteraceae family such as *Cichorium intybus*(Velayutham et al. 2006) and *C. pumilum* (Al Khateeb et al. 2012). Application of IBA induces adventitious roots, due to an increase in the biosynthesis of IAA (Woodward and Bartel 2005).

It was noted that the type of regeneration of *R. carthamoides* shoots (indirectly from root-derived callus and directly from leaf explants) affects the time of root initiation and the rooting efficiency in terms of percentage of shoots forming roots, root branching and root length. On shoots obtained from leaf explants through direct organogenesis the roots started to initiate after 7 days and within 5 weeks of culture on optimal medium (½ MS supplemented with 0.5 mg L^{-1} IBA) 97 % of the shoots formed an average 2.8 roots with an average length of 3.7 cm. The plants developed an adventitious roots system with numerous lateral branching and with high adsorption capacity (Fig. 6f). In the case of shoots derived from callus, root initiation commenced after 14 days and at the end of culture period (5 weeks) a maximum 58 % of the shoots formed an average 2.7 roots with an average length of 1.7 cm. The root system of these plants lacked lateral roots (Fig. 6b). As shown in Fig. 1, the presence of IBA was necessary for rooting of shoots derived from callus.

The rooted shoots were transplanted into small pots containing soil, sand, peat and perlite for further growth in the greenhouse. Three months after transfer to ex vitro conditions all, the DSO-plants survived (plants regenerated directly from leaf explants), while the survival rate for IDSO-plants (plants regenerated from root-derived callus) reached only 44 %. The lack of lateral roots in IDSO-plants (Fig. 6b) adversely impacts their adsorptive capacity, contributing to a lower survival rate under ex vitro conditions. In many plant species, high mortality has been observed upon the transfer of rooted shoots to soil as the cultured plants have weak root system (Mathur et al. 2008; Chandra et al. 2010).

Morphological Characterization of Plants Grown in Greenhouse

Observation of growth in greenhouse revealed that the morphology of regenerated plants varied and was dependent on whether they are derived through direct or indirect organogenesis (Table 3; Figs. 6, 7). After 3 months, DSO-plants were taller and their total fresh weight was almost twice higher than that in IDSO-plants (Table 3). The former had more leaves and longer, thicker and more branched roots (Fig. 6g–h). Significant differences exist also between both plants in respect to shape of leaves (Fig. 7). This phenomenon may be caused by somaclonal variation as a result of callus culture. No significant differences in any measuring growth parameters were observed between *R. carthamoides* plants developed through direct organogenesis and those propagated from seeds, which were the same age and grew under identical conditions as in vitro developed plants (Table 3).

Figure. 7: Alteration in the leaf morphology of *R. carthamoides* plants regenerated through direct (**a–c**) and indirect (**d–f**) shoot organogenesis. *Bars* 1 cm (**e**)

Analysis of Genetic Stability by RAPD and ISSR and Flow Cytometry

To determine the genetic stability of in vitro regenerated plants of *R. carthamoides*, we compared the genetic fidelity of the plants derived through direct (from lamina—DSO-L-plants and petiole—DSO-P-plants) and indirect organogenesis (IDSO-plants) using RAPD and ISSR markers. In addition, the RC callus line (callus derived from a root explant which was used for the induction of shoot organogenesis) was also included in the analysis, while the mother plant (MP) and plants regenerated in vitro from axillary buds (CP-plants) were used as control materials. The plants propagated from meristematic tissues exhibited the lowest tendency for genetic variation (Mallón et al. 2010; Bhatia et al. 2011). The initial RAPD and ISSR analyses of all 10 randomly-selected plants of each group (DSO-, IDSO- and CP-plants) for every tested primers showed that no genetic variation was present within them.

In RAPD–PCR analysis, the nine primers were selected for DNA amplification (Table 4). The molecular size of the RAPD fragments ranged from 100 to 2500 bp. The number of amplified fragments varied from 5 (OPA 13) to 14 (OPA 9) (Table 4). The 9 primers, produced a total of 77 bands with an average of 8.5 bands per primer. Among these 9 amplified primers, 7 generated polymorphic bands. Out of the 77 scorable bands, 19 were polymorphic with an average number of polymorphic bands per primer as 2.1. The levels of polymorphism were varied according to primers. The highest number (3) of polymorphic bands was obtained by primer OPA 1, OPA 10 and OPA 17 and the lowest (0) by primer OPA 4 and OPA 13 (Table 4). The similarity matrix calculated from the RAPD profiles was subsequently subjected by UPGMA cluster analysis using NTSYS software in order to evaluate the genetic fidelity (Fig. 3).

The Jaccard's similarity coefficient of dendrogram among MP-, CP-plants (used as the control material) and DSO-L-, DSO-P-, IDSO-plants and RC line callus ranged from 0.588 to 0.941 (Fig. 3). DSO-L- and DSO-P-plants showed maximum closeness between together, with a similarity coefficient of 1.000 and a similarity coefficient of 0.941 to MP- and CP-plants (Fig. 3). IDSO-plants were closely related to MP-, CP-, DSO-P- and DSO-L-plants but the similarity values were low and ranged from 0.706 to 0.765 (Fig. 3).

During ISSR–PCR analysis, the molecular size of amplified products ranged from 100 to 2500 bp (Table 5). The highest number of bands (12) was obtained in case of primer UBC 807, whereas the lowest number of bands (5) was recorded for the primer UBC 835 (Table 5). The eight selected ISSR primers generated a total of 65 bands, with an average of 8.1 bands per primer. Among these amplified primers, only one generated monomorphic bands (UBC 835; Table 5). Of the 65 fragments scored from ISSR primers, 42 were monomorphic and 23 were polymorphic with an average number of 2.9 polymorphic bands per primer (Table 5). The primer UBC 834 gave the highest number of polymorphic bands (4) while UBC 835 gave the lowest (0; Table 5).

The Jaccard's similarity coefficient calculated on the basis of the ISSR data revealed different levels of genetic variation than the RAPD–PCR analysis. The analysis revealed that the genetic identity of DSO-L-, DSO-P- and IDSO-plants and the callus with the MP- and CP-plants (control materials), ranged from 0.647 to 0.947 (Fig. 4). Maximum genetic similarity to the MP- and CP-plants was observed for DSO-L- and DSO-P-plants, similarly as to the coefficient calculated of the RAPD data (0.947) (Figs. 3, 4). As shown in the dendrogram (Fig. 4), the IDSO-plants were clustered to RC callus (0.824).

In order to test the genetic variability of the plants regenerated through direct (DSO-plants) and indirect organogenesis (IDSO-plants), clonal propagation (CP-plants) and RC callus line, the flow cytometry was also used. Flow cytometry is widely used to establish the nuclear DNA content, genome size or ploidy level in in vitro cultures (Ochatt et al.2011; Makowczyńska et al. 2008; Slazak et al. 2015; Peng et al. 2015). The genome size of diploid mother plant of *R. carthamoides* was 2.27 pg/2C (Table 6). To our knowledge, this is the first report on genome size in this species. The 2C DNA content in CP-, DSO-L-, DSO-P-plants was similar (2.33–2.38 pg; no statistical difference) to the mother plant (Table 6). The plants regenerated through indirect organogenesis and RC callus line showed higher nuclear DNA content (2.47 and 3.19 pg/2C, respectively) than the mother plant (a significant difference; $p < 0.05$).

Table 6: Nuclear DNA content in callus and regenerated in vitro plants of *R. carthamoides*

Plant material	Nuclear DNA content (pg/2C ± SE)
MP	2.27 ± 0.05[a]
CP	2.33 ± 0.01[a]
DSO	2.38 ± 0.01[ab]
IDSO	2.47 ± 0.03[b]
RC	3.19 ± 0.01[c]

Each value represent the mean ± standard error (SE) of nine independent samples. Values with the same letter are not significantly different from each other at a 5 % level according to the Kruskal–Wallis test and multiple comparisons of average ranks

MP, mother plant; CP, plants obtained from clonal propagation; DSO, plants regenerated by direct shoot organogenesis; IDSO, plants regenerated by indirect shoot organogenesis; RC, root-derived callus

In this study, the genetic identity of DSO-plants of *R. carthamoides* to the mother plant and axillary buds-derived plants was high and ranged from 0.941 to 0.947. Flow cytometry confirmed similar ploidy level in the mother plant and DSO- and axillary buds-derived plants (the nuclear DNA content was 2.33–2.38 pg/2C). Our results suggested that the direct plant regeneration from the explant tissues might ensure the low genetic variation.

Somaclonal variation has been frequently reported in in vitro regenerated plants (Larkin and Scowcroft 1981). The occurrence of variation during in vitro propagation depends upon the source of explants and the pathway of plant regeneration (Goto et al. 1998). Somaclonal variation is quite often associated with indirect regeneration involving an intervening callus stage (Shen et al. 2007; Yuan et al. 2009). Similar results were observed in this study for *R. carthamoides* plants regenerated through indirect shoot organogenesis from root-derived callus. The plants showed a low similarity coefficient to the mother plant and axillary bud-derived plants, which ranged from 0.647 to 0.706. Also, the nuclear DNA content in IDSO-plants was higher (2.47 pg/2C) than in the mother plant (2.27 pg/2C) (a significant difference; $p < 0.05$). Since single seedling was used to initiate shoot regenerative callus, the genetic polymorphism of IDSO-plants of *R. carthamoides*can not be due to the genetic variability of seeds but rather somaclonal variation induced by in vitro propagation. The major reason for increase in somaclonal variation upon indirect regeneration may be increase in genetic heterogeneity of callus cells. The similarity coefficient between IDSO-plants and callus was found to be relatively lower in RAPD analysis (0.588) than those assessed in ISSR analysis (similarity coefficient was 0.824).

The in vitro tissue culture condition (particularly callus induction) induces physiological stress, which may be characterized by disruption of normal developmental controls, and the mechanism for the occurrence of somaclonal variation can be largely explained in the context of pre-adapted cellular responses to stress (Cassells and Curry 2001). The responses may include activation of an otherwise cryptic mutator system like transposition, and/or malfunction of a normally functional anti-mutational system such as the mismatch repair (Guo et al. 2006). Somaclonal variation can be manifested as cytological abnormalities, various kinds of aberrations at the nucleotide sequence level (substitution, deletion, insertion), and gene activation and silencing (Kaeppler et al. 2000; Guo et al. 2006).

Although, somaclonal variation is undesirable in micropropagation but it can be an important source of new variability as for example new cultivar development. The overall results presented that direct organogenesis is good method for generation of genetically stable clones of *R. carthamoides* identical to mother plant.

Quantification of Chlorogenic Acid and 20-Hydroxyecdysone

In order to evaluate the effect of the type of regeneration pathway on secondary metabolite production, UHPLC-DAD method was used to quantify the chlorogenic acid and 20-hydroxyecdysone (20-HE) in methanolic extracts of leaves and roots of 3-month-old DSO- and IDSO-plants of *R. carthamoides* and compared with those in leaves and roots of corresponding seed-derived plants (SP). A typical chromatogram of methanolic extracts is shown in Fig. 5 where peak 1 is related to chlorogenic acid and peak 2 to 20-HE. Both compounds are typical of *R. carthamoides* species (Kokoska and Janovska 2009; Biskup et al. 2013). Chlorogenic acid accounts for most of the content of caffeic acid derivatives in *R. carthamoides* and we used the compound as a marker of caffeic acid derivatives. As shown in Table 7, the DSO- and IDSO-plants demonstrated significantly different levels of chlorogenic acid, as did plant leaves and roots. The chlorogenic acid content in DSO-plant leaves was found to be 13-times

(12.2 mg g^{-1} DW) greater than in leaves of IDSO-plants (0.9 mg g^{-1} DW). In contrast, the average content of chlorogenic acid in roots of DSO-plants was 2.5 mg g^{-1} DW: 1.5 times lower than in the roots of IDSO-plants (3.9 mg g^{-1}DW).

Table 7: Quantitative analysis of chlorogenic acid and 20-hydroxyecdysone in plants of *R. carthamoides* regenerated in vitro and in the seed-derived plants, growing in greenhouse for 3 months

Plant origin	Plant organ	Content (mg g^{-1} DW) ± SE	
		Chlorogenic acid	20-HE
DSO	Leaves	12.23 ± 0.13[d]	2.46 ± 0.07[b]
	Roots	2.49 ± 0.03[b]	2.12 ± 0.04[b]
IDSO	Leaves	0.94 ± 0.03[a]	7.38 ± 0.05[c]
	Roots	3.9 ± 0.17[c]	3.04 ± 0.07[a]
SP	Leaves	11.58 ± 0.37[d]	2.38 ± 0.11[b]
	Roots	3.31 ± 0.19[c]	2.02 ± 0.06[b]

Each value represent the mean ± standard error (SE) of three independent samples. Each sample was repeated three times (n = 9). Values within a column followed by the same letter are not significantly different from each other at a 5 % level according to the Kruskal–Wallis test and multiple comparisons of average ranks

DSO, plants regenerated by direct shoot organogenesis from leaf explants; IDSO, plants regenerated from root-derived callus; SP, seed-derived plants

The other class of bioactive compounds in *R. carthamoides* are phytoecdysteroids. Among them, 20-hydroxyecdysone is dominant (Kokoska and Janovska 2009). The results presented in Table 7 show that the average content of 20-HE in leaves and roots of IDSO-plants were found to be 3- and 1.5-times higher than those in leaves and roots of DSO-plants (7.4 mg g^{-1} DW and 3.04 mg g^{-1} DW vs. 2.5 and 2.1 mg g^{-1} DW, respectively). Chavan et al. (2014) also note the presence of quantitative differences between *Ceropegia santapaui* plants regenerated through indirect and direct organogenesis, with the former exhibiting increased levels of total phenolics and flavonoids. So far, the biosynthetic differences have not been well clarified. Chavan et al. (2014) suggest that these might be due to the exogenous supply of the plant growth regulators during in vitro multiplication.

In comparison to SP-plants, DSO-plants exhibited similar levels of both chlorogenic acid and 20-HE. A significant difference ($p < 0.05$) was observed only in terms of the chlorogenic acid content in roots of these plants, which was 33 % higher in SP-plant roots. The leaves and roots of the IDSO-plants contained significantly higher amounts of 20-HE than the analogous parts of SP-plants (Table 7). In respect to chlorogenic acid, leaves of IDSO-plants accumulated only one tenth of the compound level found in the leaves of the SP-plants, while the roots of both plants contained similar levels of the caffeic acid derivative (Table 7).

This is the first report comparing the chlorogenic acid and 20-HE contents in in vitro- and in vivo-derived plants of *R. carthamoides*. The levels of chlorogenic acid in methanolic extracts from the leaves of mature *R. carthamoides* plants were found to be 0.63 mg g^{-1}DW in those growing in Russia and 2.0 mg g^{-1} DW in those growing in Poland (Biskup et al. 2013). Timofeev et al. (2006) report that the level of 20-HE HE varied from 0.15 to 0.55 % on the basis of dry weight in 8- to 11-year-old *R. carthamoides* plants, depending on organ and stage of plant development.

In summary, *R. carthamoides* plants regenerated trough direct organogenesis (from leaf explants) and indirect organogenesis via callus tissue (from seedling explants) were obtained in this work. The pathway of regeneration affected the rooting of in vitro derived shoots, the survival of plantlets under ex vitro conditions, their morphology and the modified metabolic activity of in vitro regenerated plants. It was observed that while the leaves of plants obtained from direct organogenesis were a better source of chlorogenic acid than the leaves of regenerants obtained from callus culture, the later exhibited a superior potential for accumulation of 20-HE. Phenotypic changes and overproduction of 20-HE in plants obtained through indirect organogenesis could be related to somaclonal variation. The RAPD and ISSR analyses indicated the heterogeneity of regenerants derived from callus culture versus *R. carthamoides* mother plant and plants regenerated directly from explant tissue.

REFERENCES

1. Al Khateeb W, Hussein E, Qouta L, Alu'datt M, Al-Shara B, Abu-zaiton A (2012) In vitro propagation and characterization of phenolic content along with antioxidant and antimicrobial activities of *Cichorium pumilum* Jacq. Plant Cell Tissue Organ Cult 110:103–110. doi:10.1007/s11240-012-0134-9
2. Bhatia R, Singh KP, Sharma TR, Jhang T (2011) Evaluation of the genetic fidelity of in vitro-propagated gerbera (*Gerbera jamesonii* Bolus) using DNA-based markers. Plant Cell Tissue Organ Cult 104:131–135. doi:10.1007/s11240-010-9806-5
3. Biskup E, Lojkowska E (2009) Evaluation of biological activities of *Rhaponticum carthamoides* extracts. J Med Plant Res 31:1092–1098
4. Biskup E, Szynklarz B, Golebiowski M, Borsuk K, Stepnowski P, Lojkowska E (2013) Composition and biological activity of *Rhaponticum carthamoides* extracts obtained from plants collected in Poland and Russia. J Med Plant Res 7:687–695. doi:10.5897/JMPR012.1061

5. Cassells AC, Curry RF (2001) Oxidative stress and physiological, epigenetic and genetic variability in plant tissue culture: implications for micropropagators and genetic engineers. Plant Cell Tissue Organ Cult 64:145–157. doi:10.1023/A:1010692104861

6. Chandra S, Bandopadhyay R, Kumar V, Chandra R (2010) Acclimatization of tissue cultured plantlets: from laboratory to land. Biotechnol Lett 32:1199–1205. doi:10.1007/s10529-010-0290-0

7. Chavan JJ, Gaikwad NB, Umdale SD, Kshirsagar PR, Bhat KV, Yadav SR (2014) Efficiency of direct and indirect shoot organogenesis, molecular profiling, secondary metabolite production and antioxidant activity of micropropagated *Ceropegia santapaui*. Plant Growth Regul 72:1–15. doi:10.1007/s10725-013-9830-7

8. Clifford MN (1999) Chlorogenic acids and other cinnamates—nature, occurrence and dietary burden. J Sci Food Agric 79:362–372. doi:10.1002/(SICI)1097-0010(19990301)79:3<362:AID-JSFA256>3.0.CO;2-D

9. Duškova J, Dušek J (1995) *Leuzea carthamoides* DC in vitro. Herba Polonica 41:165–169

10. Goto S, Thakur RC, Ishii K (1998) Determination of genetic stability in long-term micropropagated shoots of *Pinus thunbergii* Parl. Using RAPD markers. Plant Cell Rep 18:193–197. doi:10.1007/s002990050555

11. Guo WL, Gong L, Ding ZF, Li YD, Li FX, Zhao SP, Liu B (2006) Genomic instability in phenotypically normal regenerants of medicinal plant *Codonopsis lanceolata* Benth. et Hook. f., as revealed by ISSR and RAPD markers. Plant Cell Rep 25:896–906. doi:10.1007/s00299-006-0131-8

12. Hiddemann W, Schaumann J, Andreeff M, Barlogie B, Herman ChJ, Leif RC, Mayall BH, Murphy RF, Sandberg AA (1984) Convention on nomenclature for DNA cytometry. Cytometry 5:445–446

13. Jaccard P (1908) Nouvelles recherches surla distribution florale. Bull Soc Vaud Sci Nat 44:223–270. doi:10.5169/seals-268384

14. Kaeppler SM, Kaeppler HF, Rhee Y (2000) Epigenetic aspects of somaclonal variation in plants. Plant Mol Biol 43:179–188. doi:10.1023/A:1006423110134

15. Kokoska L, Janovska D (2009) Chemistry and pharmacology of *Rhaponticum carthamoides*: a review. Phytochemistry 70:842–855. doi:10.1016/j.phytochem.2009.04.008

16. Kokoska L, Polesny Z, Rada V, Nepovim A, Vanek T (2002) Screening of some Siberian medicinal plants for antimicrobial activity. J Ethnopharmacol 82:51–53

17. Larkin P, Scowcroft W (1981) Somaclonal variation—a novel source of variability from cell cultures for plant improvement. Theor Appl Genet 60:197–214. doi:10.1007/BF02342540

18. Larsen JK, Munch-Petersen B, Christiansen J, Brgensen KJ (1984) Flow cytometry for discrimination of mitotic cells: resolution of M, as well as G1, S and G2 phase nucei with mitharamycin, propidium iodide, and ethidium bromide after fixation with formaldehyde. Cytometry 7:54–63

19. Li Q, Deng M, Zhang J, Zhao W, Song Y, Li Q, Huang Q (2013) Shoot organogenesis and plant regeneration from leaf explants of *Lysionotus serratus* D. Don. Sci World J:1–7. doi:10.1155/2013/280384

20. Liu F, Huang L-L, Li Y-L, Reinhoud P, Jongsma MA, Wang C-Y (2011) Shoot organogenesis in leaf explants of *Hydrangea macrophylla* 'Hyd1' and assessing genetic stabiltity of regenerants using ISSR markers. Plant Tissue Organ Cult 104:111–117. doi:10.1007/s11240-010-9797-2

21. Makowczyńska J, Andrzejewska-Golec E, Sliwinska E (2008) Nuclear content in different plant materials of *Plantago asiatica* L. cultured in vitro. Plant Cell Tissue Organ Cult 94:65–71. doi:10.1007/s11240-008-9387-8

22. Mallón R, Rodríguez-Oubiña J, González ML (2010) In vitro propagation of the endangered plant *Centaurea ultreiae*: assessment of genetic stability by cytological studies, flow cytometry and RAPD analysis. Plant Cell Tissue Organ Cult 101:31–39. doi:10.1007/s11240-009-9659-y

23. Mathur A, Mathur AK, Verma P, Yadav S, Gupta ML, Darokar MP (2008) Biological hardening and genetic fidelity testing of micro-cloned progeny of *Chlorophytum borivilianum* Sant. et Fernand. Afr J Biotechnol 7:1046–1053. doi:10.5897/AJB07.760

24. Mohat S, Parthasarathy U, Asish GR, Babu KN (2012) Evaluation of genetic stability of micropropagated plants of three species of *Garcinia* using random amplified Polymorphic DNA (RAPD) and inter simple sequence repeat (ISSR) markers. Indian J Biotechnol 11:341–343

25. Murashige T, Skoog F (1962) A revised medium for rapid growth and bioassays with tobacco tissue cultures. Physiol Plant 15:473–497. doi:10.1111/j.1399-3054.1962.tb08052.x

26. Nayak SA, Kumar S, Satapathy K, Moharana A, Behera B, Barik DP, Acharya L, Mohapatra PK, Jena PK, Naik SK (2013) In vitro plant regeneration from cotyledonary nodes of*Withania somnifera* (L.) Dunal and assessment of clonal fidelity using RAPD and ISSR markers. Acta Physiol Plant 35:195–203. doi:10.1007/s11738-012-1063-2

27. Ochatt SJ, Patat-Ochatt EM, Moessner A (2011) Ploidy level determination within the context of in vitro breeding. Plant Cell Tissue Organ Cult 104:329–341. doi:10.1007/s11240-011-9918-6

28. Olthof MR, Hollman PC, Katan MB (2001) Chlorogenic acid and caffeic acid are absorbed in humans. J Nutr 131:66–71

29. Palla KJ, Pijut PM (2011) Regeneration of plants from *Fraxinus americana* hypocotyls and cotyledons. In Vitro Cell Dev Biol Plant 47:250–256. doi:10.1007/s11627-011-9360-9

30. Peng X, Zhang T-T, Zhang J (2015) Effect of subculture times on genetic fidelity, endogenous hormone level and pharmaceutical potential of *Tetrastigma hemsleyanum*callus. Plant Cell Tissue Organ Cult 122:67–77. doi:10.1007/s11240-015-0750-2

31. Rohlf FJ (2000) NTSYS-pc: numerical taxonomy and multivariate analysis system. Exeter Software, New York

32. Ronot X, Hecquet C, Larno S, Hainque B, Adolphe M (1986) G_2 Arrest, binucleation and single parameter DNA flow cytometry analysis. Cytometry 7:286–290

33. Shen X, Chen J, Kane ME, Henny RJ (2007) Assessment of somaclonal variation in*Dieffenbachia* plants regenerated through indirect shoot organogenesis. Plant Cell Tissue Organ Cult 91:21–27. doi:10.1007/s11240-007-9273-9

34. Sláma K, Lafont R (1995) Insect hormones-ecdysteroids: their presence and actions in vertebrate. Eur J Entomol 92:355–377

35. Slazak B, Sliwinska E, Saługa M, Ronikier M, Bujak J, Słomka A, Göransson U, Kuta E (2015) Micropropagation of *Viola uliginosa* (Violaceae) for endangered species conservation and for somaclonal variation-enhanced cyclotide biosynthesis. Plant Cell Tissue Organ Cult 120:179–190. doi:10.1007/s11240-014-0592-3

36. Thiem B, Śliwińska E (2003) Flow cytometric analysis of nuclear DNA content in cloudberry (*Rubus chamaemorus* L.) in vitro cultures. Plant Sci 164:129–134

37. Timofeev NP, Lapin AA, Zelenkov VN (2006) Quality assessment of *Rhaponticum carthamoides* (Willd.) Iljin as medicinal raw material by the bromic antioxidant capacity. Chem Comput Simul Butlerov Commun 8:35–40

38. Velayutham P, Ranjithakumari BD, Baskaran P (2006) An efficient in vitro plant regeneration system for *Cichorium intybus* L.—an important medicinal plant. J Agric Tech 2:287–298

39. Woodward AW, Bartel B (2005) Auxin: regulation, action, and interaction. Ann Bot 95:707–735

40. Yuan XF, Dai ZH, Wang XD, Zhao B (2009) Assessment of genetic stability in tissue-cultured products and seedlings of *Saussurea involucrate* by RAPD and ISSR markers. Biotechnol Lett 31:1279–1287. doi:10.1007/s10529-009-9984-6

41. Zand A, Babaei A, Omidbaigi R, Daneshfar E (2014) Study on callus induction and plant regeneration of *Leuzea carthamoides* via tissue culture system. J Med Plant Res 8:260–268. doi:10.5897/JMPR11.610

42. Zhou L, Cheng Z, Chen D (2012) Simultaneous determination of six steroidal saponins and one ecdysone in *Asparagus filicinus* using high performance liquid chromatography coupled with evaporative light scattering detection. Acta Pharm Sin B 2:267–273. doi:10.1016/j.apsb.2012.04.003

CHAPTER 9

ACCUMULATION OF BIOLOGICALLY ACTIVE PHENOLIC ACIDS IN AGITATED SHOOT CULTURES OF THREE HYPERICUM PERFORATUM CULTIVARS: 'ELIXIR', 'HELOS' AND 'TOPAS'

Inga Kwiecien´[1] Aleksandra Szydłowska[1] Beata Kawka[1] Ludger Beerhues[2] Halina Ekiert[1]

[1].Chair and Department of Pharmaceutical Botany, Jagiellonian University, Collegium Medicum, Medyczna Street 9, 30-688 Krako´w, Poland

[2] Institut fu¨r Pharmazeutische Biologie, Technische Universita¨t Braunschweig, Mendelssohnstrasse 1, 38106 Braunschweig, Germany

INTRODUCTION

Hypericum perforatum L. (St. John's Wort) is one of the best known medicinal plant species used in both traditional and modern medicine. It is characterized by a rich chemical composition, and the bioactive compounds of the herb (*Herba Hyperici*) represent biogenetically different groups of plant secondary metabolites. The main constituents are derivatives of anthraquinones (hypericins), flavonoids, tannins, phloroglucinols (hyperforins), xanthones, terpenes (essential oil), and phenolic acids. The phenolic acids are represented by caffeic, *p*-coumaric, ferulic, *p*–hydroxybenzoic and vanillic acids as well as the depside—chlorogenic acid (Barnes et al. 2007).

Plants originating from natural habitats in different regions of Europe have been found to contain mostly chlorogenic (Cirak et al. 2007) and neochlorogenic acids (Silva et al. 2008), caffeic and protocatechuic acids (Dias et al. 1999; Tatsis et al. 2007), and isomers of *p*-coumaroylquinic acid (Jurgenliemk and Nahrstedt 2002). By comparison, preliminary studies conducted at our department in Kraków have revealed that, in addition to caffeic, chlorogenic, *p*–hydroxybenzoic, protocatechuic and vanillic acids, syringic acid is also present in the herbal raw material supplied by two Polish herbal companies.

Phenolic acids have valuable biological properties, including antioxidant, immunostimulant, anticancer, antiaggregatory and anti-inflammatory activities. In particular, the antioxidant properties of phenolic acids attract an enormous amount of interest from both the pharmaceutical and cosmetics industries (Khadem and Marles2010).

To date, the quite extensive biotechnological research on *H. perforatum*, and other species of the genus *Hypericum*, has primarily focused on studying the biosynthesis and accumulation of hypericins, hyperforins, xanthones and flavonoids (Beerhues 2011; Charchoglyan et al. 2007; Don Palmer and Keller 2011; Karting et al. 1996; Liu et al.2007). In addition, numerous micropropagation protocols have been developed (Goel et al.2009; Savio et al. 2012). Some of them can be useful in the production of elite varieties with predictable phytochemical profiles for pharmaceutical purposes (Murch and Saxena2006). The biosynthesis and accumulation of phenolic acids in in vitro cultures of *H. perforatum* have so far been the subject of isolated biotechnological studies. In adventitious root biomass cultivated in a balloon-type airlift bioreactor, high amounts of chlorogenic acid (0.80 mg g^{-1}) and, among flavonoids, of quercetin and hyperoside have been reported (Cui et al. 2010). Other investigations have revealed the accumulation of neochlorogenic and ferulic acids (Dias et al. 1999; Pasqua et al. 2003) in the biomass of *H. perforatum* cultured in vitro.

Citation: Inga Kwiecień, Aleksandra Szydłowska, Beata Kawka, Ludger Beerhues, Halina Ekiert, Accumulation of biologically active phenolic acids in agitated shoot cultures of three Hypericum perforatum cultivars: 'Elixir', 'Helos' and 'Topas', DOI: 10.1007/s11240-015-0830-3.

Preliminary studies of the accumulation of free phenolic acids in agar cultures of *H. perforatum* were conducted earlier at our department in Kraków, and similar investigations have recently been performed with agar cultures of three cultivars—'Elixir', 'Helos' and 'Topas'. The preliminary results were sufficiently encouraging to continue and conduct these studies within a wider scope. An additional incentive to undertake further research were the biotechnological achievements concerning the accumulation of this particular group of plant secondary metabolites, especially of rosmarinic acid (Makri and Kintzios 2004; Ekiert et al. 2013).

The aim of the present study was to establish agitated shoot cultures of three *H. perforatum* cultivars ('Elixir', 'Helos', 'Topas') and to optimize the conditions for the growth of biomass and accumulation of free phenolic acids. A precursor for one subgroup of these compounds, cinnamic acid, was included. Altogether, seventeen compounds were analyzed by HPLC. The study aimed at nominating the best 'growth' and 'production' media for the three cultivars. The cultures were performed on four variants each of Linsmaier and Skoog (LS) medium and Murashige and Skoog (MS) medium containing varying concentrations (0.1–3.0 mg l^{-1}) of plant growth regulators, namely the auxin—NAA and the cytokinin—BA. In order to assess the biosynthetic potential of the tested cultivars, the cultures were additionally analyzed for hypericin content. Our previous studies of the accumulation of hypericin with densitometric detection and of indole compounds, such as serotonin, melatonin, indole-3-acetic acid, l-tryptophan and 5-hydroxy-l-tryptophan, in agar shoot cultures of the three *H. perforatum* cultivars had indicated a remarkable biosynthetic potential, which, however, differed significantly between the cultivars (Muszyńska et al. 2014).

MATERIALS AND METHODS

Plant Material

The in vitro shoot cultures of the three *Hypericum perforatum* L. cultivars: 'Elixir', 'Helos', and 'Topas' were established in 2007 at the Institut für Pharmazeutische Biologie, Technische Universität Braunschweig (Germany). Seeds had been purchased from Richters Herbs (GooDMood, Canada). They were surface-sterilized with 70 % ethanol for 2 min and 6 % sodium hypochlorite for 5 min. After being rinsed three times with sterile water (for 10 min each time), the seeds were placed on a hormone-free Murashige and Skoog (MS) medium (Murashige and Skoog 1962) at 25 °C. Seeds germination started within 3 weeks under a 16/8 h light/dark period (16 µmol m^{-2} s^{-1}).

The resulting shoot cultures of 'Elixir', 'Helos', and 'Topas' were maintained on solid MS medium supplemented with 0.5 mg l^{-1} NAA and 0.5 mg l^{-1} BA (pH 5.7). The cultures were grown under constant artificial light (16 µmol m^{-2} s^{-1}, LF-40 W lamp, daylight, Piła) at 25 ± 2 °C. Shoots were subcultured every 6 weeks (Fig. 1a, b).

Figure. 1: In vitro cultures of *Hypericum perforatum* 'Helos' cultivar. **a**, **b** Initial agar shoot culture (MS, 0.5 mg l^{-1} NAA and 0.5 mg l^{-1} BA), **c**, **d** experimental shoot agitated culture (LS, 0.1 mg l^{-1} NAA and 0.1 mg l^{-1} BA)

Experimental cultures

Agitated shoot cultures of the three *H. perforatum* cultivars: 'Elixir', 'Helos' and 'Topas' were performed on two types of medium: MS and Linsmaier and Skoog (LS) medium (Linsmaier and Skoog 1965), which contained different amounts of the plant growth regulators NAA and BA (mg l^{-1}): 0.1 and 0.1, 1.0 and 1.0, 2.0 and 2.0, and 3.0 and 3.0, respectively. Erlenmeyer flasks (500 ml) were filled with 150 ml of medium. The agitated cultures were initiated with 6-week-old shoot clusters. The inoculum for the cultures was 1 g of fresh biomass per flask. Three subculture cycles (3 weeks each) were carried out before the main experiment. The cultures (three series) were maintained under the same light and temperature conditions as the initial cultures, on an Altel rotary shaker operating at 140 rpm with a vibration amplitude of 35 mm for 3 weeks (Fig. 1c, d).

Extraction and HPLC Analysis

The biomass was dried in fresh air at 25 ± 2 °C, ground, and extracted with boiling methanol (50 ml) for 3 h. The methanol was evaporated and the residue was dissolved in HPLC-grade methanol. The methanolic extracts were analyzed using modified HPLC method (Ellnain-Wojtaszek and Zgórka 1999). An HPLC-system (Merck–Hitachi) and a Purospher RP-18e analytical column (4 × 250 mm, 5 µm; Merck) were used. The mobile phase consisted of methanol (A) and 0.5 % acetic acid (B) (gradient elution). The flow rate was 1 ml min^{-1}. Compounds were estimated using a DAD detector. Qualification and quantification analyses were based on comparison with reference substances of caffeic, cinnamic, chlorogenic, *o*-coumaric, *p*-coumaric, 3,4-dihydoxyphenylacetic, ferulic, gallic, gentisic, *p*-hydroxybenzoic, neochlorogenic, protocatechuic, rosmarinic, salicylic, sinapic, syringic and vanillic acids (Sigma-Aldrich). Additionally, an analysis of hypericin content was performed using the Sigma-Aldrich standard.

Statistical Analysis

The obtained results are expressed as the mean ± SD of three independent experiments and determinations.

RESULTS

Accumulation of Metabolites on the Tested Variants of LS Medium

The biomass from the agitated shoot cultures of *H. perforatum* ('Elixir', 'Helos' and 'Topas'), which were grown on variants of LS medium containing different concentrations of plant growth regulators, was characterized by a varying morphological appearance. The cultures of all three cultivars on the medium with the lowest level of plant growth regulators (0.1 mg l^{-1} NAA and 0.1 mg l^{-1} BA) produced numerous green leafy shoots. With an increase in the concentration of plant growth regulators, the biomass became more compact, the shoots were shorter and formed callus tissue at the base. During the 3-week-culture cycles on the tested variants of LS medium, different increases in dry biomass were obtained, which ranged from 7.2 to 8.5 for 'Elixir', from 8.5 to 9.7 for 'Helos', and from 7.8 to 9.1 for 'Topas'. The largest increases were obtained for the cultivar 'Helos' grown on LS media containing 0.1 mg l^{-1} NAA and 0.1 mg l^{-1} BA (9.5-fold) and 2 mg l^{-1}NAA and 2 mg l^{-1} BA (9.7-fold)(Fig. 1c, d).

The chromatographic analysis of the extracts from the in vitro biomass of the three cultivars proved the presence of the same seven compounds out of the seventeen that were tested—chlorogenic, *p*-coumaric, 3,4-dihydroxyphenylacetic, neochlorogenic, protocatechuic, syringic and vanillic acids (Fig. 2; Table 1). None of the extracts was found to contain the following acids: caffeic, cinnamic, *o*-coumaric, ferulic, gallic, gentisic, *p*-hydroxybenzoic, rosmarinic, salicylic and sinapic acids.

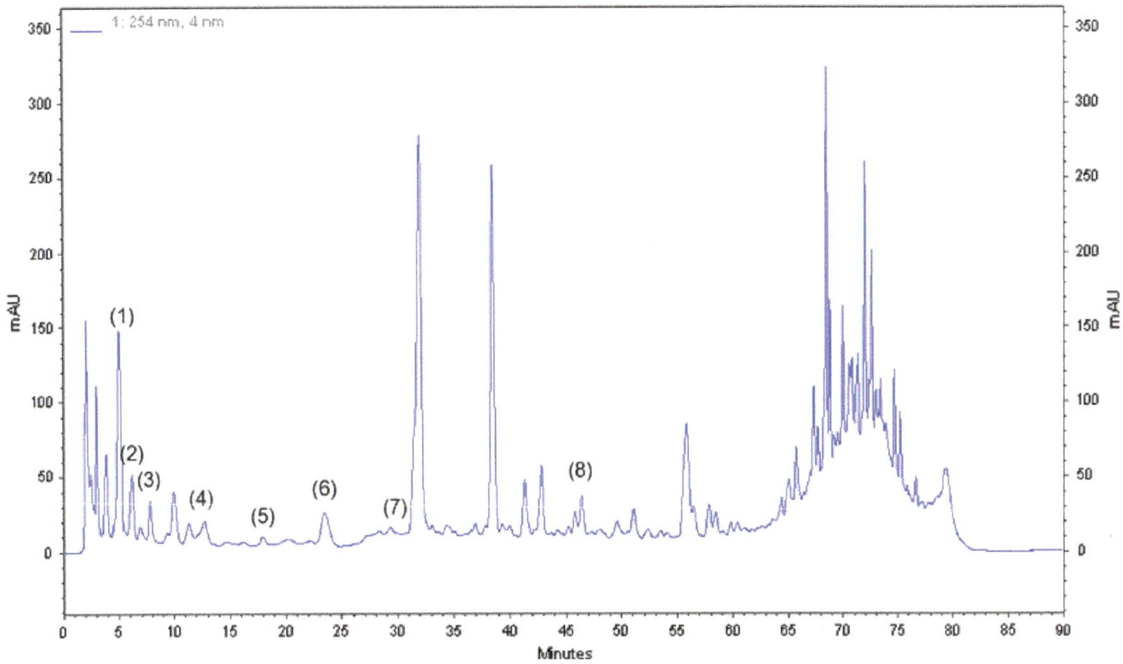

Figure. 2: Chromatogram of methanolic extract of *Hypericum perforatum* 'Helos'cultivar grown on LS medium—1.0 mg l^{-1} NAA and 1.0 mg l^{-1} BA; (*1*) protocatechuic acid, (*2*) neochlorogenic acid, (*3*) 3,4-dihydoxyphenylacetic acid, (*4*) chlorogenic acid, (*5*) vanillic acid, (*6*) syryngic acid, (*7*) *p*-coumaric acid, (*8*) hypericin

Table 1: Content (mg 100 g^{-1} DW ± SD) of free phenolic acids and hypericin in biomass extracts from in vitro cultures of *Hypericum perforatum* cultivars grown on different LS medium variants (three series)

H. perforatum cultivar	NAA/BA (mg l^{-1})	Phenolic acids [mg 100 g^{-1} DW]							Hypericin (mg 100 g^{-1} DW)
		Protocatechuic acid	Neochlorogenic acid	3,4-Dihydoxy-phenylacetic acid	Chlorogenic acid	Vanillic acid	Syryngic acid	*p*-Coumaric acid	
Elixir	0.1/0.1	18.73 ± 1.03	98.80 ± 3.33	55.99 ± 1.40	11.64 ± 1.36	0.70 ± 0.13	13.43 ± 2.33	12.59 ± 0.80	7.76 ± 0.85
	1.0/1.0	15.74 ± 0.77	32.40 ± 2.13	30.48 ± 2.13	12.01 ± 1.28	0.52 ± 0.21	19.19 ± 1.54	2.45 ± 1.93	12.75 ± 0.96
	2.0/2.0	14.23 ± 0.92	26.94 ± 1.57	41.08 ± 1.41	11.81 ± 1.53	0.45 ± 0.04	13.68 ± 3.99	2.67 ± 3.11	14.46 ± 1.32
	3.0/3.0	10.46 ± 1.01	19.02 ± 1.63	53.11 ± 3.03	8.46 ± 1.03	0.44 ± 0.07	12.91 ± 1.85	4.42 ± 2.57	13.67 ± 1.72
Helos	0.1/0.1	18.66 ± 1.22	118.81 ± 6.60	47.49 ± 2.82	10.81 ± 2.95	0.00 ± 0.00	16.74 ± 2.10	8.12 ± 1.61	7.53 ± 0.74
	1.0/1.0	19.13 ± 1.38	30.49 ± 0.68	75.07 ± 2.83	11.40 ± 2.28	0.01 ± 0.00	16.28 ± 2.77	4.86 ± 1.02	6.22 ± 0.64
	2.0/2.0	18.49 ± 2.00	35.22 ± 3.37	84.71 ± 2.53	11.94 ± 0.99	0.02 ± 0.00	23.63 ± 2.15	9.17 ± 1.33	5.78 ± 0.52
	3.0/3.0	15.68 ± 1.11	32.67 ± 0.83	75.18 ± 2.60	12.33 ± 1.62	0.03 ± 0.05	21.17 ± 1.18	6.67 ± 1.30	3.08 ± 0.23
Topas	0.1/0.1	16.66 ± 1.20	47.37 ± 4.32	32.35 ± 2.36	18.79 ± 1.19	0.00 ± 0.00	16.57 ± 1.5	4.06 ± 1.87	3.57 ± 0.19
	1.0/1.0	16.47 ± 1.16	51.49 ± 3.16	86.70 ± 2.38	14.74 ± 1.33	0.00 ± 0.00	18.44 ± 1.86	3.63 ± 2.71	5.01 ± 0.43
	2.0/2.0	13.80 ± 1.26	21.61 ± 1.20	59.71 ± 2.33	5.98 ± 2.23	0.21 ± 0.03	12.68 ± 2.41	3.33 ± 2.43	7.43 ± 0.75
	3.0/3.0	16.92 ± 0.91	21.81 ± 1.79	53.13 ± 1.68	5.49 ± 1.21	0.00 ± 0.00	7.91 ± 2.54	3.07 ± 1.35	3.59 ± 0.49

The amounts of the individual compounds as well as their total amount in the extracts from the biomass of all three cultivars was clearly dependent on the concentrations of the plant growth regulators in the tested LS medium variants. The total phenolic acids content in the shoots increased, depending on the LS variant, 1.37-fold for 'Elixir', 1.79-fold for 'Helos', and 1.22-fold for 'Topas' (Fig. 3). The maximum total phenolic acids content for 'Elixir' (161.6 mg 100 g^{-1} DW) was found on LS medium containing auxin and cytokinin at 0.1 mg l^{-1} each. An almost identical maximum total amount of metabolites (160.5 mg 100 g^{-1} DW) was found in the shoots of 'Topas' growing on LS medium containing 1 mg l^{-1}NAA and 1 mg l^{-1} BA. The highest maximum total amount of the compounds analyzed (222.4 mg 100 g^{-1} DW) was present in extracts from the shoots of 'Helos', which also grew on LS medium containing NAA and BA at 1.0 mg l^{-1} each (Fig. 3).

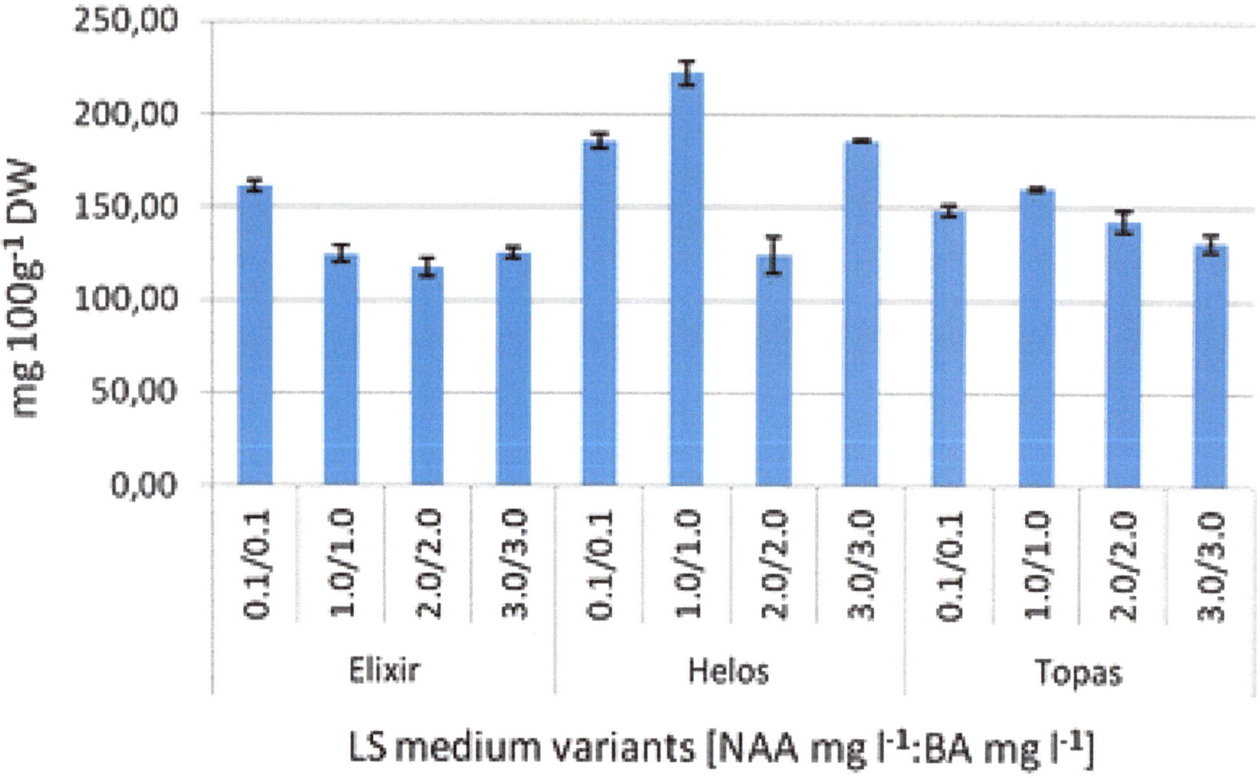

Figure. 3: Content (mg 100 g^{-1} DW ± SD) of total free phenolic acids in biomass extracts from in vitro cultures of *Hypericum perforatum* cultivars grown on LS medium variants (three series)

The largest amounts accumulating in the biomass of all three cultivars were those of 3,4-dihydroxyphenylacetic acid (31.66–29.29 mg 100 g^{-1} DW) and neochlorogenic acid (24.22–87.77 mg 100 g^{-1} DW). Acids that accumulated in lower amounts were chlorogenic (5.35–15.34 mg 100 g^{-1} DW), protocatechuic (16.60–21.43 mg 100 g^{-1} DW) and syringic (10.23–20.37 mg 100 g^{-1} DW). Even lower were the amounts of *p*-coumaric acid (3.51–12.56 mg 100 g^{-1} DW). Vanillic acid was present in very small amounts (0.01–1.56 mg 100 g^{-1} DW) and was not detectable in two extracts (Table 1).

The amount of hypericin in the biomass of *H. perforatum* in vitro cultures grown on variants of LS medium ranged from 3.06 to 10.98 mg 100 g^{-1} DW. The highest amounts were present in extracts from the shoots of 'Elixir'. The lowest amount of this metabolite was in 'Helos' (max. 5.88 mg 100 g^{-1} DW) (Table 1).

Accumulation of Metabolites on the Tested Variants of MS Medium

Like on the tested variants of LS medium, the biomass from the agitated cultures of 'Elixir', 'Topas' and 'Helos' was also morphologically diverse on the tested variants of MS medium. On the medium containing auxin and cytokinin at 0.1 mg l^{-1} each, the biomass consisted of numerous green leafy shoots. An increase in the concentrations of the plant growth regulators in the medium resulted in morphological changes. The shoots became shorter, more compact, with callus tissue forming at their base. The increases in dry biomass during 3-week-growth cycles on the tested variants of MS medium also varied, namely from 6.4 to 8.6 for 'Elixir', from 8.5 to 9.2 for 'Helos', and from 5.9 to 8.2 for 'Topas'.

As in the case of the LS variants, the highest increases were obtained for the cultivar 'Helos'. They were particularly pronounced (9.2- and 8.8-fold) on two variants of MS medium containing auxin and cytokinin 0.1 mg l^{-1} and 1 mg l^{-1} each. A high increase (8.6-fold) was also recorded for the biomass of 'Elixir' on MS medium containing 1.0 mg l^{-1}NAA and 1.0 mg l^{-1} BA.

The analyzed extracts from the biomass growing on the tested variants of MS medium had the same qualitative composition of phenolic acids as in the case of the variants of LS medium (Table 2). Of the seventeen compounds analyzed, the presence of seven was confirmed: chlorogenic, *p*-coumaric, 3,4-dihydroxyphenylacetic, neochlorogenic, protocatechuic, syringic, and vanillic acids. The amounts of the individual phenolic acids as well as their total amount in the analyzed extracts from the biomass of the cultivars studied varied greatly, depending on the MS medium variant (Table 2).

Table 2: Content (mg 100 g^{-1} DW ± SD) of free phenolic acids and hypericin in biomass extracts from in vitro cultures of *Hypericum perforatum* cultivars grown on different MS medium variants (three series)

H. perforatum cultivar	NAA/BA (mg l^{-1})	Phenolic acids [mg 100 g^{-1} DW]							Hypericin (mg 100 g^{-1} DW)
		Protocatechuic acid	Neochlorogenic acid	3,4-Dihydoxy-phenylacetic acid	Chlorogenic acid	Vanillic acid	Syryngic acid	*p*-Coumaric acid	
Elixir	0.1/0.1	18.73 ± 1.03	98.80 ± 3.33	55.99 ± 1.40	11.64 ± 1.36	0.70 ± 0.13	13.43 ± 2.33	12.59 ± 0.80	7.76 ± 0.85
	1.0/1.0	15.74 ± 0.77	32.40 ± 2.13	30.48 ± 2.13	12.01 ± 1.28	0.52 ± 0.21	19.19 ± 1.54	2.45 ± 1.93	12.75 ± 0.96
	2.0/2.0	14.23 ± 0.92	26.94 ± 1.57	41.08 ± 1.41	11.81 ± 1.53	0.45 ± 0.04	13.68 ± 3.99	2.67 ± 3.11	14.46 ± 1.32
	3.0/3.0	10.46 ± 1.01	19.02 ± 1.63	53.11 ± 3.03	8.46 ± 1.03	0.44 ± 0.07	12.91 ± 1.85	4.42 ± 2.57	13.67 ± 1.72
Helos	0.1/0.1	18.66 ± 1.22	118.81 ± 6.60	47.49 ± 2.82	10.81 ± 2.95	0.00 ± 0.00	16.74 ± 2.10	8.12 ± 1.61	7.53 ± 0.74
	1.0/1.0	19.13 ± 1.38	30.49 ± 0.68	75.07 ± 2.83	11.40 ± 2.28	0.01 ± 0.00	16.28 ± 2.77	4.86 ± 1.02	6.22 ± 0.64
	2.0/2.0	18.49 ± 2.00	35.22 ± 3.37	84.71 ± 2.53	11.94 ± 0.99	0.02 ± 0.00	23.63 ± 2.15	9.17 ± 1.33	5.78 ± 0.52
	3.0/3.0	15.68 ± 1.11	32.67 ± 0.83	75.18 ± 2.60	12.33 ± 1.62	0.03 ± 0.05	21.17 ± 1.18	6.67 ± 1.30	3.08 ± 0.23
Topas	0.1/0.1	16.66 ± 1.20	47.37 ± 4.32	32.35 ± 2.36	18.79 ± 1.19	0.00 ± 0.00	16.57 ± 1.5	4.06 ± 1.87	3.57 ± 0.19
	1.0/1.0	16.47 ± 1.16	51.49 ± 3.16	86.70 ± 2.38	14.74 ± 1.33	0.00 ± 0.00	18.44 ± 1.86	3.63 ± 2.71	5.01 ± 0.43
	2.0/2.0	13.80 ± 1.26	21.61 ± 1.20	59.71 ± 2.33	5.98 ± 2.23	0.21 ± 0.03	12.68 ± 2.41	3.33 ± 2.43	7.43 ± 0.75
	3.0/3.0	16.92 ± 0.91	21.81 ± 1.79	53.13 ± 1.68	5.49 ± 1.21	0.00 ± 0.00	7.91 ± 2.54	3.07 ± 1.35	3.59 ± 0.49

The total amount of metabolites increased for the biomass of individual cultivars as follows: 1.4-fold for 'Helos', 1.77-fold for 'Topas', and 1.95-fold for 'Elixir' (Fig. 4). The highest total amount of the compounds (220.65 mg 100 g^{-1} DW) was confirmed in the biomass of the shoots of 'Helos' growing on MS medium containing 0.1 mg l^{-1} NAA and 0.1 mg l^{-1} BA. Notable amounts of the compounds (211.89 mg 100 g^{-1} DW) were also found in the extracts from the shoots of 'Elixir' growing on the same MS variant (Fig. 4).

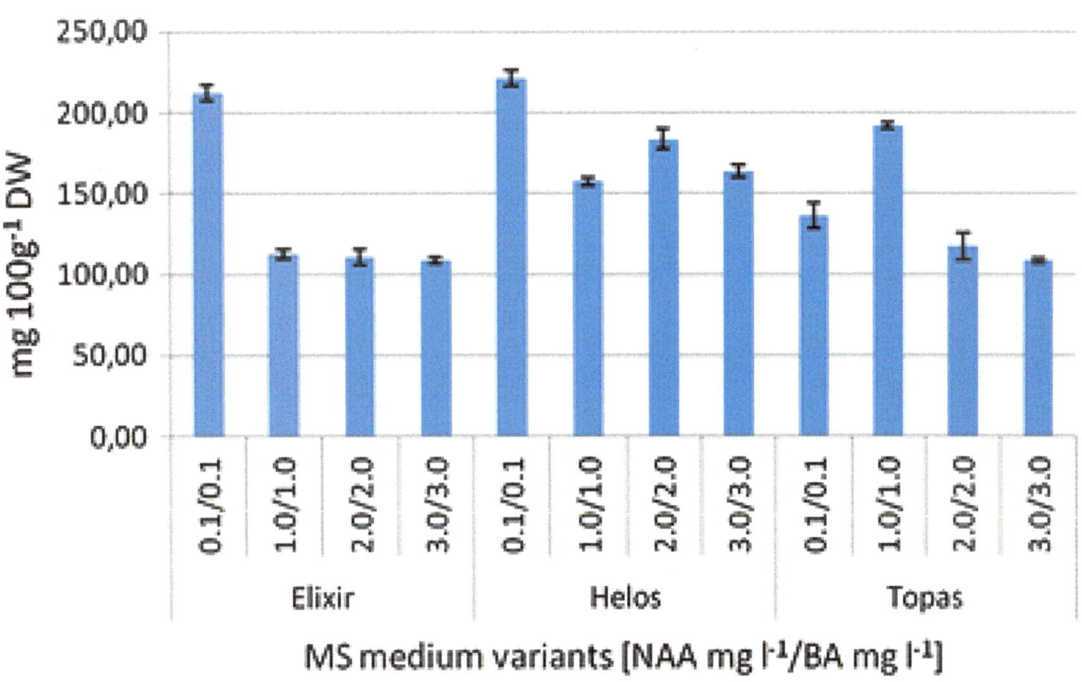

Figure. 4: Content (mg 100 g^{-1} DW ± SD) of total free phenolic acids in biomass extracts from in vitro cultures of *Hypericum perforatum* cultivars grown on MS medium variants (three series)

Neochlorogenic acid and 3,4-dihydroxyphenylacetic acid were also among the quantitatively dominant compounds in the biomass of all the cultivars. Their amounts changed from 19.02 to 118.81 mg 100 g^{-1} DW and from 30.48 to 86.70 mg 100 g^{-1} DW, respectively. Three other phenolic acids were likewise accumulated in considerable amounts: chlorogenic (5.49 to 18.79 mg 100 g^{-1} DW), protocatechuic (10.46 to 19.13 mg 100 g^{-1} DW) and syringic (7.91 to 23.63 mg 100 g^{-1} DW). Like on the variants of LS medium, the amounts of *p*-coumaric acid in the analyzed extracts were much lower (2.45 to 12.59 mg 100 g^{-1} DW). The vanillic acid content did not exceed 0.70 mg 100 g^{-1} DW and the presence of this compound was not confirmed in four of the analyzed extracts, which were from 'Topas' and 'Helos' (Table 2).

The amount of hypericin in extracts of the biomass grown on MS medium variants ranged from 3.08 to 14.45 mg 100 g^{-1} DW. Also on LS medium variants, the cultivar 'Elixir' was the richest source of this metabolite. The highest hypericin content was confirmed in the biomass of the shoots of 'Elixir' growing on MS medium containing 2.0 mg l^{-1} NAA and 2.0 mg l^{-1} BA. In 'Helos' shoot extracts the maximum amount of hypericin was 7.53 mg 100 g^{-1} DW (Table 2).

Discussion

The biomass of shoots of all three *H. perforatum* cultivars grown in agitated cultures increased considerably. A particularly rapid increase in shoot dry biomass, more than ninefold, during 3-week-growth cycles on the three tested media variants was observed for 'Helos'.

The analysis of extracts from the shoots of the three cultivars studied revealed a qualitatively different composition of phenolic acids in comparison with open-air grown *H. perforatum* of different origin (Dias et al. 1999; Jurgenliemk and Nahrstedt 2002; Gorzkiewicz and Ekiert 2004, unpublished; Cirak et al. 2007; Tatsis et al. 2007; Silva et al.2008). The biomass of all the cultivars grown in vitro was notable for the presence of high amounts of 3,4-dihydroxyphenylacetic acid (Tables 1, 2).

The concentrations of the tested plant growth regulators, NAA and BA, in culture medium affected considerably the amounts of individual compounds and consequently their total amount in the cultured biomass. On the tested variants of LS medium, the amounts of individual compounds in the shoots increased from 1.08 to 15.0-fold (Table 1). The total amount of the compounds increased 1.37-, 1.79- and 1.22-fold for 'Elixir', 'Helos' and 'Topas', respectively (Fig. 3). On the tested variants of MS medium, the amounts of individual metabolites increased from 1.14 to 5.19-fold (Table 2). The differences in the total amount of the compounds were also considerable: 1.95-, 1.40- and 1.77-fold, respectively (Fig. 4).

The impact of the concentrations of plant growth regulators in culture media on the accumulation of metabolites is a well-known phenomenon. Our previous studies had demonstrated a clear influence of the concentrations of plant growth regulators on the accumulation of biogenetically different groups of metabolites—coumarins (Ekiert 2004), lignans (Szopa and Ekiert 2011, 2013, 2015), and phenolic acids (Ekiert et al. 2009, 2014; Szopa and Ekiert 2012, 2014; Szopa et al. 2013).

Our studies of in vitro cultures of *Aronia melanocarpa* have recently also demonstrated the importance of the basic composition of the culture media (LS and MS) on the accumulation of free phenolic acids. Variants of MS medium with the same composition of plant growth regulators as in variants of LS medium allowed us to obtain from 1.61- to 2.00-fold higher amounts of free phenolic acids (Szopa et al. 2013; Szopa and Ekiert2014). In the present study with the three cultivars, the results were different. Only in four out of twelve extracts from the shoots grown on MS variants, the total amount of the compounds was from 1.19 to 1.47 times higher than in the extracts from the shoots grown on identical variants of LS medium.

The amounts of individual compounds as well as their total amounts obtained by us are significant due to the high degree of differentiation of the in vitro cultured biomass. Cultures with a high degree of differentiation usually provide significantly higher amounts of secondary metabolites than undifferentiated callus cultures. Our previous studies on the accumulation of free phenolic acids have shown significant amounts of these compounds in shoot cultures of *Ruta graveolens* (Ekiert et al. 2009), shoot-differentiating callus cultures of *Ruta graveolens* ssp. *divaricata* (Ekiert et al. 2014), and shoot cultures of *Aronia melanocarpa* (Szopa et al. 2013; Szopa and Ekiert 2014). Furthermore, our studies have demonstrated the importance of the type of culture on the accumulation of phenolic acids. In our preliminary studies, agar cultures of the three cultivars of *H. perforatum* grown on MS medium variants accumulated several times lower amounts of free phenolic acids (unpublished). These results confirmed the observations of other research teams, which demonstrated that liquid media are better for the accumulation of phenolic compounds in *H. perforatum* in vitro systems (Savio et al. 2012).

Based on the results of the present study, agitated shoot cultures of the cultivar 'Helos' can be nominated as a rich, potential biotechnological source of the bioactive phenolic acids analyzed. As a good 'production' medium, we can propose the variant of LS medium containing 1.0 mg l^{-1} NAA and 1.0 mg l^{-1} BA, and/or the variant of MS medium containing 0.1 mg l^{-1} NAA and 0.1 mg l^{-1} BA. The total amount of the compounds of interest on these media was approx. 220 mg 100 g^{-1} DW (Figs. 3, 4). Both nominated media are also good 'growth' media, which guarantee about 9-fold increases in the dry weight of shoots during 3-week-culture cycles. Of the three cultivars studied, the cultivar 'Helos' appears to have the highest enzymatic potential for the biosynthesis of phenolic acids. The shoots of this cultivar grown in vitro accumulate significant amounts of 3,4-dihydroxyphenylacetic acid and neochlorogenic acid (max. 129.29 mg 100 g^{-1} DW and 118.81 mg 100 g^{-1} DW, respectively), and also large quantities of other phenolic acids, such as chlorogenic, protocatechuic, and syringic (max. 12.97, 21.40, and 23.63 mg 100 g^{-1} DW, respectively) (Tables 1, 2). Protocatechuic acid and chlorogenic acid stand out because of their exceptional biological activities. Their antioxidant, antiradical, antibacterial and anticancer properties are proven (Rice-Evans et al. 1996; Sánchez-Maldonado et al. 2011). Moreover, antiproliferative and apoptotic effects of 3,4-dihydroxyphenylacetic acid have been demonstrated (Kampa et al. 2004).

As shown by the qualitative and quantitative analyses of hypericin (performed for comparative purposes), 'Helos' was characterized by a moderate capacity for the biosynthesis of this compound. The highest hypericin content was found in extracts from the shoots of 'Elixir'. Our earlier studies on the accumulation of nonhallucinogenic indole compounds and hypericin in agar cultures of the same three cultivars of *H. perforatum*had also indicated the shoots of 'Helos' to accumulate the highest amounts of indole compounds (serotonin and 5-hydroxy-l-tryptophan) and 'Elixir' as the richest source of hypericin (Muszyńska et al. 2014). The results of ongoing studies confirm the different enzymatic potential of cells from in vitro cultures of the cultivars being analyzed and consequently a different phytochemical profile of the cultured biomass. Our present study has confirmed the substantial amounts of 3,4-dihydroxyphenylacetic acid in the biomass grown in vitro. This is the first report of the presence of this compound in plant in vitro cultures.

The results obtained encourage the continuation of the biotechnological research related to the influence of different in vitro culture conditions on the biosynthetic ability of the investigated *H. perforatum* cultivars.

REFERENCES

1. Barnes J, Anderson LA, Phillipson D (2007) St John's Worth. In: Herbal medicines. Pharmaceutical Press, London, pp 549–569
2. Beerhues L (2011) Biosynthesis of the active *Hypericum perforatum* constituents. In: Odabas MS, Çırak C (eds) *Hypericum*. Med Arom Plant Sci Biotechnol 5:70–77
3. Charchoglyan A, Abrahamyan A, Fujii I, Boubakir Z (2007) Differential accumulation of hyperforin and secohyperforin in *Hypericum perforatum* tissue cultures. Phytochemistry 68:2670–2677
4. Cirak C, Radusiěnė J, Karabük BS, Janulis V, Ivanauskas L (2007) Variation of bioactive compounds in *Hypericum perforatum* growing in Turkey during its phenological cycle. J Integr Plant Biol 49:615–620
5. Cui XH, Chakrabarty D, Lee EJ, Paek KY (2010) Production of adventitious roots and secondary metabolites by *Hypericum perforatum L.* in a bioreactor. Bioresour Technol 101:4708–4716
6. Dias ACP, Seabra RM, Andrade PB, Fernandes-Ferreira M (1999) The development and evaluation of an HPLC-DAD method for the analysis of the phenolic fractions from in vivo and in vitro biomass of *Hypericum* species. J Liq Chromatogr Relat Technol 22:215–227
7. Don Palmer C, Keller WA (2011) Plant regeneration from petal explants of *Hypericum perforatum* L. Plant Cell Tiss Organ Cult 105:129–134
8. Ekiert H (2004) Accumulation of biologically active furanocoumarins within in vitro cultures of medicinal plants. In: Ramawat KG (ed) Biotechnology of medicinal plants. Vitalizer and therapeutic. Science Publishers, Enfield, pp 267–296
9. Ekiert H, Szewczyk A, Kuś A (2009) Free phenolic acids in *Ruta graveolens* L. in vitro culture. Pharmazie 64:694–696
10. Ekiert H, Kwiecień I, Szopa A (2013) Rosmarinic acid production in plant in vitro cultures. Pol J Cosmet 16:49–58
11. Ekiert H, Piekoszewska A, Muszyńska B, Baczyńska S (2014) Accumulation of *p*-cumaric acid and other bioactive phenolic acids in in vitro culture of *Ruta graveolens* ssp.*divaricata* (Tenore) Gams. Med Int Rev 26:24–31
12. Ellnain-Wojtaszek M, Zgórka G (1999) High-performance liquid chromatography and thin-layer chromatography of phenolic acids from *Ginkgo biloba L.* leaves collected within vegetative period. J Liq Chromatogr Relat Technol 22:1457–1471
13. Goel MK, Kukreja AK, Bisht NS (2009) In vitro manipulations in St. John's wort (*Hypericum perforatum* L.) for incessant and scale up micropropagation using adventitious roots in liquid medium and assessment of clonal fidelity using RAPD analysis. Plant Cell Tiss Organ Cult 96:1–9
14. Jurgenliemk G, Nahrstedt A (2002) Phenolic compounds from *Hypericum perforatum*. Planta Med 68:88–91
15. Kampa M, Alexaki VI, Notas G, Nifli AP, Nistikaki A, Hatzoglou A, Bakogeorgou E, Kouimtzoglou E, Blekas G, Boskou D, Gravanis A, Castanas E (2004) Antiproliferative and apoptotic effects of selective phenolic acids on T47D human breast cancer cells: potential mechanisms of action. Breast Cancer Res 6:63–74
16. Karting T, Gobel I, Heydel B (1996) Production of hypericin, pseudohypericin and flavonoids in cell cultures of various *Hypericum* species and their chemotypes. Planta Med 62:51–53
17. Khadem S, Marles R (2010) Monocyclic phenolic acids; hydroxy- and polyhydroxybenzoic acids: occurrence and recent bioactivity studies. Molecules 15:7985–8005
18. Linsmaier EM, Skoog F (1965) Organic growth factor requirements of tobacco tissue cultures. Physiol Plant 18:100–127
19. Liu XN, Zhang XQ, Zhang SX, Sun JS (2007) Regulation of metabolite production by precursors and elicitors in liquid cultures of *Hypericum perforatum*. Plant Cell Tiss Organ Cult 91:1–7
20. Makri O, Kintzios S (2004) In vitro rosmarinic acid production: an update. In: Ramawat KG (ed) Biotechnology of medicinal plants. Vitalizer and therapeutic. Science Publishers, Enfield, pp 19–31
21. Murashige T, Skoog F (1962) A revised medium for rapid growth and bioassays with tobacco tissue cultures. Physiol Plant 15:473–497
22. Murch SJ, Saxena PK (2006) St. John's wort (*Hypericum perforatum* L.); challenges and strategies for production of chemically consistent plant. Can J Plant Sci 86:765–771
23. Muszyńska B, Ekiert H, Kwiecień I, Maślanka A, Zodi R, Beerhues L (2014) Comparative analysis of therapeutically important indole compounds in in vitro cultures of *Hypericum perforatum* cultivars by HPLC and TLC analysis coupled with densitometric detection. Nat Prod Commun 9:1437–1440

24. Pasqua G, Agato P, Monacelli P, Santamaria AR, Argentieri AP (2003) Metabolites in cell suspension cultures, calli, and in vitro regenerated organs of *Hypericum perforatum* cv. Topas. Plant Sci 165:977–982

25. Rice-Evans CA, Miller NJ, Paganga G (1996) Structure-antioxidant activity relationships of flavonoids and phenolic acids. Free Radic Biol Med 20:933–956

26. Sánchez-Maldonado AF, Schieber A, Gänzle MG (2011) Structure-function relationships of the antibacterial activity of phenolic acids and their metabolism by lactic acid bacteria. J Appl Microbiol 111:1176–1184

27. Savio LEB, Astarita LV, Santarem ER (2012) Secondary metabolism in micropropagated*Hypericum perforatum* L. grown in non-aerated liquid medium. Plant Cell Tiss Organ Cult 108:465–472

28. Silva BA, Malva JO, Dias A (2008) St. John's Wort (*Hypericum perforatum*) extracts and isolated phenolic compounds are effective antioxidants in several in vitro models of oxidative stress. Food Chem 110:611–619

29. Szopa A, Ekiert H (2011) Lignans in *Schisandra chinensis* in vitro cultures. Pharmazie 66:633–634

30. Szopa A, Ekiert H (2012) *In vitro* cultures of *Schisandra chinensis* (Turcz.) Baill. (Chinense Magnolia Vine)—a potential biotechnological rich source of therapeutically important phenolic acids. Appl Biochem Biotechnol 166:1941–1948

31. Szopa A, Ekiert H (2013) Production of deoxyschizandrin and γ-schizandrin in shoot-differentiating and undifferentiating callus cultures of *Schisandra chinensis* (Turcz.) Baill. (Chinese magnolia vine). J Biotechnol 165:209–213

32. Szopa A, Ekiert H (2014) Production of biologically active phenolic acids in *Aronia melanocarpa* (Michx.) Elliott in vitro cultures cultivated on different variants of the Murashige and Skoog medium. Plant Growth Regul 72:51–58

33. Szopa A, Ekiert H (2015) Production of schisantherin A and gomisin G in in vitro cultures of *Schisandra chinensis*. Phytochem Lett 11:440–444

34. Szopa A, Ekiert H, Muszyńska B (2013) Accumulation of hydroxybenzoic acids and other biologically active phenolic acids in shoot and callus cultures of *Aronia melanocarpa*(Michx.) Elliot (black chokeberry). Plant Cell Tiss Organ Cult 113:323–329

35. Tatsis EC, Boeren S, Exarchou V, Troganis AN, Vervoort J, Gerothanassis IP (2007) Identification of the major constituents of *Hypericum perforatum* by LC/SPE/NMR and/or LC/MS. Phytochemistry 68:383–393

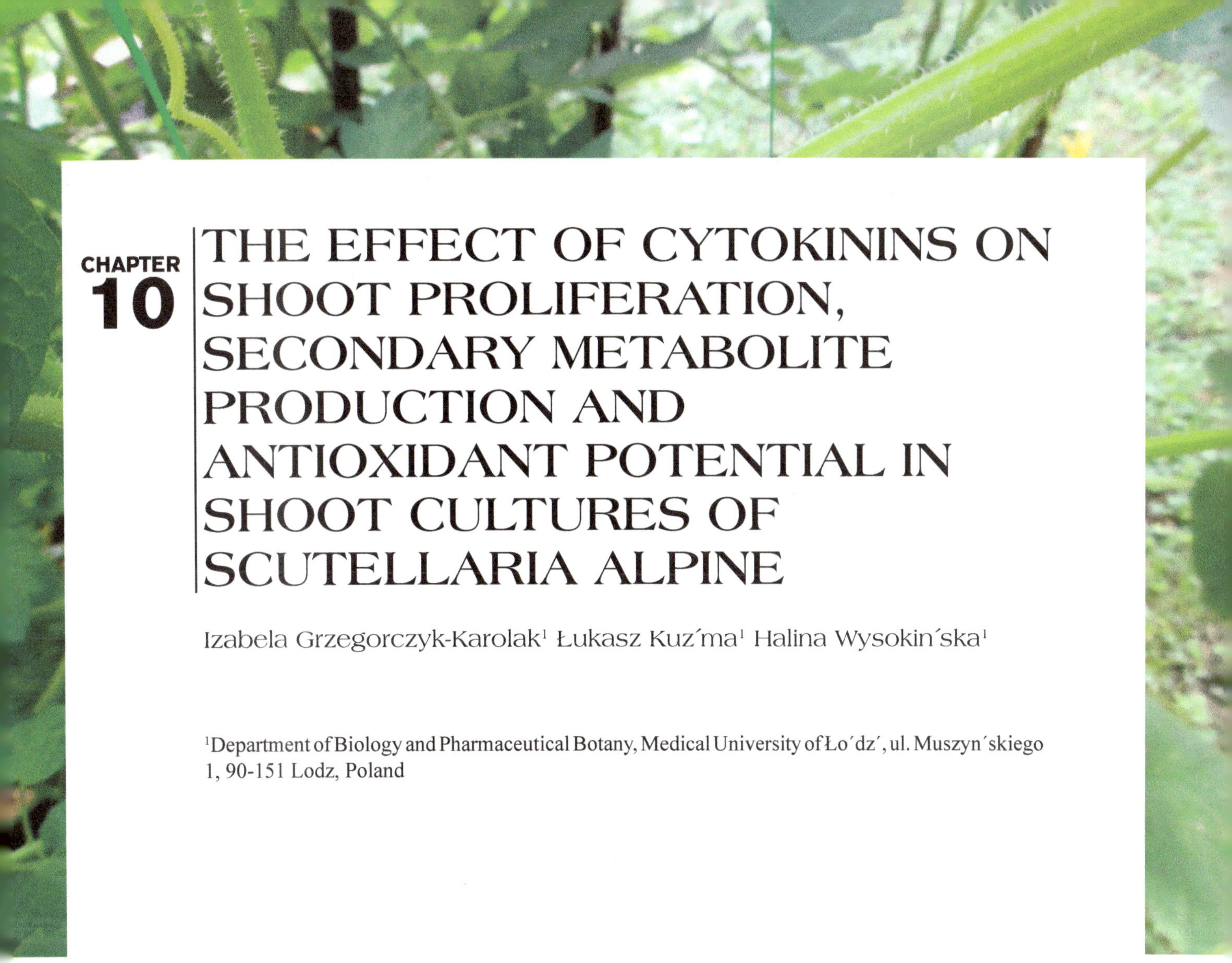

CHAPTER 10

THE EFFECT OF CYTOKININS ON SHOOT PROLIFERATION, SECONDARY METABOLITE PRODUCTION AND ANTIOXIDANT POTENTIAL IN SHOOT CULTURES OF SCUTELLARIA ALPINE

Izabela Grzegorczyk-Karolak[1] Łukasz Kuz´ma[1] Halina Wysokin´ska[1]

[1]Department of Biology and Pharmaceutical Botany, Medical University of Ło´dz´, ul. Muszyn´skiego 1, 90-151 Lodz, Poland

ABSTRACT

The present study evaluates the effects of various cytokinins on Scutellaria alpina shoot proliferation and production of polyphenolic metabolites (baicalin, wogonoside, luteolin, luteolin 7-O-glucoside, verbascoside). The shoots were induced from shoot tips on MS medium supplemented with IAA (indole-3-acetic acid, 0.57 lM) and various concentrations of 6-benzylaminopurine (BAP), kinetin, zeatin (1, 2, 4, 8 lM) or tidiazuron (TDZ) (0.2, 0.5, 1 lM). Among the cytokinins tested, BAP was the most effective for shoot induction, and the highest number of shoots (25 per explant) was achieved with 2 and 4 lM BAP. Maximum biomass production was also achieved on these media. Significantly higher baicalin, wogonoside and verbascoside contents were recorded in treatments containing cytokinins combined with 0.57 lM IAA, when compared to cytokinin-free medium. TDZ at a concentration of 0.5 lM was the most effective for polyphenol production. However, supplementation with cytokinins often results in the reduction of luteolin and its 7-O-glucoside production in the shoot culture of S. alpina. ABTS [2,20 -azino-bis(3-ethylbenzothiazoline-6-sulphonic acid)] and ferric reducing antioxidant power assays were used to identify the antioxidant potential of methanolic extracts from shoots cultured in the presence of different types and concentrations of cytokinins. In both tests, the shoots from medium supplemented with 0.5 lM TDZ demonstrated the strongest antioxidant activity. The results indicate that higher polyphenolic content correlated with greater reducing power and antiradical efficiency.

INTRODUCTION

The genus Scutellaria (Lamiaceae) comprises about 300 species, which are distributed in Europe, the United States and East Asia. Many of the plants belonging to the Scutellaria genus have been used for centuries in traditional Chinese medicine for the treatment of hyperlipidemia, arteriosclerosis, allergy, antibacterial and inflammatory diseases (Shang et al. 2010). The pharmacological properties of the plants are mainly due to the presence of flavonoids such as baicalein, baicalin, wogonin and wogonoside (Shang et al. 2010). Baicalein is known to have a neuroprotective effect against MPTP (1-methyl-4-phenyl-1, 2, 3, 6-tetrahydropyridine) neurotoxicity (Cheng et al. 2008). The flavone protects rat brain mitochondria against chronic cerebral hyperfusion-induced oxidative damage (He et al. 2009). Its glucoside, baicalin, is a potent anti-inflammatory and antitumor agent (Chang et al. 2002; Woo et al. 2006; Shang et al. 2010). Animal experimentation has shown that baicalin can ameliorate the intestinal cytotoxicity caused by the antitumor agent Cpt-11 (Takasuna et al. 1995). It inhibits

Citation: Izabela Grzegorczyk-Karolak, Łukasz Kuźma, Halina Wysokińska, The effect of cytokinins on shoot proliferation, secondary metabolite production and antioxidant potential in shoot cultures of Scutellaria alpine, DOI: 10.1007/s11240-015-0804-5.

HIV-1 infection and replication by inhibiting HIV-1 reverse transcriptase (Kitamura et al. 1998). Wogonoside inhibits lipopolysaccharide-induced angiogenesis (Chen et al. 2009). It demonstrates strong activity against lipid peroxidation and an inhibitory effect on histamine and IgE production (Lim 2003).

Besides flavonoids, the phenylethanoid glycosides, verbascoside and martynoside have been isolated from some species of Scutellaria (Shang et al. 2010). Verbascoside has been found to have a wide spectrum of biological properties, demonstrating antibacterial (Shoyama et al. 1986), antiviral (Kernan et al. 1998), antioxidant (Gao et al. 1999) and cytotoxic activities (Pettit et al. 1990). Due to the pharmacological importance of the secondary metabolites of Scutellaria, interest has grown in the use of biotechnological methods to increase production. Plant tissue culture offers a valuable alternative method for plant micropropagation and the production of biologically active substances under controlled conditions, regardless of season. In recent years, various strategies have been developed which could increase the accumulation of biomass and biosynthesis of secondary metabolites in plant cell and organ cultures (Cole et al. 2007). Attempts to produce phytochemicals of Scutellaria species through tissue culture have been reported earlier for S. baicalensis, S. columnae, S. lateriflora and S. racemosa (Seo et al. 1993; Nishikawa and Ishimura 1997; Stojakowska and Kisiel 1998; Cole et al. 2008). Previous studies describe the successful establishment of shoot regenerating calli and whole plants of S. altissima and these have been found to produce important secondary metabolites (GrzegorczykKarolak et al. 2013, 2015). Following on from these studies, the present study focuses on shoot cultures of S. alpina, a perennial herb distributed throughout the mountain regions of central and southern Europe. The species is characterized by the presence of both valuable flavonoids and high levels of verbascoside (Nishikawa et al. 1999), making it particularly suitable for the present study. This paper reports the effect of different cytokinines (BAP, zeatin, kinetin, TDZ) on the growth of shoot cultures of S. alpina, their capacity to produce flavonoids (baicalin, wogonoside, luteolin, luteolin 7-glucoside) and verbascoside, and their antioxidant activity.

MATERIALS AND METHODS

Establishment of in Vitro Culture

Scutellaria alpina seeds provided by the Botanical Garden of the Institute of Ecology and Botany in Vacratot (Hungary) were used to initiate in vitro culture. The seeds were washed in running tap water, surface sterilized using sodium hypochlorite (1 %) for 10 min and then rinsed three times in sterile distilled water. They were placed on MS (Murashige and Skoog 1962) agar (0.7 %) medium supplemented with kinetin 0.02 mg l^{-1} and gibberellic acid 1 mg l^{-1} for germination. The shoot culture was initiated from shoot tips of 3-week-old aseptically germinated seedlings. The explants were grown on MS agar medium supplemented with 0.57 lM 3-inoliloacetic acid (IAA) and 2 lM 6-benzylaminopurine (BAP), and were subcultured every 5 weeks (Grzegorczyk-Karolak et al. 2013). All cultures were kept at 26 ± 2 C under a 16 h photoperiod provided by cool white fluorescent lamps (approximately 40 lmol m^{-2} s^{-1}).

Shoot Proliferation and Biomass Production

The shoot tips (0.5 cm) of in vitro growing shoots of S. alpina were used as starting material in the experiments. The explants were incubated in tubes containing 25 ml of MS agar medium supplemented with 0.57 lM IAA and various types and concentrations of cytokinin (1, 2, 4, 8 lM BAP, zeatin or kinetin and 0.2, 0.5 or 1 lM TDZ). Shoot tips grown on MS medium without cytokinin, only with 0.57 lM IAA, were used as a control. After 5 weeks, shoot fresh and dry weights [g/tube], the average number of shoots formed per explant (shoot multiplication rate), the length of the main shoots [cm] and shoot morphology in terms of hyperhydricity were recorded. Experiments were repeated three times, and each series consisted of 10–20 explants.

Preparation of Extracts

This extraction was performed as described previously (Grzegorczyk-Karolak et al. 2015). Briefly, lyophilized plant material (250 mg for phytochemical analysis and 500 mg for antioxidant assay) was pulverized and pre-extracted with chloroform overnight to remove chlorophyll. After filtration, the remaining plant material was extracted three times with 30 ml 70 % methanol for 15 min using an ultrasonic bath. The extracts were combined and evaporated under reduced pressure.

Determination of Secondary Metabolites

Ultra-high performance liquid chromatography (UPLC) was used for quantitative evaluation of the verbascoside, baicalin, wogonoside, luteolin and luteolin-7-O-glucoside content of shoots propagated on media with different types and concentrations of cytokinins. Chromatographic analysis was performed as described earlier for Scutellaria altissima (Grzegorczyk-Karolak et al. 2013) using an Agilent Technologies 1290 Infinity UPLC apparatus equipped with a diode array detector on a Zorbax Eclipse Plus C18 column (100 9 3 mm; 1.8 lm Agilent Technologies). The mobile phase consisted of 0.1 % formic acid in acetonitrile (v/v) (solvent A) and 0.1 % formic acid in water (v/v) (solvent B). A gradient program was applied as follows: 0–15 min 20–30 % solvent A, 15.1–17 min 99 % solvent A at flow 0.3 ml min^{-1} . The detection wavelength was set at 320 nm. In the analyzed samples, baicalin, wogonoside, luteolin, 7-O-glucoside of luteolin and verbascoside were identified by comparing their retention times (8.6, 13.6, 12.4, 4.1, 3.6 min., respectively), UV adsorption profiles and MS data with corresponding standard compounds. Liquid chromatography–mass spectrometry (LC–MS/MS) data of the compounds has been previously published (Grzegorczyk-Karolak et al. 2013, 2015). Calibration curves were used to quantify the compounds. Linear standard curves were obtained by plotting standard concentration (baicalin: 220–1100 lg ml^{-1} , wogonoside: 320–1600 lg ml^{-1} , luteolin: 220–1000 lg ml^{-1} , luteolin-7-glucoside: 200–1000 lg ml^{-1} and verbascoside: 200–1000 lg ml^{-1}) as a function of peak area. The mean recovery values for the two fortification levels of the individual components were as follows: baicalin—112.8 and 100.63 %, wogonoside—95.11 and 99.44 %, luteolin—105.4 and 89.45 %, its glucoside—100.42 and 94.29 %, verbascoside—95.48 and 93.07 %. The contents of the compounds in the tested plant materials were expressed as mg g-1 of dry weight (wt).

Antioxidant Assays

The antioxidant potential of methanolic extracts from S. alpina shoot cultures was evaluated using ABTS [2,20 - azino-bis(3-ethylbenzothiazoline-6-sulphonic acid)] cation radical and ferric reducing antioxidant power (FRAP) assays.

FRAP Assay

The FRAP assay was performed according to Pulido et al. (2000). Briefly, 3 ml of freshly prepared FRAP reagent was mixed with 300 ll redistilled water and 100 ll of the methanol extract. The FRAP reagent was prepared by adding 2.5 ml 10 mM TPTZ (2,4,6-tris(2-pyridyl)-s-triazine) solution in 40 mM HCl, 2.5 ml of 20 mM aqueous FeCl3x6H2O solution and 45 ml of 0.3 M acetate buffer at pH 3.6. The reaction mixture was incubated at 37 C for 15 min. The absorbance was measured after 15 min at 595 nm relative to a blank (a sample containing methanol instead extract). The antioxidant activity was determined against a standard of known FRAP value: ferrous sulphate calculated from a calibration curve of 0–2000 lM.

ABTS Radical Scavenging Assay

The antioxidant activity was also determined using the ABTS radical cation decolourisation test according to Re et al. (1999). An ABTS stock solution was prepared by mixing 7 mM ABTS (Sigma Aldrich) with 2.45 mM potassium persulfate (Sigma Aldrich). This mixture was kept in the dark at room temperature for 12–16 h before use. ABTS solution was diluted with methanol until an absorbance of 0.7 ± 0.05 was obtained at 734 nm for the negative control, which was prepared by mixing 2 ml ABTS solution with 2 ml of methanol. The assays were performed by adding 2 ml of extracts (at concentrations of 2.0, 20.0, 100.0, 200.0, 500.0 and 1000.0 lg ml^{-1}) to 2 ml of ABTS solution. The absorbance was measured after 10 min at 734 nm. The results were expressed as EC50: the concentration of sample at which 50 % of maximum scavenging activity was recorded.

Statistical Analysis

The results were expressed as the means ± (standard error) SE of three repeated independent experiments. The significance of the differences between the mean culture growth values was calculated using the Kruskal–Wallis test (p B 0.05) test with Statistica 10.0 software (Statsoft Poland). Correlation analysis was also performed in order to determine relationships among variables.

RESULTS

Effect of Cytokinins on Shoot Multiplication and Biomass Production

The shoot tips of S. alpina were cultured on MS agar solidified medium supplemented with IAA (0.57 lM) and a cytokinin: either BAP, kinetin, zeatin or TDZ. BAP, kinetin and zeatin were added to the medium at the concentrations of 1, 2, 4 or 8 lM. However, as the biological activity of TDZ is known to be generally higher than that of adenin-type cytokinins (Guo et al. 2011), it was used at lower concentrations in the present study (0.2, 0.5 and 1 lM). Since the shoot tip explants did not grow on the hormone-free medium, shoots grown on MS medium supplemented with auxin (0.57 lM IAA) were used as control samples. The effect of cytokinins on axillary shoot proliferation of S. alpina and biomass, in terms of fresh and dry weight, were recorded after 5 weeks of culture. The results are presented in Figs. 1 and 2. For all types and concentrations of cytokinins, the mean number of axillary shoots per S. alpina explant was higher, in most cases significantly, than those of the control MS medium supplemented only with 0.57 lM IAA (Fig. 1). The most effective cytokinin was BAP, with the highest mean number of axillary shoots (25.0 per explant) being induced by BAP at 2 lM, and the number of shoots not being significantly different (at p B 0.05) compared with BAP at 4 lM (23.5 per explant). However, increasing the BAP concentration to 8 lM or decreasing it to 1 lM resulted in a significant reduction in the number of shoots to 17.2 and 18.5 per explant, respectively. Most of the shoots grown in the presence of BAP demonstrated normal morphology and had well developed leaves; a maximum 9 % of proliferating shoots were hyperhydrated with a reduced number of leaves when the highest BAP concentration (8 lM) was used. In the presence of other cytokinins, the multiplication rate of S. alpina shoots was significantly lower and ranged from 6.4 to 8.2 for TDZ, from 3.0 to 6.8 for kinetin and from 5.3 to 6.9 for zeatin (Fig. 1). The quantitative response of the different cytokinins on shoot length is presented in Fig. 1. Shoot length was the greatest on the control medium (8.05 cm), followed by 1–4 lM zeatin supplementation (5.4–5.5 cm). The shortest shoots (1.8–1.9 cm) were obtained on medium supplemented with higher concentrations of BAP (4–8 lM).

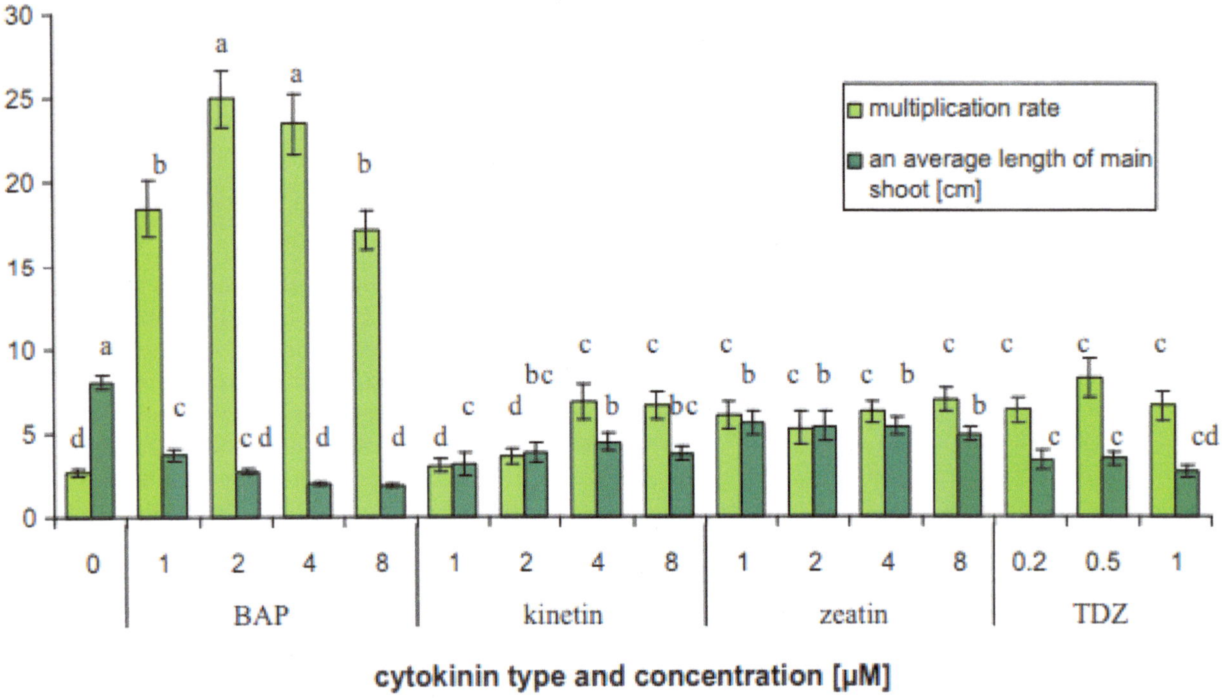

Figure. 1: The effect of various cytokinins and their concentrations on shoot proliferation of S. alpina and their length after 5 weeks of culture on MS medium with IAA (0.57 lM). The results are expressed as means of three replicates ± SE. The means with the same letter for the same parameter do not differ significantly according to the Kruskal–Wallis test (p B 0.05)

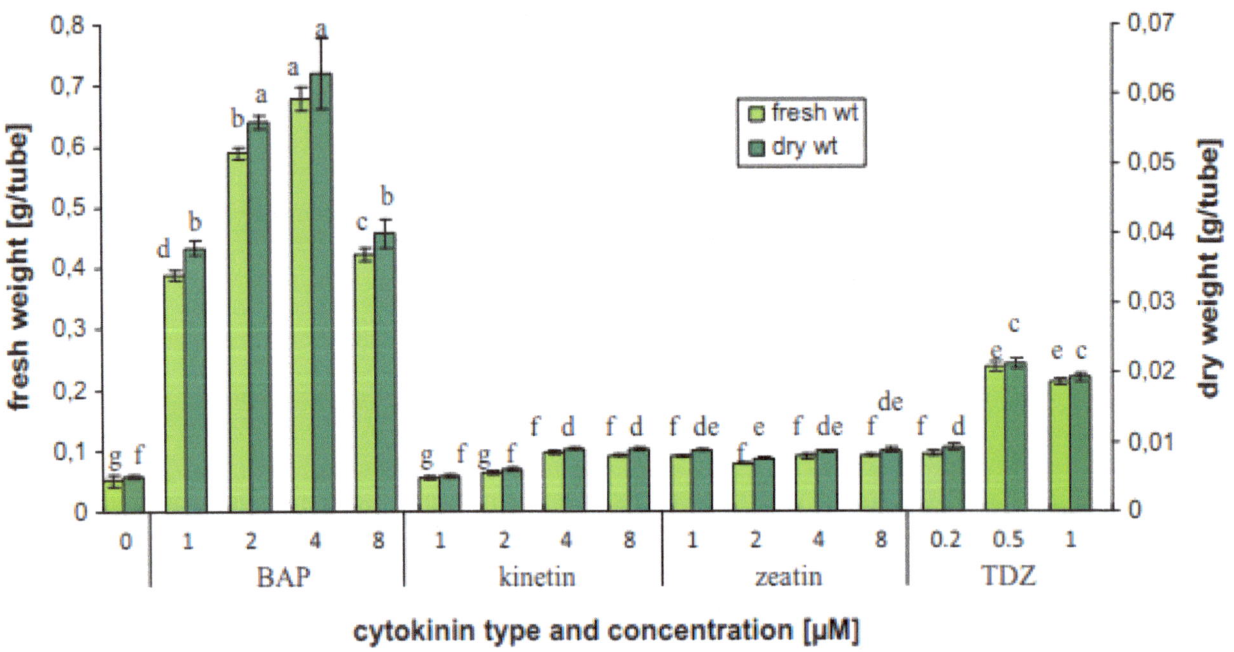

Figure. 2: The effect of various cytokinins and their concentrations on fresh weight and dry weight of S. alpina shoots after 5 weeks of culture on MS medium with IAA (0.57 lM). The results are expressed as means of three replicates ± SE. The means with the same letter for the same parameter do not differ significantly according to the Kruskal–Wallis test (p B 0.05)

The addition of cytokinin to the auxin-containing medium (0.57 lM IAA) also significantly increased the mean dry and fresh weights of cultures, except in the presence of 1 and 2 lM kinetin (Fig. 2). In these cases, no significant differences were observed in the number of shoots produced per explant when compared to those on the control medium (Fig. 1). The highest biomass (dry and fresh weight per culture vessel) was achieved for S. alpina shoots cultured in the presence of 4 lM BAP: a 78-fold increase in fresh weight (0.678 g per tube from an initial inoculum of 0.0087 g per tube) and a 74-fold increase in dry weight (0.063 g per tube from an initial inoculum of 0.00085 g per tube) were observed after 5 weeks of culture on the medium. Similar results were obtained in the shoot culture treated with 2 lM BAP (Fig. 2).

Effect of Cytokinins on Polyphenolic Compound Production

Verbascoside, baicalin, wogonoside, luteolin and luteolin 7-O-glucoside were identified in all S. alpina shoot extracts by a comparison of their retention times, UV spectra and mass spectra with those of standard compounds (Grzegorczyk-Karolak et al. 2013). However, the level of the polyphenolic compounds was quantitatively modified by the type and concentration of cytokinin present in the culture medium (MS with 0.57 lM IAA) (Fig. 3). The predominant metabolite was baicalin. Its content in S. alpina shoot biomass ranged from 4.53 to 18.79 mg g^{-1} dry weight (wt) (Fig. 3b). The maximum amount of the metabolite was found in the extract from the biomass of shoots growing on the medium containing 0.5 lM TDZ. Baicalin content fell to around 75 % of the amount when lower (0.2 lM) and higher (1 lM) concentrations of TDZ were used (Fig. 3b). The levels of baicalin in shoots grown on the media containing TDZ were 3–4 times higher than those seen in the control medium, i.e. MS supplemented only with auxin. Also, the addition of other cytokinins (BAP, kinetin, zeatin) to a medium supplemented with 0.57 lM IAA stimulated the accumulation of baicalin in proliferating shoots, especially when they were added at higher concentrations (4 or 8 lM). For example, baicalin production by shoots grown on media containing 0.57 lM IAA and 8 lM BAP or zeatin was approximately 12.5 mg g^{-1} dry wt, while that of shoots grown at an equimolar concentration of kinetin was 15.5 mg g-1 dry wt (Fig. 3b). The levels were about 20–30 % less than in the shoots grown in the presence of 0.5 lM TDZ.

A similar tendency was noted with respect to wogonoside, although the flavone was found in S. alpina shoot cultures in lower amounts than baicalin. The greatest wogonoside content (7.04 mg g^{-1} dry wt) was observed in the presence of 0.5 lM TDZ, this value being about 1.5 times more than on media containing 0.2 or 1.0 lM of the cytokinin and up to 9 times more compared to the control medium (0.78 mg g^{-1} dry wt) (Fig. 3c). High levels of wogonoside (above 4 mg g^{-1} dry wt) were also found in the shoots grown on media supplemented with 8 lM BAP, 8 lM zeatin or 4 lM kinetin (Fig. 3c). TDZ was also more effective than the other cytokinins tested (BAP, zaeatin, kinetin) for verbascoside production. Its quantity was highest in the shoots multiplied on media containing 0.5 or 1 lM TDZ combined with 0.57 lM IAA (9.76 mg g-1 dry wt and 9.36 mg g^{-1} dry wt, respectively). The levels were about 40 % higher than that found on the control medium (MS without cytokinin). However, supplementation with zeatin (2 and 4 lM), kinetin (1 lM) and BAP (1 and 8 lM) resulted in lower verbascoside production compared to treatment with auxin only (0.57 lM IAA). The lowest content of the phenylethanoid (3.75 mg g^{-1} dry wt) was reached on MS medium supplemented with 1 lM kinetin (Fig. 3a).

The quantity of luteolin was reduced below that observed in the control shoots following treatment with cytokinin (Fig. 3d). Only cultures treated with 2 lM BAP displayed amounts of luteolin similar to those of the control (about 3.5 mg g^{-1} dry wt). These cultures also revealed significantly higher luteolin 7-glucoside levels than controls (1.68 vs 1.39 mg g^{-1} dry wt). In the other hormonal variants, the luteolin 7-glucoside levels were either lower or identical to those recorded in the control (Fig. 3d). It should also be noted that although the different types and concentrations of cytokinins have varying degrees of influence on the production of luteolin and its glucoside, the ratios between these compounds tend to be similar in the S. alpina shoots, ranging from approximately 1.9 to 2.3:1. Exceptionally, treatment with 1 lM TDZ resulted in a luteolin content only 50 % more than that of its glucoside (Fig. 3d).

Antioxidant Activity

The ABTS and FRAP in vitro assays were used to evaluate the antioxidant activity of methanolic extracts from S. alpina shoot cultures multiplied on MS medium supplemented with BAP, kinetin, zeatin or TDZ. Extract of shoots from MS medium containing 0.57 lM IAA, without cytokinin, was used as control. The ABTS test was used to characterize antiradical scavenging properties, while the FRAP assay was employed to measure the capacity of the extracts to reduce ferric ions (Fe3? to Fe2?). The results are presented in Fig. 4.

Similarly to secondary metabolites, an exogenous supply of a cytokinin was found to significantly affect the antioxidant properties of analyzed shoot cultures (Fig. 4). The ABTS assay revealed the free radical scavenging activity of shoot cultures, expressed as EC50, to range from 28.27 to 118.56 lg ml^{-1} , while the FRAP test of the same extracts found reducing power to range from 775.62 Fe(II) g-1 extract dry wt to 265.73 Fe(II) g^{-1} extract dry wt. For both assays, the highest activity was found in shoots grown with 0.5 lM TDZ, which was also characterized by the highest level of secondary metabolites analyzed. The EC50 values indicated that the weakest free radical scavenging properties were found for plant material cultured on medium supplemented only with auxin (control), and shoots kept in the presence of auxin combined with 2 and 4 lM zeatin (Fig. 4). The extracts which exhibited the weakest effects on reducing ferric ions were those from shoots grown on medium supplemented with 2 lM zeatin and those treated with 1 lM kinetin.

The analysis of correlation coefficient values revealed a strong relationship between the antioxidant activity of shoot cultures treated with different cytokinins and the total polyphenolic compound content, calculated as the sum of verbascoside, baicalin, wogonoside, luteolin, luteolin 7-glucoside, ($r = 0.97$ and -0.93 for FRAP and ABTS assays, respectively). Some correlations were also found with the content of some individual compounds (Table 1).

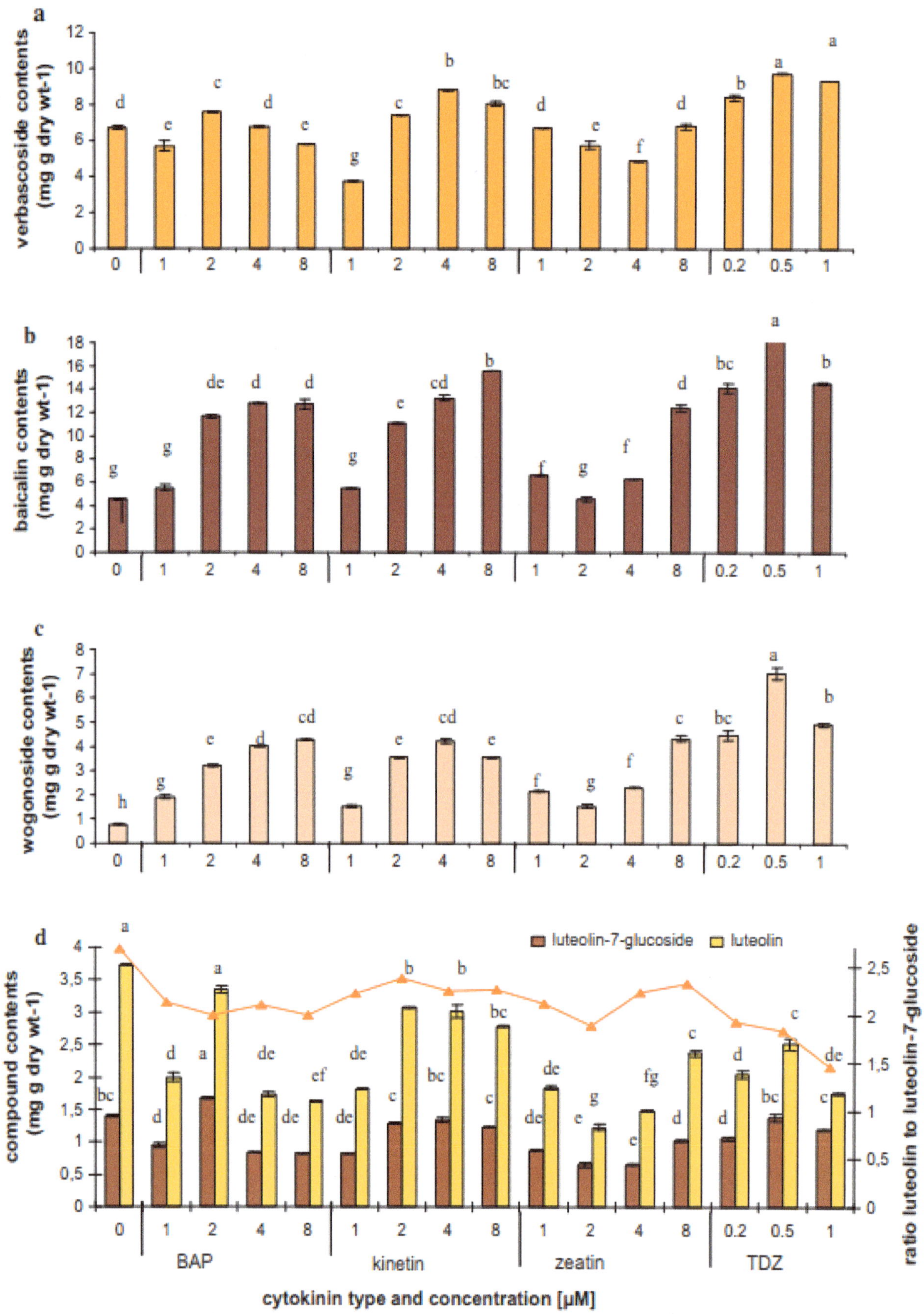

Figure. 3: The effect of various cytokinins and their concentrations on polyphenolic compound a verbascoside, b baicalin, c wogonoside, d luteolin and luteolin-7-glucoside) production in S. alpina shoots after 5 weeks of culture on MS medium with IAA (0.57 lM). The results are expressed as means of three replicates ± SE. The means with the same letter for the same parameter do not differ significantly according to the Kruskal–Wallis test (p B 0.05)

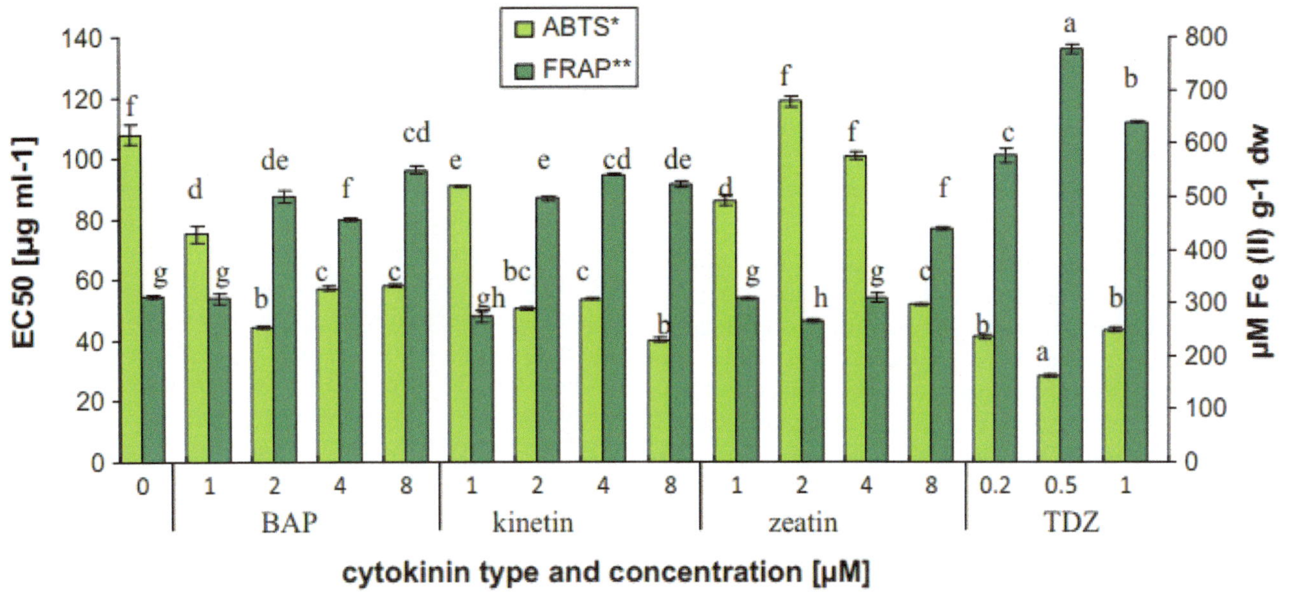

Figure. 4: Antioxidant activities of S. alpina shoot extracts determined by the ABTS radical scavenging and reducing power (FRAP) assays. The results are expressed as mean ± SE. The means with the same letter for the same test do not differ significantly according to the Kruskal–Wallis test (p B 0.05). Single asterisk EC50 the concentration of sample (lg ml-1) showing 50 % of maximal radical scavenging activity. Double asterisk Ferric reducing antioxidant power expressed in lM Fe(II) g-1 dw

Table 1: Comparison of correlation coefficient (r) between different parameters (biomass, proliferation rate, antioxidant assays) and metabolite (verbascoside, baicalin, wogonoside, luteolin-7-glucoside, luteolin, total polyphenolic) contents in S. alpina shoot cultures grown on MS media with IAA (0.57 lM) and different type and concentrations of cytokinins

	Shoot proliferation	Fresh wt (g tube^{-1})	Dry wt (g tube^{-1})	ABTS EC$_{50}$ (µg ml^{-1})	FRAP (µM Fe(II) g^{-1} dry wt)
Verbascoside	−0.02	0.04	0.02	**−0.74**	**0.84**
Baicalin	0.17	0.23	0.22	**−0.93**	**0.95**
Wogonoside	0.15	0.23	0.21	**−0.86**	**0.94**
Luteolin-7-glucoside	0.10	0.10	0.10	−0.56	0.54
Luteolin	−0.04	−0.07	−0.06	−0.29	0.22
Total polyphenols[a]	0.13	0.19	0.17	**−0.93**	**0.97**

Bold values indicate strong or very strong correlation

[a] Total polyphenolic compound content calculated as the sum of verbascoside, baicalin, wogonoside, luteolin and luteolin 7-glucoside

DISCUSSION

The current study is the first to evaluate the role of different types of cytokinin, i.e. BAP, kinetin, zeatin and TDZ, in combination with 0.57 lM IAA on S. alpina shoot culture: its growth, shoot proliferation, secondary metabolite production and antioxidant activity. Almost all cytokinins increased shoot multiplication, apart from kinetin at concentrations of 1 and 2 lM. Among these, 2 and 4 lM BAP were the most effective treatments, producing 23–25 shoots per explant within 5 weeks of culture. BAP has been reported to be more efficient at inducing shoot proliferation than other cytokinins in many other plants, including some Scutellaria species. Joshee et al. (2007) found the maximum number of shoots from shoot tips of S. integrifolia (23 per explant) to be obtained on MS medium supplemented with 2.2 lM BAP, while for S. discolor the best results were obtained in the presence of 4.4 lM BAP (Sinha et al. 1999). Also, the increase in fresh and dry biomass of S. alpina shoots grown for 5 weeks on media containing BAP was largest, although the shoots were shorter than those grown in the presence of other tested cytokinins, especially kinetin and zeatin, and those cultured on control medium (MS without cytokinin). These findings suggest that shoot biomass could be affected more by proliferation rate than the growth of individual shoots. The results of the present study confirm that changes in the type and concentration of cytokinin applied can drastically affect product accumulation. For example, cytokinins have been reported to increase production of alkaloids in cell cultures of Catharanthus roseus (Decendit et al. 1992), lignans in Phyllanthus amarus shoots (Nitnaware et al. 2011) and anthocyanins in Oxalis linearis callus (Meyer and van Staden 1995). Conversely, they have also been observed to inhibit secoiridoid production in Blackstonia perfoliata shoots (Sabovljevic et al. 2006) and rutin in the callus and adventitious root culture of Morus alba (Lee et al. 2011). The findings of the present study demonstrate that total polyphenolic content (calculated as the sum of verbascoside, baicalin, wogonoside, luteolin and luteolin 7-glucoside) in S. alpina shoot cultures varied from 13.37 to 39.49 mg g^{-1} dry wt, which was found to be much higher under optimal cytokinin conditions than in cytokinin-free controls (17.15 mg g^{-1} dry wt). This difference is a result of the positive effect of suitable concentrations of cytokinins on the accumulation of baicalin, wogonoside and verbascoside. A different tendency was observed regarding luteolin and luteolin 7-O-glucoside: luteolin concentration was significantly lower in S. alpina shoots grown in the presence of cytokinins, except 2 lM BAP, than those grown on control medium.

Although BAP was more effective in improving S. alpina shoot proliferation and biomass production than TDZ, TDZ was more effective than BAP, kinetin and zeatin in inducing baicalin, wogonoside and verbascoside production. These findings indicate that polyphenolic compound production in S. alpina shoot culture is not associated with biomass accumulation, as confirmed by a correlation coefficient (r) lower than 0.23 between the concentrations of analyzed compounds and the fresh and dry weights of S. alpina shoot cultures treated with the cytokinins (Table 1). Biomass-independent production of total phenolic content was also recently reported by Ali et al. (2013) in callus cultures of Artemisia absinthium.

The highest amounts of baicalin, wogonoside and verbascoside were observed in the S. alpina shoots grown in the presence of 0.5 lM TDZ. Under these conditions, baicalin content (18.8 mg g^{-1} dry wt) was 8.5 times higher than that observed in the aerial parts of field-grown plants of S. alpina (Zgo´rka 2006). The present study represents the first description of the wogonoside content of S. alpina. Comparing these findings and those of Islam et al. (2011) with other closely related and pharmacological important Scutellaria species, the following observations can be made. The accumulation of wogonoside in S. alpina shoots grown on MS medium supplemented with 0.5 lM TDZ can be seen to be greater than that found in the leaves, stems and roots of S. racemosa (0–2.05 mg g^{-1} dry wt) and S. tomentosa (0–5.68 mg g^{-1} dry wt), as well as the stems of S. baicalensis (5.6 mg g-1 dry wt). However, these levels are lower than those observed in the roots of S. baicalensis, which were found to contain 12 mg g^{-1} dry wt of wogonoside. Although the roots of S. baicalensis also contained the highest level of verbascoside of the Scutellaria plants, (2.35 mg g^{-1} dry wt) (Islam et al. 2011), this value was significantly lower than that previously observed by Nishikawa et al. (1999) in the roots of S. alpina, which produced about 2.8 % dry weight of verbascoside. This level was three times higher than that obtained in our shoot culture of S. alpina grown under optimized cytokinin conditions: i.e. on MS medium in the presence of TDZ (0.5 or 1 lM) (about 9.5 mg g^{-1} dry wt).

The stimulatory effect of TDZ on axillary shoot proliferation and adventitious shoot induction is well documented. However, the effect of TDZ on the biosynthesis of secondary metabolites has rarely been investigated and it is not clear how TDZ increases the production of flavones (baicalin and wogonoside) or phenylethanoid glycoside (verbascoside) in S. alpina shoot culture. According to Ramawet and Merrilion (1999), the plant growth regulators do not react with the intermediate compounds of the biosynthetic pathway, but appear to change cytoplasmatic conditions for product formation. Sakakibara (2006) speculate that cytokinin signifi- cantly suppresses the transport of macronutrients such as nitrate, ammonium, sulfate and phosphate, while nitrate regulates the expression of genes involved in the phenylpropanoid and flavonoid pathways. Previous studies report that while TDZ is believed to be the cytokinin responsible for elevated production of hypericin in Hypericum perforatum shoot culture (Liu et al. 2007), it is also responsible for reducing the levels of hypericin and pseudohypericin in H. maculatum and H. sampsonii shoots (Coste et al. 2011; Liu et al. 2007). In our study, TDZ decreased luteolin and luteolin 7-glucoside levels in shoot culture of S. alpina in a similar way to other cytokinins, such as kinetin, zeatin and BAP (with the exception of 2 lM BAP). The similarity between the biochemical pathways of luteolin and its glucoside in shoot cultures treated with cytokinins was con- firmed by the high positive correlation between their ratio obtained (0.88). The high positive correlations between baicalin, wogonoside and to an extent, verbascoside biosynthesis, and the weak correlation between these polyphenols and the biosynthesis of luteolin and its glucoside suggest that accumulation of these phytochemicals in the presence of cytokinins can occur via different pathways. The present study is the first to evaluate the antioxidant potential of S. alpina. The antioxidant activity of S. alpina, as assessed by the ABTS and FRAP tests, was found to have a strong correlation with total polyphenolic levels in shoot extracts. Such a relationship is not surprising and has been previously published for many other phenolic-rich species of the Lamiaceae family such as Salvia officinalis (Grzegorczyk-Karolak et al. 2007), Lycopus lucidus (S´lusarczyk et al. 2009) or Dracocephalum moldavica (Weremczuk-Jezyna et al. 2013).

CONCLUSION

The present study demonstrates that S. alpina can be successfully propagated in vitro using shoot tip explants. Of the cytokinins tested, BAP was the most effective for shoot proliferation and biomass production. The type of cytokinin and its concentration was also found to significantly influence the accumulation of bioactive metabolites S. alpina shoot cultures and their antioxidant activity. These findings indicate that shoots regenerated from shoot tips on MS medium containing auxin (0.57 lM IAA) and 0.5 lM TDZ can be a rich source of baicalin, wogonoside and verbascoside, compounds with many therapeutic properties. Further research is needed to evaluate whether production of the metabolites in S. alpina shoots depends on the type and concentration of auxin combined with TDZ, and how this can be the case.

REFERENCES

1. Ali M, Abbasi BH, ul-haq I (2013) Production of commercially important secondary metabolites and antioxidant activity in cell suspension cultures of Artemisia absinthium L. Ind Crops Prod 49:400–406. doi:10.1016/j.indcrop.2013.05.033
2. Chang WH, Chem CH, Lu FJ (2002) Different effects of baicalein, baicalin and wogonin on mitochondrial function, glutathione content and cell cycle progression in human hepatoma cell lines. Planta Med 68:128–132. doi:10.1055/s-2002-20246
3. Chen Y, Lu N, Ling Y, Gao Y, Wang L, Sun Y, Qi Q, Feng F, Liu W, You Q, Gao Q (2009) Wogonoside inhibits lipopolysaccharide-induced angiogenesis in vitro and in vivo via toll-like receptor 4 signal transduction. Toxicology 259:10–17. doi:10.1016/j.tox.2009.01.010

4. Cheng Y, He G, Mu X, Zhang T, Li X, Hu J, Xu B, Du G (2008) Neuroprotective effect of baicalein against MPTP neurotoxicity: behavioral, biochemical and immunohistochemical profile. Neurosci Lett 441:16–20. doi:10.1016/j.neulet.2008.05.116

5. Cole IB, Praveen KS, Murch SJ (2007) Medicinal biotechnology in the genus Scutellaria. In Vitro Cell Dev Biol Plant 43:318–327. doi:10.1007/s11627-007-9055-4

6. Cole IB, Cao J, Alan AR, Saxena PK, Murch SJ (2008) Comparisons of Scutellaria baicalensis, Scutellaria lateriflora and Scutellaria racemosa: genome size, antioxidant potential and phytochemistry. Planta Med 74:474–481. doi:10.1055/s-2008-1034358

7. Coste A, Vlase L, Halmagyi A, Deliu C, Coldea G (2011) Effects of plant growth regulators and elicitors on production of secondary metabolites in shoot cultures of Hypericum hirsutum and Hypericum maculatum. Plant Cell Tiss Organ Cult 106:279–288. doi:10.1007/s11240-011-9919-5

8. Decendit A, Liu D, Ouelhazi L, Doireau P, Mérillon JM, Rideau M (1992) Cytokinin-enhanced accumulation of indole alkaloids in Catharanthus roseus cell cultures—the factors affecting the cytokinin response. Plant Cell Rep 11:400–403. doi:10.1007/BF00234369

9. Gao Z, Huang K, Yang X, Xu H (1999) Free radical scavenging and antioxidant activities of flavonoids extracted from the radix of Scutellaria baicalensis Georgi. Biochim Biophys Acta 1472:643–650. doi:10.1016/S0304-4165(99)00152-X

10. Grzegorczyk I, Matkowski A, Wysokińska H (2007) Antioxidant activity of extracts from in vitro cultures of Salvia officinalis L. Food Chem 104:536–541. doi:10.1016/j.foodchem.2006.12.003

11. Grzegorczyk-Karolak I, Kuźma Ł, Wysokińska H (2013) The use of long-term Scutellaria altissima callus cultures for shoot regeneration, production of bioactive metabolites and micropropagation. J Med Plants Res 7:3003–3313. doi:10.5897/JMPR2013.5226

12. Grzegorczyk-Karolak I, Kuźma Ł, Wysokińska H (2015) Study on the chemical composition and antioxidant activity of extracts from shoot culture and regenerated plants of Scutellaria altissima L. Acta Physiol Plant 37:1736–1744. doi:10.1007/s11738-014-1736-0

13. Guo B, Abbasi BH, Zeb A, Xu LL, Wei YH (2011) Thidiazuron: a multi-dimensional plant growth regulator. Afr J Biotechnol 10:8984–9000. doi:10.5897/AJB2013.12456

14. He XL, Wang YH, Gao M, Li XX, Zhang TT, Du GH (2009) Baicalein protects rat brain mitochondria against chronic cerebral hypoperfusion-induced oxidative damage. Brain Res 1249:212–221. doi:10.1016/j.brainres.2008.10.005

15. Islam MN, Downey F, Ng CY (2011) Comparative analysis of bioactive phytochemicals from Scutellaria baicalensis, Scutellaria lateriflora, Scutellaria racemosa, Scutellaria tomentosa and Scutellaria wrightii by LC-DAD-MS. Metabolomics 7:446–453. doi:10.1007/s11306-010-0269-9

16. Joshee N, Mentreddy SR, Yadav AK (2007) Mycorrhizal fungi and growth and development of micropropagated Scutellaria integrifolia plants. Ind Crops Prod 25:169–177. doi:10.1016/j.indcrop.2006.08.009

17. Kernan MR, Amarquaye A, Chen J, Chan LDFJ, Sesin N, Parkinson ZJ, Barrett C, Ye M, Stoddart CA, Sloan B, Blanc P, Limbach C, Mrisho S, Rozhon EJ (1998) Antiviral phenylpropanoid glycosides from the medicinal plant Markhamia lutea. J Nat Prod 61:564–570. doi:10.1021/np9703914

18. Kitamura K, Honda M, Yoshizaki H, Yamamoto S, Nakane H, Fukushima M, Ono K, Tokunaga T (1998) Baicalin, an inhibitor of HIV-1 production in vitro. Antiviral Res 37:131–140. doi:10.1016/S0166-3542(97)00069-7

19. Lee Y, Lee DE, Lee HS, Kim KS, Lee WS, Kim SH, Kim MW (2011) Influence of auxins, cytokinins, and nitrogen on production of rutin from callus and adventitious roots of the white mulberry tree (Morus alba L.). Plant Cell Tiss Organ Cult 105:9–19. doi:10.1007/s11240-010-9832-3

20. Lim BO (2003) Effects of wogonin, wogonoside, and 3,5,7,2′,6′-pentahydroxyflavone on chemical mediator production in peritoneal exduate cells and immunoglobulin E of rat mesenteric lymph node lymphocytes. J Ethnopharmacol 84:23–29. doi:10.1016/S0378-8741(02)00257-X

21. Liu XN, Zhang XQ, Sun JS (2007) Effects of cytokinins and elicitors on the production of hypericins and hyperforin metabolites in Hypericum sampsonii and Hypericum perforatum. Plant Growth Regul 53:207–214. doi:10.1007/s10725-007-9220-0

22. Meyer HJ, van Staden J (1995) The in vitro production of an anthocyanin from callus cultures of Oxalis linearis. Plant Cell Tiss Organ Cult 40:55–58. doi:10.1007/BF00041119

23. Murashige T, Skoog F (1962) A revised medium for rapid growth and bioassays with tobacco tissue cultures. Physiol Plant 15:473–497. doi:10.1111/j.1399-3054.1962.tb08052.x

24. Nishikawa K, Ishimaru K (1997) Flavonoids in root cultures of Scutellaria baicalensis. J Plant Physiol 151:633–636

25. Nishikawa K, Furukawa H, Fujioka T, Fujii H, Mihashi K, Shimomura K, Ishimaru K (1999) Phenolics in tissue cultures of Scutellaria. Nat Med 53:209–213

26. Nitnaware KM, Naik DG, Nikam TD (2011) Thidiazuron-induced shoot organogenesis and production of hepatoprotective lignan phyllanthin and hypophyllanthin in Phyllanthus amarus. Plant Cell Tiss Organ Cult 104:101–110. doi:10.1007/s11240-010-9796-3

27. Pettit RG, Numata A, Takemura T, Ode RH, Nurula AS, Schmidt JM, Cragg GM, Pase CP (1990) Isolation of acteoside and isoacteoside from Castilleja linariaefolia. J Nat Prod 53:456–458. doi:10.1021/np50068a026

28. Pulido R, Bravo L, Saura-Calixto F (2000) Antioxidant activities of dietary phenols are determined by a modified ferric reducing/antioxidant power assay. J Agric Food Chem 48:3396–3402. doi:10.1021/jf9913458

29. Ramawet K, Merrilion J (1999) Biotechnology – secondary metabolites. Mohan Primlani for Oxford and YBH, Delhi, pp 243–247

30. Re R, Pellegrini N, Proteggente A, Pannala A, Yang M, Rice-Evans C (1999) Antioxidant activity applying an improved ABTS radical cation decolorizing assay. Free Radical Biol Med 26:1231–1237. doi:10.1016/S0891-5849(98)00315-3

31. Sabovljevic A, Rosic N, Jankovic T, Grubisic D (2006) Secoiridoid content of Blackstonia perfoliata in vivo and in vitro. In Vitro Cell Dev Biol. Plant 42:427–431. doi:10.1079/IVP2006759

32. Sakakibara H (2006) Cytokinins: activity, biosynthesis and translocation. Annu Rev Plant Biol 57:431–449

33. Seo WT, Park YH, Choe TB (1993) Identification and production of flavonoids in a cell-suspension culture of Scutellaria baicalensis Georgi. Plant Cell Rep 12:414–417. doi:10.1007/BF00234703

34. Shang XF, He XR, He XY, Li MX, Zhang RX, Fan PC, Zhang QL, Jia ZP (2010) The genusScutellaria an ethnopharmacological and phytochemical review. J Ethnopharmacol 128:279–313. doi:10.1016/j.jep.2010.01.006

35. Shoyama Y, Matsumoto M, Nishioka I (1986) Four caffeoyl glycosides from callus tissue ofRehmannia glutinosa. Phytochem 25:1633–1636. doi:10.1016/S0031-9422(00)81224-8

36. Sinha S, Pokhrel S, Vaidya BN, Joshee N (1999) In vitro micropropagation and callus induction in Scutellaria discolor Colebr.- A medicinally important plant of Nepal. Indian J Pl Genet Resources 12:219–223

37. Ślusarczyk S, Hajnos M, Skalicka-Woźniak K, Matkowski A (2009) Antioxidant activity of polyphenols from Lycopus lucidus Turcz. Food Chem 113:134–138. doi:10.1016/j.foodchem.2008.07.037

38. Stojakowska A, Kisiel W (1998) Secondary metabolites from a callus culture of Scutellaria colomnae. Fitoterapia 70:324–325. doi:10.1016/S0367-326X(99)00022-2

39. Takasuna K, Kasai Y, Kitano Y, Mori K, Kobayashi R, Hagiwara T, Kakihata K, Hirohashi M, Nomura M, Nagai E et al (1995) Protective effects of kampo medicines and baicalin against intestinal toxicity of a new anticancer camptothecin derivative, irinotecan hydrochloride (CPT-11), in rats. Jpn J Cancer Res 86:978–984

40. Weremczuk-Jeżyna I, Grzegorczyk-Karolak I, Frydrych B, Królicka A, Wysokińska H (2013) Hairy roots of Dracocephalum moldavica: rosmarinic acid content and antioxidant potential. Acta Physiol Plant 35:2095–2103. doi:10.1007/s11738-013-1244-7

41. Woo KJ, Lim JH, Suh SI, Kwon YK, Shin SW, Kim SC, Choi YH, Park JW, Kwon TK (2006) Differential inhibitory effects of baicalein and baicalin on LPS-induced cyclooxygenase-2 expression through inhibition of C/EBP β DNA-binding activity. Immunobiology 211:359–368. doi:10.1016/j.imbio.2006.02.002

42. Zgórka G (2006) Retention behavior of silica bonded and novel polymeric reversed phase sorben ts in studies on flavones as chemotaxonomic markers of Scutellaria L. genus. J Chromatogr A 1120:230–236.

CHAPTER 11 LIGHT AND TEMPERATURE CONDITIONS AFFECT BIOFLAVONOID ACCUMULATION IN CALLUS CULTURES OF CYCLOPIA SUBTERNATA VOGEL (HONEYBUSH)

A. Kokotkiewicz M. Luczkiewicz[1], A. Bucinski[2]

[1]Department of Pharmacognosy, Faculty of Pharmacy, Medical University of Gdansk, al. gen. J. Hallera 107, 80-416 Gdan´sk, Poland

[2]Department of Biopharmacy, Faculty of Pharmacy, Ludwik Rydygier Collegium Medicum in Bydgoszcz, Nicolaus Copernicus University in Torun, ul. dr A. Jurasza 2, 85-089 Bydgoszcz, Poland

ABSTRACT

Callus cultures of the endemic South-African legume Cyclopia subternata were cultivated under varying light and temperature conditions to determine their influence on biomass growth and bioflavonoids accumulation. Experimental modifications of light included complete darkness, light of different spectral quality (white, red, blue and yellow) and ultraviolet C (UVC) irradiation. The calli were also subjected to elevated temperature or cold stress. Among the tested light regimes, cultivation under blue light resulted in the highest levels of hesperidin (H)—118.00 mg 100 g−1 dry weight (DW) on 28 days of experiment, as well as isoflavones: 7-O-β-glucosides of calycosin (CG), pseudobaptigenin (PG) and formononetin (FG)—28.74, 19.26 and 10.32 mg 100 g−1 DW, respectively, in 14-days old calli. UVC irradiation applied on 20 days stimulated the accumulation of H (204.14 mg 100 g−1 DW), CG (31.84 mg 100 g−1 DW) and PG (18.09 mg 100 g−1 DW) in 28 days culture by 140, 46 and 165 %, respectively, without negatively influencing callus growth. Low temperature (13 °C) increased CG content by over 1,500 % (235.29 mg 100 g−1 DW) when applied during the whole 28-days growth cycle, at the same time causing 95 % decrease in culture growth in comparison to reference calli maintained at 24 °C. On the contrary, elevated temperature (29 °C) applied during the second half of the culture period resulted in over 300 and 500 % increase in CG and PG content (61.76 and 58.89 mg 100 g−1, respectively) while maintaining relatively high biomass yield.

The South-African shrubs of the genus Cyclopia (Fabaceae) are used to manufacture the traditional, sweet-scented herbal tea, commonly known as honeybush. They contain a range of biologically active polyphenols, including xanthones, benzophenones, flavanones, flavones, dihydrochalcones and isoflavones. Among the last group are the methoxy-substituted derivatives, represented by calycosin and formononetin (Louw et al. 2013). These compounds, and/or their corresponding glucosides, were shown to exhibit multidirectional biological effects including estrogenic (Chen et al. 2013; Louw et al. 2013) and antitumor (Chen et al. 2013; Zhang et al. 2013). However, unlike soybean phytoestrogens (i.e. genistein and daidzein), methoxylated isoflavones like calycosin are less available for biological activity studies. Chinese plants of the genus Astragalus, considered a major source of free and glucosidated calycosin, are threatened by overexploitation because of slow growth of the roots combined with high market demand for natural medicines (Wu et al. 2011; Xu et al. 2011).

Citation: Adam Kokotkiewicz, Adam Bucinski, Maria Luczkiewicz, Light and temperature conditions affect bioflavonoid accumulation in callus cultures of Cyclopia subternata Vogel (honeybush), DOI: 10.1007/s11240-014-0502-8.

Cell cultures of Cyclopia subternata were previously shown to accumulate 7-O-β-glucosides of calycosin, pseudobaptigenin and formononetin, absent in intact plant material (Kokotkiewicz et al. 2012, 2013), and can thus be utilized for the production of these derivatives independently of their natural resources. In the present work, the effect of varying light and temperature regimes on biomass growth and accumulation of bioflavonoids in C. subternata callus cultures (Kokotkiewicz et al. 2009) was investigated.

For each experiment, 1.5 g portions of callus (grown on MS medium containing 3.0 % w/v sucrose, 20.19 µM 4-CPPU, 1.96 µM 2,4,5-T and solidified with 0.7 % w/v agar, taken on 20 days of the growth cycle) were transferred into baby food jars containing 25 ml of the growth medium with the same composition (culture containers and reagents from Sigma-Aldrich, St. Louis, US-MO) and closed with polypropylene caps. The cultures were maintained in complete darkness or under continuous light (88 ± 8 µmol m^{-2} s^{-1}, TLD 35W white fluorescent tubes, Philips, Amsterdam, Netherlands), applied directly or through colour filters (106 primary red, 101 yellow, 119 dark blue, Lee Filters, Andover, UK). For ultraviolet C (UVC) exposure, polypropylene lids of growth vessels were removed in aseptic conditions and the calli were irradiated for 5 min from 0.15 m distance (TUV 30W/G30 T8 lamp, Philips, Amsterdam, the Netherlands). Except for temperature modification experiments, the cultures were grown at 24 ± 1 °C. The specific experimental schemes are given in Figs. 1, 2, 3 and 4.

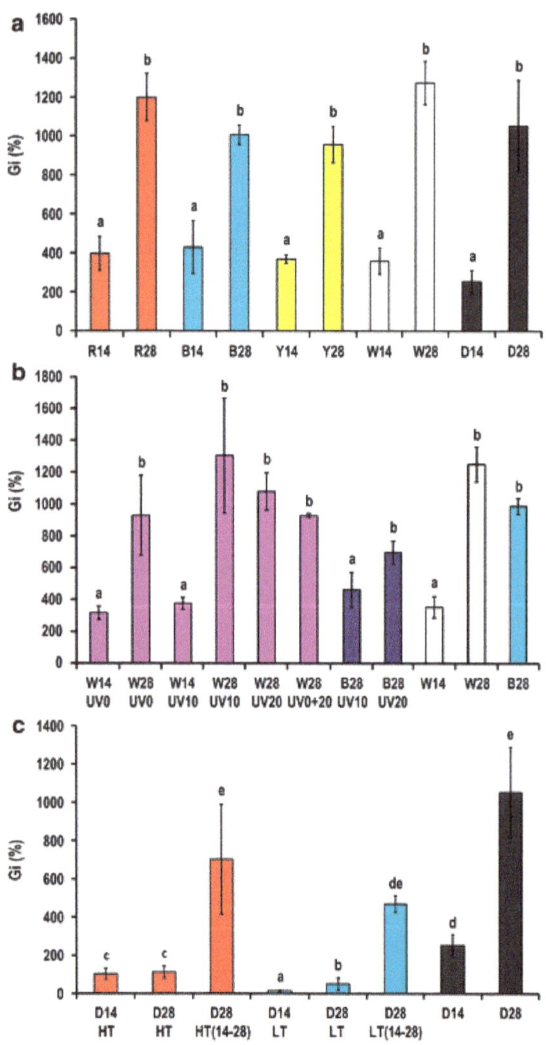

Figure. 1: The effect of different light conditions (**a**), UVC irradiation (**b**) and temperature regimes (**c**) on the growth of C. subternata callus. Experimental modifications applied: R red light;B blue light; Y yellow light; W white light; D darkness; 14, 28 callus grown for 14 or 28 days, respectively; UV0, UV10, UV20, UV0+20 UVC exposure on 0 (start), 10, 20 or 0 and 20 days (double irradiation) of the experiment, respectively; HT, LT experiment maintained at 29 or 13 °C, respectively; 14–28, elevated/lowered temperature applied only during a second half of the experiment. Growth indices were calculated as follows: $Gi = [(G_1 - G_0)/G_0] \times 100$ % where Gi is the growth index, G_1 is the callus fresh weight at the end of a cultivation period and G_0 is the fresh weight of the inoculum. Different lettersindicate significant differences between means (n = 3) based on Tukey's range test ($p < 0.05$). (Color figure online)

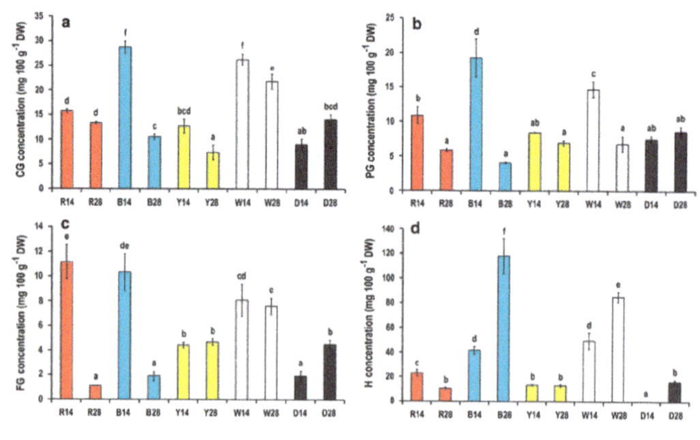

Figure. 2: The effect of different light conditions on the accumulation of calycosin 7-O-β-glucoside (**a**), pseudobaptigenin 7-O-β-glucoside (**b**), formononetin 7-O-β-glucoside (**c**) and hesperidin (**d**), in C. subternata callus. Abbreviations and statistical designations as defined in Fig. 1. (Color figure online)

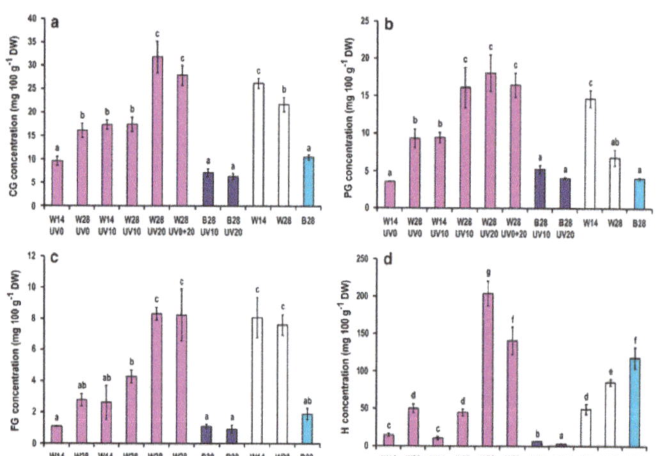

Figure. 3: The effect of UVC irradiation on the accumulation of calycosin 7-O-β-glucoside (**a**), pseudobaptigenin 7-O-β-glucoside (**b**), formononetin 7-O-β-glucoside (**c**) and hesperidin (**d**), in C. subternata callus. Abbreviations and statistical designations as defined in Fig. 1. (Color figure online)

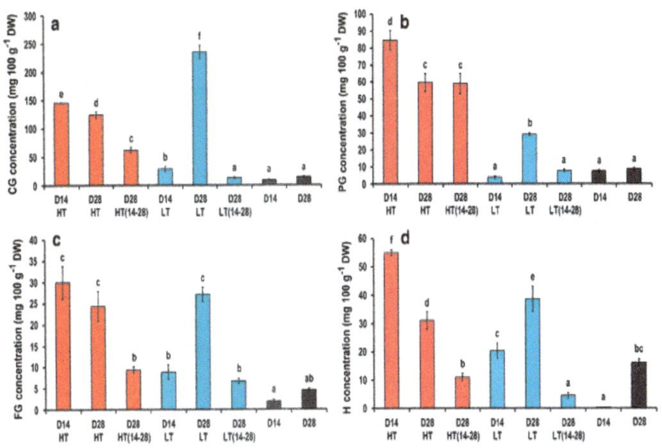

Figure. 4: The effect of different temperature regimes on the accumulation of calycosin 7-O-β-glucoside (**a**), pseudobaptigenin 7-O-β-glucoside (**b**), formononetin 7-O-β-glucoside (**c**) and hesperidin (**d**), in C. subternata callus. Abbreviations and statistical designations as defined in Fig. 1. (Color figure online)

The calli were harvested after 14 or 28 days and their growth indices calculated (Fig. 1). The samples were freeze-dried, extracted and assessed for the production of isoflavonoids (CG, PG, FG) and flavonoids (H), identified as major phenolic metabolites in the investigated cultures (Kokotkiewicz et al. 2009, 2012), using the previously described HPLC method (Kokotkiewicz et al. 2009, 2013).

Lighting conditions are well-known factors affecting primary and secondary metabolism in plant cell cultures (Ramakrishna and Ravishankar 2011). The current experiments demonstrated that the light regime did not significantly influence the growth of C. subternata callus. All biomasses were characterized by fast growth, with Gi values exceeding 1,000 % after the 28-days experiment (Fig. 1a). On the other hand, light conditions proved to significantly affect bioflavonoids accumulation. The presence of light showed to be beneficial, but not necessary, for the production of the examined compounds. The highest amounts of H (28 days) and isoflavones (14 days) were recorded in the calli maintained under blue light (Fig. 2). This is in agreement with previous reports, indicating the stimulatory effect of blue light on the accumulation of plant phenolics in in vitro cultures, with examples including flavonoids in Saussurea medusa calli (Guo et al.2007) and phenolic acids in shoots of Ruta graveolens (Szopa et al. 2012). Unfortunately, in the present work the production of the most valuable metabolites (i.e. isoflavones) was not correlated with high biomass yield, thus limiting the usefulness of the above described experimental scheme.

Further part of the study included irradiation of the calli using UV light, which was previously shown to increase the accumulation of isoflavones in legume plants—the examples include genistein in Genista tinctoria callus (Tůmová and Tůma 2011) and methoxylated isoflavones (calycosin, formononetin and CG) in A. membranaceous leaves (Xu et al. 2011). It was decided to subject C. subternata callus to highly energetic UVC radiation, which so far has been scarcely studied with respect to its influence on isoflavone biosynthesis. Due to the high energy of UVC, it can be conveniently used in plant cell culture experiments as a stress factor, applied in short bursts at the desired moments of the growth period.

As presented in Fig. 1b, there were no statistical differences between the Gi values of calli grown without UVC treatment and subjected to various irradiation schemes. To the contrary, the effects of UVC on bioflavonoid accumulation were clearly dependent on the moment of exposure (Fig. 3). Irradiation on 0 or 10 days of the experiment resulted in statistically unchanged (CG and PG) or lowered (FG and H) secondary metabolite concentrations on 28 days, as compared to the control. On the other hand, application of UVC on 20 days significantly increased the levels of CG, PG and H by 46, 165 and 140 %, respectively, in the cultures maintained under white light, yielding the best results in the described series of experiments. As opposed to the experiments depicted in Fig. 2, this strategy also enables to match the high biomass yield with high secondary metabolite accumulation (both achieved on 28 days).

The parallelly conducted experiments involved cultivation of C. subternata calli under elevated (29 °C) and lowered (13 °C) temperatures (as compared to standard 24 °C), as both these strategies were effective in increasing isoflavone accumulation in the selected legume species. For instance, Thanonkeo and Panichajakul (2006) reported over twofold increase in biomass growth and threefold higher isoflavone content in Pueraria candolleicallus grown at 32 °C. On the other hand, cold stress was shown to cause a significant increase in calycosin and its glucoside content in A. membranaceus seedlings (Pan et al.2007).

As shown in Fig. 1c, elevated temperature almost completely inhibited callus growth, which is indicative of its high sensitivity to environmental changes. Thus, despite the significantly increased isoflavone levels (Fig. 4), the above approach has no practical value for the production of the examined compounds. In order to stimulate isoflavone biosynthesis without compromising biomass growth, the calli were subjected to elevated temperature only during the second half of the culture period. This strategy proved partially successful—the relatively high growth rate was maintained but increased accumulation was recorded only for CG and PG (over 300 and 500 %, respectively).

Cold stress was shown to significantly inhibit the growth of C. subternata callus. However, low temperature applied during the whole culture period resulted in elevated bioflavonoid accumulation. The highest (>1,500 %) increase was recorded for CG, thus confirming the results by Pan et al. (2007). Unfortunately, the attempt to correlate isoflavone production with high biomass yield, by applying cold stress only during the 14–28-days period, proved unsuccessful and resulted in secondary metabolite concentrations comparable to the control group (Fig. 4).

Summing up, bioflavonoid accumulation in C. subternata callus was shown to be strongly affected by light and temperature regime applied. The best results, in terms of biomass growth and secondary metabolite content, were obtained by maintaining the cultures at 29 °C during the second half of the growth cycle, or by UVC exposure of the callus on 20 days of experiment. These strategies may be useful for biotechnological production of methoxylated isoflavoned based on C. subternata cell cultures.

ACKNOWLEDGMENTS

This study was supported by Grant No. NN302 041936 from the Polish Ministry of Education and Science.

REFERENCES

1. Chen J, Zhao X, Ye Y, Wang Y, Tian J (2013) Estrogen receptor beta-mediated proliferative inhibition and apoptosis in human breast cancer by calycosin and formononetin. Cell Physiol Biochem 32:1790–1797. doi:10.1159/000356612
2. Guo B, Liu Y-G, Yan Q, Liu C-Z (2007) Spectral composition of irradiation regulates the cell growth and flavonoid biosynthesis in callus cultures of Saussurea medusa Maxim. Plant Growth Regul 52:259–263. doi:10.1007/s10725-007-9192-0
3. Kokotkiewicz A, Wnuk M, Bucinski A, Luczkiewicz M (2009) In vitro cultures of Cyclopiaplants (honeybush) as a source of bioactive xanthones and flavanones. Z Naturforsch 64c:533–540
4. Kokotkiewicz A, Luczkiewicz M, Sowinski P, Glod D, Gorynski K, Bucinski A (2012) Isolation and structure elucidation of phenolic compounds from Cyclopia subternataVogel (honeybush) intact plant and in vitro cultures. Food Chem 133:1373–1382. doi:10.1016/j.foodchem.2012.01.114
5. Kokotkiewicz A, Luczkiewicz M, Kowalski W, Badura A, Piekus N, Bucinski A (2013) Isoflavone production in Cyclopia subternata Vogel (honeybush) suspension cultures grown in shake flasks and stirred-tank bioreactor. Appl Microbiol Biotechnol 97:8467–8477. doi:10.1007/s00253-013-5099-z
6. Louw A, Joubert E, Visser K (2013) Phytoestrogenic potential of Cyclopia extracts and polyphenols. Planta Med 79:580–590. doi:10.1055/s-0032-1328463
7. Pan H, Fang C, Zhou T, Wang Q, Chen J (2007) Accumulation of calycosin and its 7-O-β-D-glucoside and related gene expression in seedlings of Astragalus membranaceus Bge. var. mongholicus (Bge.) Hsiao induced by low temperature stress. Plant Cell Rep 26:1111–1120. doi:10.1007/s00299-006-0301-8
8. Ramakrishna A, Ravishankar GA (2011) Influence of abiotic stress signals on secondary metabolites in plants. Plant Signal Behav 6:1720–1731. doi:10.4161/psb.6.11.17613
9. Szopa A, Ekiert H, Szewczyk A, Fugas E (2012) Production of bioactive phenolic acids and furanocoumarins in in vitro cultures of Ruta graveolens L. and Ruta graveolens ssp.divaricata (Tenore) Gams. under different light conditions. Plant Cell Tiss Organ Cult 110:329–336. doi:10.1007/s11240-012-0154-5
10. Thanonkeo S, Panichajakul S (2006) Production of isoflavones, daidzein and genistein in callus cultures of Pueraria candollei Wall. ex Benth. var. mirifica. Songklanakarin J Sci Technol 28:45–53
11. Tůmová L, Tůma J (2011) The effect of UV light on isoflavonoid production in Genista tinctoria culture in vitro. Acta Physiol Plant 33:635–640. doi:10.1007/s11738-010-0566-y
12. Wu SQ, Lian ML, Gao R, Park SY, Piao XC (2011) Bioreactor application on adventitious root culture of Astragalus membranaceus. In Vitro Cell Dev Biol Plant 47:719–724. doi:10.1007/s11627-011-9376-1
13. Xu R-Y, Nan P, Yang Y, Pan H, Zhou T, Chen J (2011) Ultraviolet irradiation induces accumulation of isoflavonoids and transcription of genes of enzymes involved in the calycosin 7-O-β-D-glucoside pathway in Astragalus membranaceus Bge. var. mongholicus(Bge) Hsiao. Physiol Plant 142:265–273. doi:10.1111/j.1399-3054.2011.01474.x
14. Zhang D, Wang S, Zhu L, Tian Y, Wang H, Zhuang Y, Li Y, Wang D (2013) Profiling of hepatocellular carcinoma cell cycle regulating genes targeted by calycosin. BioMed Res Int 2013:317926. doi:10.1155/2013/317926

CHAPTER 12

EFFECT OF PLANT GROWTH REGULATORS ON THE ACCUMULATION OF INDOLIZIDINE ALKALOIDS IN SECURINEGA SUFFRUTICOSA CALLUS CULTURES

Danuta Raj[1] Adam Kokotkiewicz[2] Andrzej Drys[3] Maria Luczkiewicz[2]

[1]Department of Pharmacognosy, Faculty of Pharmacy, Wroclaw Medical University, ul. Borowska 211 A, 50-556 Wrocław, Poland

[2] Department of Pharmacognosy, Faculty of Pharmacy, Medical University of Gdansk, al. gen. J. Hallera 107, 80-416 Gdan´sk, Poland

[3] Department of Physical Chemistry, Faculty of Pharmacy, Wroclaw Medical University, ul. Borowska 211 A, 50-556 Wrocław, Poland

INTRODUCTION

Securinine (S), the main pharmacologically active constituent of *Securinega suffruticosa*(Pall.) Rehd., Phyllanthaceae, is an indolizidine alkaloid and γ-aminobutyric acid$_A$receptor (GABA$_A$R) competitive antagonist. S is considered a promising agent for the prevention and treatment of neurodegenerative diseases, including Alzheimer's disease, amyotrophic lateral sclerosis and post-alcohol degeneration (Raj and Luczkiewicz 2008; Gupta et al. 2011). Apart from the above properties, S was shown to exhibit cytotoxic activity against colon (Rana et al. 2010) and breast cancer cell lines (Li et al. 2014), as well as inhibit toxoplasma growth (Holmes et al. 2011). The second main alkaloid of *S. suffruticosa*, allosecurinine (AS), acts as a cholinergic agent (Raj and Luczkiewicz 2008).

In previous years, in vitro cultures of *S. suffruticosa* were established in order to provide an efficient and reliable source of securinega alkaloids (Wysokińska 1979; Ide et al. 1986; Ide 1991; Yuan et al. 2005, 2007; Raj et al. 2009, 2015). Despite promising results in terms of alkaloid accumulation, these reports provided limited data on the influence of phytohormones on the accumulation of S and related compounds in the investigated biomasses. Given that the biosynthesis of alkaloids in plant cell cultures is strongly affected by both the type and concentration of PGRs used (Verpoorte et al. 1991), modifications of phytohormone composition can be considered a simple, yet effective tool for improving the productivity of an in vitro system. However, contrary to other alkaloid groups such as indole (Verpoorte et al. 1991; van der Heijden et al. 2004; Pietrosiuk et al.2007) and diterpene derivatives (Sabater-Jara et al. 2010; Malik et al. 2011), reports concerning the influence of exogenous PGRs on indolizidine alkaloid accumulation are rather scarce and limited to the most commonly used phytohormones like 2,4-dichlorophenoxyacetic acid (2,4-D), 1-naphthaleneacetic acid (NAA), and kinetin (KIN) (Wysokińska 1979; Ide 1991; Yuan et al. 2007).

The aim of the presented work was to investigate the influence of different combinations of PGRs (39 modifications in total) on biomass growth and the accumulation of S and AS in*S. suffruticosa* callus cultures. Consequently, the experiment led to the selection a phytohormone composition providing the highest concentrations of the investigated compounds. Correlations between callus growth and the accumulation of indolizidine alkaloids were established and possible relationships between the chemical structure of the PGRs used and the observed effects were also discussed. To the authors' knowledge, this is the first such comprehensive study on the influence of phytohormones on alkaloid accumulation in a single in vitro model.

Citation: Danuta Raj, Adam Kokotkiewicz, Andrzej Drys, Maria Luczkiewicz, Effect of plant growth regulators on the accumulation of indolizidine alkaloids in Securinega suffruticosa callus cultures, DOI: 10.1007/s11240-015-0811-6.

MATERIALS AND METHODS

Plant Material

The experiments were conducted with the use of previously established (Raj et al. 2009) *S. suffruticosa* callus cultures, grown on solidified (0.7 % w/v agar) Schenk–Hildebrandt (SH) medium (Schenk and Hildebrandt 1972) containing 5.0 mg l^{-1} 2,4-D, 5.0 mg l^{-1} KIN and 3 % w/v sucrose, and subcultured at 4-week intervals.

Experimental Procedures and Culture Conditions

For the experiment, 1 g portions of *S. suffruticosa* callus, taken on 30 d of the growth cycle, were transferred to baby food jars (Sigma-Aldrich, St. Louis, US-MO) filled with 25 ml of the solidified (0.7 % w/v agar) SH media supplemented with 3 % (w/v) sucrose and the respective phytohormones (10 replicates per modification, 2 series of 5 repetitions). The cultures were maintained at 24 ± 1 °C under a 16/8 (light/dark) photoperiod (88 ± 8 μmol m^{-2} s^{-1}, Philips TLD 35 fluorescent lamps). After 4-weeks, the calli were harvested and their growth indices (Gi) calculated as described by Yuan et al. (2007). After freeze-drying (Lyovac GT2, Fin-Aqua-Santasalo-Sohlberg Co, Tuusula, Finland), the samples were evaluated for S and SA content using the previously developed HPTLC-densitometric method (Raj et al. 2009).

PGRs used in the study were either sterile-filtered [forchlorfenuron (CPPU), diphenylurea (DPU), indole-3-acetic acid (IAA), 2-isopentenyladenine (2iP), thidiazuron (TDZ), 2,3,5-triiodobenzoic acid (TIBA); Schleicher & Schuell 0.2 μm CA filters, GE Healthcare, Little Chalfont, UK] into the pre-autoclaved media or co-autoclaved (1 bar, 120 °C, 20 min) with other media components (all reagents from Sigma-Aldrich, purified water produced by double distillation). The pH of the media was adjusted to 5.8 prior to autoclaving.

Experimental Scheme

In order to test the effect of auxins and anti-auxins on the growth rate and accumulation of S and AS, the KIN-containing media were supplemented interchangeably with 2,4-D, IAA, NAA, TIBA, 4-chlorophenoxyacetic acid (4-CPA), indole-3-butyric acid (IBA), indole-3-propionic acid (IPA), β-naphthoxyacetic acid (BNOA), 2,4,5-trichlorophenoxyacetic acid (2,4,5-T) and picloram (PCL). The respective PGRs and KIN were applied at both 1:10 and 10:1 ratio—the media containing 5.0 or 0.5 mg l^{-1} KIN were supplemented with 0.5 or 5.0 mg l^{-1} auxin/auxin inhibitor, respectively. *S. suffruticosa* calli grown on SH media supplemented solely with 0.5 or 5.0 mg l^{-1} KIN (C1 and C2, respectively) and the callus maintained on PGR-free SH medium (C0) were the reference groups.

Analogously, the influence of cytokinins on callus growth and the accumulation of the indolizidine alkaloids was investigated by enriching the 2,4-D-supplemented media interchangeably with CPPU, DPU, 2iP, KIN, TDZ, 6-benzylaminopurine (BAP) and zeatin (ZEA). The cytokinins and 2,4-D were applied at both 1:10 and 10:1 ratio—the media containing 5.0 or 0.5 mg l^{-1} 2,4-D were supplemented with 0.5 or 5.0 mg l^{-1} auxins, respectively. *S. suffruticosa* calli grown on SH media supplemented solely with 0.5 or 5.0 mg l^{-1} 2,4-D (C3 and C4, respectively) and the callus maintained on PGR-free SH medium (C0) were the reference groups.

In the next part of the study, the PGRs which provided the best results in terms of indolizidine alkaloid content (IAA and ZEA of auxin and cytokinin group, respectively) were applied jointly at 1:10 and 10:1 ratio (2 modifications: 0.5 mg l^{-1} IAA + 5.0 mg l^{-1}ZEA and 5.0 mg l^{-1} IAA + 0.5 mg l^{-1} ZEA). *S. suffruticosa* calli grown in the presence of 0.5 mg l^{-1} IAA and 5.0 mg l^{-1} KIN (modification providing the highest alkaloid concentration in both variable auxin and variable cytokinin groups) and on the PGR-free medium were the reference groups.

Statistical Analysis

The differences in Gi values and S and AS concentrations observed between the calli grown on media with different PGR composition were tested for statistical significance ($p < 0.05$) using the analysis of variance, followed by Tukey's range test (Statistica 10.0 software, StatSoft, Tulsa, US-OK). Correlations between Gi and accumulation of S and A were also investigated for each of the experiments ($p < 0.05$).

RESULTS

The effects of different PGR compositions on the growth rate (Gi value) and indolizidine alkaloid content of *S. suffruticosa* callus were presented in Fig. 1. The biomass maintained on the PGR-free medium (C0 modification—Fig. 2a) for 30 d was characterized by relatively low Gi value (218 %). AS concentration in the biomass (1.42 mg g^{-1} DW) was over twofold higher than that of S (0.61 mg g^{-1} DW).

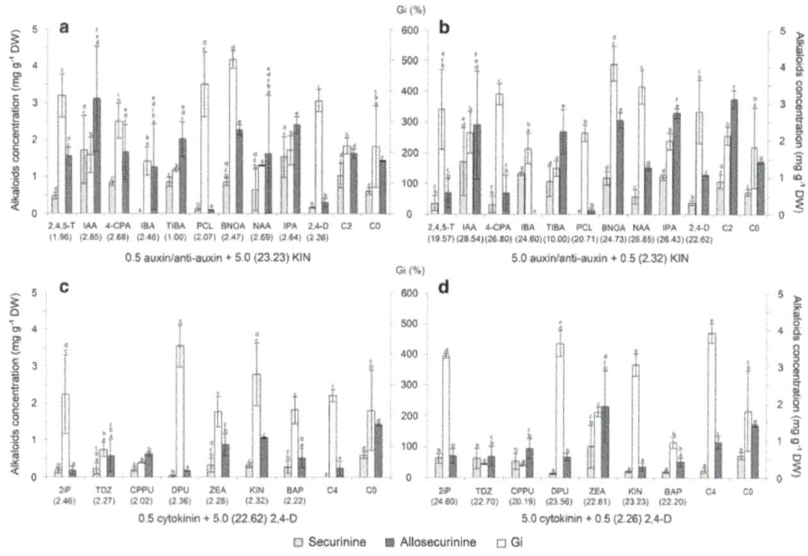

Figure. 1: The effect of different combinations of auxins/anti-auxins and cytokinins on the growth rate (Gi) and indolizidine alkaloid content of *S. suffruticosa* callus: **a** 0.5 mg l^{-1}auxin/anti-auxin + 5.0 mg l^{-1} KIN, **b** 5.0 mg l^{-1} auxin/anti-auxin + 0.5 mg l^{-1} KIN, **c**0.5 mg l^{-1} cytokinin + 5.0 mg l^{-1} 2,4-D, **d** 5.0 mg l^{-1} cytokinin + 0.5 mg l^{-1} 2,4-D (micromolar concentrations are given in *parentheses*). Control groups: *C0*phyto-hormone-free medium, *C1* 0.5 mg l^{-1} KIN, *C2* 5.0 mg l^{-1} KIN, *C3* 0.5 mg l^{-1} 2,4-D,*C4* 5.0 mg l^{-1} 2,4-D. *Different letters* indicate significant differences between means ($p < 0.05$)

Figure. 2: *S. suffruticosa* calli grown using different PGR modifications: **a** PGR-free medium, **b**0.5 mg l^{-1} BNOA + 5.0 mg l^{-1} KIN, **c** 5.0 mg l^{-1} BNOA + 0.5 mg l^{-1} KIN, **d** 0.5 mg l^{-1}TDZ + 5.0 mg l^{-1} 2,4-D, **e** 5.0 mg l^{-1} TDZ + 0.5 mg l^{-1} 2,4-D, **f** 0.5 mg l^{-1} IAA + 5.0 mg l^{-1}KIN

2,4-D added at 0.5 mg l^{-1} as the sole PGR (Fig. 1d, C3) significantly stimulated callus growth (Gi = 471 %, over twofold increase compared to C0 group) but this effect was not observed when 5.0 mg l^{-1} concentration was applied (Fig. 1c, C4). 2,4-D decreased the accumulation of S and AS in a dose-dependent manner, with stronger inhibitory effect at higher concentration.

The addition of 0.5 or 5.0 mg l^{-1} KIN (C1 and C2, respectively, Fig. 1a, b) had no significant effect on callus growth, but significantly increased the accumulation of both alkaloids. No clear dose-related effects were observed, however, lower KIN concentration (C1) favoured AS production (3.12 mg g^{-1}, ca. twofold increase as compared to C0 (Fig. 1b).

Auxins/anti-auxins at 0.5 mg l^{-1}, applied jointly with 5.0 mg l^{-1} KIN, exhibited variable effects on callus growth, ranging from inhibitory (TIBA, NAA) through neutral (IAA, CPA, IBA, IPA) to stimulatory (2,4,5-T, PCL, BNOA, 2,4-D). The combination of 0.5 BNOA and 5.0 KIN provided the highest growth rate (Gi = 502 %) of all PGR modifications (callus depicted in Fig. 2b). None of the auxins supplemented at 0.5 mg l^{-1} had statistically significant stimulatory effect on S accumulation, however, IAA, BNOA and IPA significantly increased AS content. 2,4,5-T and IBA had significant negative effect on S production, whereas PCL and 2,4-D significantly decreased the accumulation of both alkaloids (Fig. 1a). The highest mean concentrations of S and AS (1.73 and 3.11 mg g^{-1}DW) were recorded for the culture supplemented with 0.5 mg l^{-1} IAA and 5.0 mg l^{-1} KIN (Fig. 1a). Statistical analysis showed negative correlations between callus growth (Gi) and AS concentration (−0.27, p = 0.007), as well as between Gi and the sum of alkaloids (−0.25, p = 0.013).

The inclusion of 5.0 mg l^{-1} auxin into the 0.5 mg l^{-1} KIN-supplemented medium either did not affect (2,4,5-T, IAA, IBA, PCL, IPA, 2,4-D) or significantly stimulated callus growth (4-CPA, BNOA, NAA). The only exception was the anti-auxin TIBA which at 5.0 mg l^{-1}significantly decreased the growth rate. As in the case of 0.5 mg l^{-1} auxin/5.0 mg l^{-1} KIN modifications, the highest Gi value (490 %) was recorded for BNOA (callus depicted in Fig. 2c). Of the PGRs tested, IAA, IBA, TIBA, BNOA and IPA had no effect on S accumulation whereas 2,4,5-T, 4-CPA, PCL, NAA and 2,4-D exerted significant inhibitory effect. Except for IAA and IPA, AS concentrations in the examined calli were significantly lowered in comparison to reference culture (C1). The combination of 5.0 mg l^{-1} IAA and 0.5 mg l^{-1} KIN provided the highest mean concentrations of S (1.42 mg g^{-1} DW) and AS (2.42 mg g^{-1} DW). The statistical analysis showed no important correlations between the growth rate (Gi) and indolizidine alkaloid accumulation.

As depicted in Fig. 1c, cytokinins at 0.5 mg l^{-1}, applied together with 5.0 mg l^{-1} 2,4-D, had inhibitory (TDZ, CPPU), neutral (2iP, ZEA, KIN, BAP) or stimulatory (DPU, Gi = 427 %) effect on callus growth, as compared to C4 reference group (2,4-D only). All the cytokinins tested at 0.5 mg l^{-1} significantly increased S accumulation. The same effect was observed for AS, except for 2iP and DPU which had no significant effect on its concentration. The highest amounts of S and AS were recorded in the calli grown in the presence of 0.5 mg l^{-1}ZEA (0.32 and 0.88 mg g^{-1} DW, respectively) or 0.5 mg l^{-1} KIN (0.32 and 1.06 mg g^{-1} DW, respectively). At the same time, all phytohormone modifications tested provided lower indolizidine alkaloid concentrations compared to the PGR-free (C0) medium (Fig. 1c). No statistically significant correlations between callus growth and alkaloid concentrations were established.

When supplemented at 5.0 mg l^{-1} and jointly with 0.5 mg l^{-1} 2,4-D, the cytokinins had significant negative effect on callus growth. The inhibitory effect varied from ca. 20 % (2iP, KIN) to over 90 % decrease in Gi value recorded for TDZ and CPPU. DPU was the only PGR that did not significantly affect biomass growth (Gi = 436 %, comparable to C3 group). Of the cytokinins examined, DPU, KIN and BAP had no significant effect on S accumulation, whereas 2iP, TDZ and CPPU caused ca. twofold increase in its concentration. AS accumulation was negatively influenced by 2iP, TDZ, DPU, KIN and BAP but remained statistically unchanged after TDZ and CPPU supplementation. The callus supplemented with 0.5 mg l^{-1} ZEA and 5.0 mg l^{-1} 2,4-D was characterized by the highest indolizidine alkaloid content (0.85 and 1.93 mg g^{-1} DW of S and AS, respectively). The obtained results, however, were not significantly different than those of control (C0) group (Fig. 1d). Statistical analysis showed negative correlation (−0.31, p = 0.009) between biomass growth (Gi) and S accumulation, however, no such relationship was established for AS and the sum of alkaloids.

As presented in Fig. 3, the combinations of IAA and ZEA (i.e. the PGRs of auxin and cytokinin groups which provided the highest indolizidine alkaloid accumulation—Fig. 1) had no significant effect on callus growth. The recorded concentrations of S (1.02–1.20 mg g^{-1} DW) and AS (2.00–2.24 mg g^{-1} DW) were significantly higher than those in the control group but also significantly lower in comparison to the combination of 0.5 mg l^{-1} IAA and 5.0 mg l^{-1} KIN which consequently proved to be most effective PGR composition in the whole experiment (1.73 and 3.11 mg g^{-1} DW of S and AS, respectively).

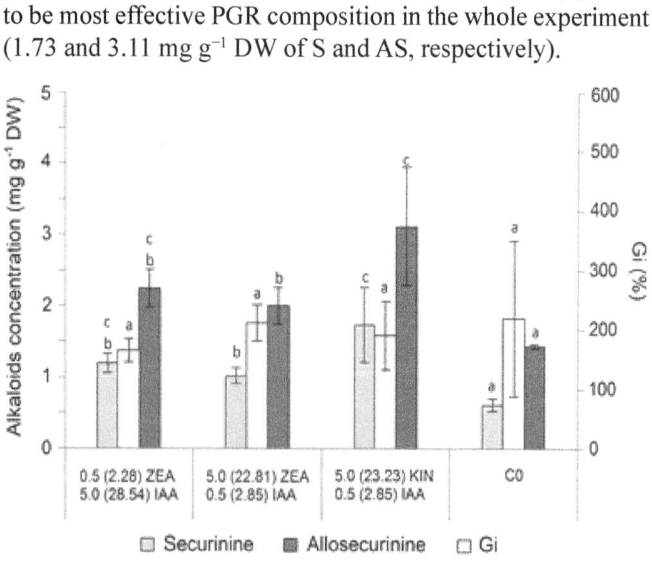

Figure. 3: The effect of the combinations of IAA with ZEA or KIN [concentrations given in mg $l^{-1}$and µM (values in *parentheses*)] on the growth rate (Gi) and indolizidine alkaloid content of *S. suffruticosa* callus. **C0**: phytohormone-free medium. *Different letters* indicate significant differences between means ($p < 0.05$)

DISCUSSION

Modifications of phytohormone composition offer wide possibilities of improving cell culture productivity with respect to therapeutically relevant constituents (Murthy et al.2014).

However, difficulties can arise from often opposite effects of auxins and cytokinins towards secondary metabolism. Consequently, in order to provide high productivity, plant cell cultures need to be maintained in two-phase systems or supplemented with a balanced, empirically optimized auxin/cytokinin composition (Verpoorte et al. 1991; Collin 2001; Luczkiewicz and Cisowski 2001; Ramawat and Mathur 2007).

In the presented work, 39 different PGR combinations were tested for their influence on the growth and indolizidine alkaloid content in *S. suffruticosa* callus cultures. Since the growth of *S. suffruticosa* callus was previously shown to be both auxin- and cytokinin dependent (Raj 2011), the experiment was conducted using auxin/cytokinin combinations. Given that the effects of phytohormones on primary and secondary metabolism are usually dose-dependent (Verpoorte et al. 1991; Ramawat and Mathur 2007), the media with opposite auxin/cytokinin ratios were tested—the experimental approach so far successfully applied in the studies concerning phytohormone-related accumulation of phenolics, including anthocyanins (Luczkiewicz and Cisowski 2001), isoflavones (Luczkiewicz et al.2014) and phenolic acids (Szopa and Ekiert 2014), in plant cell cultures.

When applied alone, 2,4-D and KIN, respectively, had inhibitory and stimulatory effect on alkaloid accumulation which is in agreement with previous studies on cell cultures of alkaloid-bearing plants (Verpoorte et al. 1991; Ramawat and Mathur 2007). The experiments also demonstrated that simultaneous application of KIN and 2,4-D was not beneficial in terms of growth rates and alkaloid concentrations obtained. The influence of different combinations of the above PGRs on the growth and alkaloid content in *S. suffruticosa* callus was also investigated by other authors. Yuan et al. (2007) found that at the constant KIN concentration, the increase in 2,4-D content resulted in slower growth rate and lower alkaloid concentration. In another study, Ide (1991) tested several compositions of 2,4-D and KIN, the best of which yielded 0.06 mg g^{-1} S (compared to 0.32 mg g^{-1} S in the present work). Because of different PGR concentrations applied, the results of previous studies are not directly comparable with the current work. Nevertheless, the general conclusion is that a combination of 2,4-D and KIN can be used for both initiation and maintenance of *S. suffruticosa* cell cultures—while preserving their ability to accumulate alkaloids.

As indicated by previous studies, the replacement of 2,4-D with other auxins often resulted in significant increase of alkaloid accumulation in plant cell cultures (Verpoorte et al.1991). In the present work, the combination of 0.5 mg l^{-1} IAA and 5.0 mg l^{-1} KIN provided the highest accumulation of S and AS among all PGR compositions tested. Positive effect on indolizidine alkaloid accumulation was also noted for NAA, confirming the results by Wysokińska (1979), Ide (1991) and Yuan et al. (2007), as well as IPA whose influence on alkaloid metabolism has so far been scarcely investigated.

Contrary to the above PGRs, most auxins included in the study negatively influenced alkaloid accumulation in *S. suffruticosa*, with PCL exhibiting the strongest inhibitory effect. Also, no positive results were found for TIBA, an anti-auxin included in the experiments because of its stimulatory effect on secondary metabolism observed in previous work by the authors (Luczkiewicz et al. 2002, 2014).

When applied at 0.5 mg l^{-1}, cytokinins were shown to promote alkaloid accumulation in *S. suffruticosa*, thus reversing the negative effect of 2,4-D on secondary metabolism. Nevertheless, the recorded concentrations of S and AS were still lower than in C0 group (PGR-free medium). In general, media compositions involving purine-type cytokinins (especially ZEA and KIN) had more beneficial effects since they significantly increased alkaloid accumulation without negatively affecting callus growth. Phenylurea derivatives, on the other hand, either inhibited culture growth (TDZ and CPPU) or strongly increased biomass yield at the expense of alkaloid production (DPU).

An important observation arising from the study is that PGR combinations of 0.5 mg l^{-1}auxin and 5.0 mg l^{-1} cytokinin employed in the study showed negative correlation between callus growth and alkaloid accumulation, indicating that the growth medium could not be optimized for both high biomass yield and secondary metabolite content. It also needs to be mentioned that several PGRs included in the study (e.g. PCL, TIBA, TDZ, CPPU, CPU) were not previously examined with respect to their influence on indolizidine alkaloid biosynthesis. Therefore, it is difficult to assess whether the observed effects follow a general rule or are rather species-specific.

In the final part of the study, *S. suffruticosa* calli were grown on the media supplemented with the combinations of auxin and cytokinin which provided the highest indolizidine alkaloid content (IAA and ZEA, respectively) in order to check for synergistic effects between both PGRs. The experiment proved unsuccessful, yielding lower concentrations of S and AS compared to IAA/KIN composition. Nevertheless, the maximum securinine content obtained in the present work (1.73 mg g^{-1} DW) was higher in comparison to the reports by Wysokińska (1979) and Ide (1991) (1.17 and 0.4 mg g^{-1} DW, respectively).

CONCLUSIONS

Cell growth and securinine concentration in *S. suffruticosa* callus cultures proved to be strongly affected by the type and amount of the PGRs applied. Of the phytohormone compositions tested, the highest indolizidine alkaloid content was achieved using the combination of 0.5 mg l^{-1} IAA and 5.0 mg l^{-1} KIN, providing higher securinine concentration in comparison to previous reports. However, negative correlation between callus growth and indolizidine alkaloid accumulation indicates that developing a two-phase production system may be necessary to further improve alkaloid productivity of *S. suffruticosa* cell cultures.

CONFLICT OF INTEREST

The authors declare that they have no conflict of interest and the presented work is compliant with ethical standards of Plant Cell, Tissue and Organ Culture. All the authors read and approved the manuscript in its final form.

REFERENCES

1. Collin H (2001) Secondary product formation in plant tissue cultures. Plant Growth Regul 34:119–134. doi:10.1023/A:1013374417961

2. Gupta K, Chakrabarti A, Rana S, Ramdeo R, Roth BL, Agarwal ML, Tse W, Agarwal MK, Wald DN (2011) Securinine, a myeloid differentiation agent with therapeutic potential for AML. PLoS ONE 6:e21203. doi:10.1371/journal.pone.0021203Central
3. Holmes M, Crater AK, Dhudshia B, Thadani AN, Ananvoranich S (2011) *Toxoplasma gondii*: inhibitory activity and encystation effect of securinine and pyrrolidine derivatives on Toxoplasma growth. Exp Parasitol 127:370–375. doi:10.1016/j.exppara.2010.09.002
4. Ide A (1991) *Securinega suffruticosa*: In vitro culture and the formation of securinega alkaloids. In: Bajaj YPS (ed) Biotechnology in agriculture and forestry, vol 15: medicinal and aromatic plants III. Springer, Berlin, pp 420–431. doi:10.1007/978-3-642-84071-5_25
5. Ide A, Fujioka N, Nagano K, Tanaka N, Iwasaki K, Yamane Y, Koga D, Yagishita K, Nakao K, Kurisu Y, Fujioka N, Kohda H, Miyagawa H, Yamasaki K (1986) Growth and alkaloid production of callus culture induced from *Securinega suffruticosa*. Phytochemistry 26:145–148. doi:10.1016/S0031-9422(00)81499-5
6. Li M, Han S, Zhang G, Wang Y, Ji Z (2014) Antiproliferative activity and apoptosis-inducing mechanism of l-securinine on human breast cancer MCF-7 cells. Pharmazie 69:217–223. doi:10.1691/ph.2014.3802
7. Luczkiewicz M, Cisowski W (2001) Optimisation of the second phase of a two phase growth system for anthocyanin accumulation in callus cultures of *Rudbeckia hirta*. Plant Cell Tiss Organ Cult 65:57–68. doi:10.1023/A:1010652507981
8. Luczkiewicz M, Zárate R, Dembińska-Migas W, Migas P, Verpoorte R (2002) Production of pulchelin E in hairy roots, callus and suspension cultures of *Rudbeckia hirta* L. Plant Sci 163:91–100. doi:10.1016/S0168-9452(02)00065-1
9. Luczkiewicz M, Kokotkiewicz A, Glod D (2014) Plant growth regulators affect biosynthesis and accumulation profile of isoflavone phytoestrogens in high-productive in vitro cultures of *Genista tinctoria*. Plant Cell Tiss Organ Cult 118:419–429. doi:10.1007/s11240-014-0494-4
10. Malik S, Cusidó RM, Mirjalili MH, Moyano E, Palazón J, Bonfill M (2011) Production of the anticancer drug taxol in *Taxus baccata* suspension cultures: a review. Process Biochem 46:23–34. doi:10.1016/j.procbio.2010.09.004
11. Murthy HN, Lee E-J, Paek K-Y (2014) Production of secondary metabolites from cell and organ cultures: strategies and approaches for biomass improvement and metabolite accumulation. Plant Cell Tiss Organ Cult 118:1–16. doi:10.1007/s11240-014-0467-7
12. Pietrosiuk A, Furmanowa M, Łata B (2007) *Catharanthus roseus*: micropropagation and in vitro techniques. Phytochem Rev 6:459–473. doi:10.1007/s11101-006-9049-6
13. Raj D (2011) Badania biotechnologiczne nad akumulacją związków alkaloidowych w kulturach in vitro *Securinega suffruticosa*, PhD Thesis, Wroclaw Medical University
14. Raj D, Luczkiewicz M (2008) *Securinega suffruticosa*. Fitoterapia 79:419–427. doi:10.1016/j.fitote.2008.02.011
15. Raj D, Kokotkiewicz A, Luczkiewicz M (2009) Densitometric HPTLC analysis of indolizidine alkaloids in the herb and in vitro cultures of *Securinega suffruticosa*. J Planar Chromatogr Mod TLC 22:371–376. doi:10.1556/JPC.22.2009.5.11
16. Raj D, Kokotkiewicz A, Luczkiewicz M (2015) Production of therapeutically relevant indolizidine alkaloids in *Securinega suffruticosa* in vitro shoots maintained in liquid culture systems. Appl Biochem Biotechnol 175:1576–1587. doi:10.1007/s12010-014-1386-0Central
17. Ramawat KG, Mathur M (2007) Factors affecting the production of secondary metabolites. In: Ramawat KG, Merillon JM (eds) Biotechnology: secondary metabolites. CRC Press, Boca Raton, pp 59–102. doi: 10.1201/b10756-4
18. Rana S, Gupta K, Gomez J, Matsuyama S, Chakrabarti A, Agarwal M, Agarwal A, Agarwal M, Wald D (2010) Securinine induces p73-dependent apoptosis preferentially in p53-deficient colon cancer cells. FASEB J 24:2126–2134. doi:10.1096/fj.09-148999
19. Sabater-Jara AB, Tudela LR, López-Pérez AJ (2010) In vitro culture of *Taxus* sp.: strategies to increase cell growth and taxoid production. Phytochem Rev 9:343–356. doi:10.1007/s11101-010-9167-z
20. Schenk RU, Hildebrandt AC (1972) Medium and techniques for induction and growth of monocotyledonous and dicotyledonous plant cell cultures. Can J Bot 50:199–204. doi:10.1139/b72-026
21. Szopa A, Ekiert H (2014) Production of biologically active phenolic acids in *Aronia melanocarpa* (Michx.) Elliott in vitro cultures cultivated on different variants of the Murashige and Skoog medium. Plant Growth Regul 72:51–58. doi:10.1007/s10725-013-9835-2
22. van der Heijden R, Jacobs DI, Snoeijer W, Hallard D, Verpoorte R (2004) The*Catharanthus* alkaloids: pharmacognosy and biotechnology. Curr Med Chem 11:607–628. doi:10.2174/0929867043455846
23. Verpoorte R, van der Heijden R, van Gulik WM, ten Hoopen HJG (1991) Plant biotechnology for the production of alkaloids: present status and prospects. In: Brossi A (ed) The alkaloids: chemistry and pharmacology, vol 40. Academic Press, San Diego, pp 1–187. doi:10.1016/S0099-9598(08)60268-0
24. Wysokińska H (1979) Wpływ niektórych regulatorów wzrostu i prekursorów biosyntezy alkaloidów na kultury tkankowe i komórkowe *Securinega suffruticosa* (Pall.) Rehd., PhD Thesis, Medical University of Lodz
25. Yuan W, Lu Z, Liu Y, Meng C, Cheng KD, Zhu P (2005) Three new podocarpane-type diterpenoids from callus of *Securinega suffruticosa*. Chem Pharm Bull 53:1610–1612. doi:10.1248/cpb.53.1610
26. Yuan W, Zhu P, Cheng K, Meng C, Wu F, Zhu H (2007) Callus of *Securinega suffruticosa*, a cell line accumulates dextro Securinega alkaloids. Nat Prod Res 21:234–242.

CHAPTER 13

COMPARISON OF THE MORPHOGENIC POTENTIAL OF FIVE GENTIANA SPECIES IN LEAF MESOPHYLL PROTOPLAST CULTURE AND PLOIDY STABILITY OF REGENERATED CALLI AND PLANTS

Karolina Tomiczak[1] Elwira Sliwinska[2] Jan J. Rybczyn´ski[1]

[1].Department of Experimental Plant Biology, Polish Academy of Sciences Botanical Garden, Center for Biological Diversity Conservation in Powsin, Prawdziwka 2, 02-973 Warsaw, Poland

[2] Laboratory of Molecular Biology and Cytometry, Department of Plant Genetics, Physiology and Biotechnology, UTP University of Science and Technology, Kaliskiego Ave. 7, 85-789 Bydgoszcz, Poland

ABSTRACT

The morphogenic potential of five Gentiana species of medicinal and ornamental value, i.e.G. cruciata, G. kurroo, G. lutea, G. septemfida, and G. tibetica, was compared using agarose-bead leaf mesophyll protoplast culture. Modified Murashige and Skoog medium containing 2.0 mg l^{-1} 1-naphthaleneacetic acid and 0.1 mg l^{-1} thidiazuron ensured the highest cell division frequency during both protoplast and protoplast-derived cell culture or callus formation. Indirect plant regeneration mainly via the somatic embryogenesis pathway was achieved for G. kurroo and G. tibetica with the greatest efficiency occurring with MS medium supplemented with 1.0 mg l^{-1} kinetin, 0.5 mg l^{-1} gibberellic acid, and 80 mg l^{-1} adenine sulfate. A considerable percentage of autopolyploid and aneuploid cells was detected in callus lines obtained from protoplasts using flow cytometry. The presence of polyploid cells in calli resulted in the regeneration of 85 % polyploid plants of G. kurrooand 14 % polyploids of G. tibetica. Chromosome counting and stomatal characteristic confirmed the ploidy variation among regenerants.

INTRODUCTION

Plants belonging to the genus *Gentiana* L. are the source of important pharmacologically active phytochemicals having anti-inflammatory, antipyretic, antimicrobial, cyto- and hepatoprotective, analgesic, stomachic and gastroprotective properties (Singh 2008), and have thus been frequently used in traditional herbal medicine and in pharmacy. Of the 360 species of gentian, *Gentiana lutea* L. is widely used for pharmaceutical purposes in Europe, *G. kurroo* Royle in India (Behera and Raina 2012) and Pakistan (Qureshi et al.2007), and *G. tibetica* King in China (Zhao et al. 2010). Conversely, because of their very decorative flowers, gentians are also used as ornamental plants, either as cut and potted flowers or in rock and naturalistic gardens (Takahata et al. 1995). *Gentiana cruciata* L. and *G. septemfida* Pall owing to their their blue and violet flowers, and ease of cultivation are popular for average garden conditions (Köhlein 1991).

The majority of *Gentiana* species are rare or even face possible extinction because of their exploitation by man. However, modern biotechnological techniques combined with in vitro cultures offer attractive alternative approaches to traditional cultivation techniques resulting both in the rapid micropropagation of valuable gentian species (Momčilović et al.1997; Hosokawa et al. 1998), and providing an efficient source of secondary metabolites (Tiwari et al. 2007; Cai et al. 2009; Hayta et al. 2011). Moreover, this technique enables modification of gentian genome for introduction of some

Citation: Karolina Tomiczak, Elwira Sliwinska, Jan J. Rybczyński, Comparison of the morphogenic potential of five Gentiana species in leaf mesophyll protoplast culture and ploidy stability of regenerated calli and plants, DOI: 10.1007/s11240-016-1000-y.

desirable characteristics (Mishiba et al. 2006; Nakatsuka et al. 2010) and the conservation of gentian biodiversity (Tanaka et al. 2004; Suzuki et al. 2005; Mikuła et al. 2008). An important prerequisite for the aforementioned use of in vitro cultures is the development of an efficient plant regeneration system from a suitable explant type that expresses a high degree of morphogenic potential.

Of the various regeneration systems available, the protoplast-to-plant system deserves special attention as it forms the basis of protoclonal variation, somatic hybridization or direct genetic transformation by DNA uptake. So far, plants of the genus *Gentiana* have been regenerated only from green leaf mesophyll protoplasts of *G. scabra* (Takahata and Jomori 1989), *G. triflora* (Jomori et al. 1995) and *G. scabra x G. triflora* (Nakano et al.1995), from callus protoplasts of *G. crassicaulis* (Meng et al. 1996) and from cell suspension protoplasts of *G. kurroo* (Fiuk and Rybczyński 2007; Wójcik and Rybczyński2015) and *G. macrophylla* (Hu et al. 2015). Although leaves are the most widely accessible plant material for protoplast isolation, cultures derived from leaf mesophyll protoplasts of gentians are usually associated with low frequency of morphogenetic processes. Some findings of Nakano et al. (1995) and Takahata et al. (1995) indicated that thidiazuron (TDZ) was a more effective cytokinin than benzylaminopurine (BAP) in cultures of mesophyll protoplasts of *G. triflora* and *G. triflora* × *G. scabra*. Recently, we succeeded in regenerating plants from leaf mesophyll protoplasts of *G. decumbens* using TDZ in media used for protoplast culture and callus proliferation (Tomiczak et al. 2015). However, only tetraploid plants were regenerated, indicating that culture conditions applied are able to disrupt cell division and promote endoreduplication (Tomiczak et al. 2015).

Thus, the aims of this study were (1) to compare the morphogenic potential of five other*Gentiana* species (*G. cruciata*, *G. kurroo*, *G. lutea*, *G. septemfida* and *G. tibetica*) in leaf mesophyll protoplast cultures and (2) to assess the ploidy stability of protocalli and plants obtained from protoplasts. Here the genome size of *G. kurroo* is estimated, to the best of our knowledge, for the first time.

MATERIALS AND METHODS

Plant Material

Seeds of *Gentiana* species were obtained from the PAS Botanical Garden-CBDC in Powsin, in Poland (*G. cruciata*) and from the Botanic Garden in Pruhonice, in the Czech Republic (*G. kurroo*, *G. lutea*, *G. septemfida*, *G. tibetica*). The seeds were sterilized with 70 % (v/v) ethanol for 30 s and then with 20 % (v/v) "Domestos" for 15 min, followed by three rinses in sterile water. Germination was carried out in a Petri dish, on half-strength Murashige and Skoog medium (MS; Murashige and Skoog 1962) supplemented with 0.5 mg l^{-1}gibberellic acid (GA_3), 10 g l^{-1} sucrose and 7 g l^{-1} agar, pH 5.8. Plantlets derived from seedlings were transferred to glass jars containing half-strength MS medium with 15 g l^{-1}sucrose and 8 g l^{-1} agar and were cultured in a phytotron at 21 ± 1 °C under 16-h day length. Illumination was provided by 100 µM m^{-2} s^{-1} daylight fluorescent tubes.

Protoplast Isolation

For protoplast isolation, young but fully expanded leaves harvested from in vitro-grown plantlets were used. The leaf abaxial epidermis was peeled off and 1 g of leaf tissue was incubated in 10 ml of cell and protoplast wash solution (CPW; Frearson et al. 1973) supplemented with 0.33 M mannitol (pH 5.8) for 30 min to induce cell plasmolysis. Cell wall digestion was carried out for 3–4 h in the dark at 26 °C in filter-sterilized enzyme solution consisting of 1.0 % (w/v) Cellulase Onozuka R-10 (Yakult Honsha Co., Ltd, Japan) and 0.5 % (w/v) Macerozyme R-10 (Yakult Honsha Co., LTD, Japan) dissolved in CPW with the addition of 0.5 M mannitol and 5 mM 2-(*N*-Morpholino)ethanesulfonic acid (MES), pH 5.8. In order to improve the release of protoplasts, gentle agitation on a rotary shaker (50 rpm) was employed. The released protoplasts were separated from the debris by filtering through a 45 µm nylon sieve and centrifuged at 180 rpm for 10 min. In the case of *G. cruciata*, *G. kurroo*, *G. lutea*, and *G. septemfida* the protoplast pellet was rinsed three times with CPW solution supplemented with 0.5 M mannitol and collected each time by centrifugation at 180 rpm for 10 min. To rinse pellet of *G. tibetica* protoplasts, the CPW solution was replaced by wash solution (WS) containing 1.0 mM $CaCl_2$, 0.5 M mannitol, and 5 mM MES, pH 5.8. The yield of protoplasts was estimated with the use of a Bürker counting chamber and expressed as the number of protoplasts obtained from 1 g of fresh weight of gentian leaves. Protoplast diameter was measured using "analySIS® FIVE" software (Olympus). To assess their viability, protoplasts were stained with 0.01 % (w/v) fluorescein diacetate (FDA) solution (Larkin 1976) and observed through an epifluorescence microscope (Vanox AHBT3, Olympus, 100× magnification). For each species, mean yield, diameter and viability were calculated on the basis of results obtained from 6 independent isolations, and 10 microscopic fields of view per isolation.

Protoplast Culture

Protoplasts were cultured at a density of 1.0×10^5 cells per ml in 100-µl agarose beads, composed of protoplast culture medium (PCM) solidified with 0.8 % (w/v) Sea Plaque®Agarose (Cambrex Bio Science Rockland Inc., USA). The PCM medium contained full MS macro- and micronutrients except for NH_4NO_3, MS vitamins, 30 g l^{-1} glucose, 3.0 g l^{-1}glutamine, 0.5 M mannitol, and appropriate plant growth regulators (PGRs), namely: α-naphthaleneacetic acid (NAA) in combination with TDZ or BAP (Table 1). From 12 to 14 beads were formed per 50-mm Petri dish and submerged in 5.0 ml of the liquid medium. Cultures were incubated at either 21 or 26 °C in the dark. Cell wall regeneration was monitored for the first week of culture by staining the protoplasts with 0.001 % (w/v) Calcofluor White (Fluorescent Brightener 28, Sigma-Aldrich) solution. Seven days after culture initiation, the plating efficiency was calculated. Protoplast-derived cells, which had divided at least once, were scored using an inverted light microscope (IMT-2, Olympus, 100× magnification) from 10 fields of view per each agarose bead culture. Plating efficiency was expressed as the number of cells that divided at least once per total number of plated protoplasts multiplied by 100. To prevent cultures from browning, liquid PCM medium was refreshed every week by replacing half the volume

(2.5 ml) of the current medium with the same volume of fresh medium. To stimulate further cell division and colony formation, the mannitol concentration of the freshly added medium was gradually reduced after 4 weeks of the protoplast culture; firstly to 0.33 M (in fifth and sixth week), then to 0.17 M (in seventh and eighth week) and finally to zero (in the following weeks).

Table 1: Combinations of plant growth regulators used in particular media for protoplast culture (PCM), callus proliferation (CPM) and plant regeneration (PRM)

Medium code	Plant growth regulators
PCM1	2.0 mg l^{-1} NAA, 0.1 mg l^{-1} TDZ
PCM2	2.0 mg l^{-1} NAA, 1.0 mg l^{-1} BAP
CPM1	2.0 mg l^{-1} NAA, 0.2 mg l^{-1} TDZ
CPM2	2.0 mg l^{-1} NAA, 1.0 mg l^{-1} BAP
CPM3	1.0 mg l^{-1} dicamba, 0.1 mg l^{-1} NAA, 2.0 mg l^{-1} BAP, 80 mg l^{-1} adenine sulfate
CPM4	0.5 mg l^{-1} 2,4-D, 1.0 mg l^{-1} kinetin
PRM1	0.1 mg l^{-1} NAA, 8.0 mg l^{-1} TDZ
PRM2	0.1 mg l^{-1} NAA, 6.0 mg l^{-1} BAP
PRM3	1.0 mg l^{-1} kinetin, 0.5 mg l^{-1} GA_3, 80 mg l^{-1} adenine sulfate

Callus Culture and Plant Regeneration

Visible protoplast-derived microcalli (about 1–2 mm in diameter) were transferred with agarose beads to 90-mm Petri dish containing 0.8 % agar-solidified MS callus proliferation medium (CPM) supplemented with 30 mg l^{-1} sucrose and various PGRs (Table 1). Cultures were maintained in the dark at 26 °C. The effect of 4 different PGR combinations on the degree of callus proliferation was estimated after 4 weeks of culture.

Calli lines obtained after 4–8 weeks were placed on 0.8 % agar-solidified MS plant regeneration medium (PRM; Table 1) and cultured under growth cabinet conditions (21 °C, 16/8 h photoperiod). The number of regenerating somatic embryos and/or shoots was estimated after 3 months of culture. Somatic embryos and/or shoots regenerating from callus tissue were transferred to glass jars containing agar-solidified half-strength MS medium supplemented with 15 g l^{-1} sucrose for further growth and rooting.

Flow Cytometry

Randomly chosen proliferating calli (7 per species) and young leaves of all regenerated plants of *G. kurroo* (20) and *G. tibetica* (70) were analyzed using flow cytometry. About 6-month-old seed-derived plants (over 20 per species) that were the source of the protoplasts constituted the control. *Petunia hybrida* cv. P × Pc6 (2.85 pg/2C; Marie and Brown 1993) and *Pisum sativum* L. 'Set' (9.11 pg/2C; Sliwinska et al. 2005) served as an internal standards for the estimation of the total nuclear DNA content of *G. kurroo* and *G. tibetica* plants, respectively. Plant material was chopped with a sharp razor blade in a Petri dish containing 1.0 ml of nuclei-isolation buffer (0.1 M Tris, 2.5 mM $MgCl_2 \times 6H_2O$, 85 mM NaCl, 0.1 % Triton X-100; pH 7.0) supplemented with 50 µg ml^{-1} of RNase A and 50 µg ml^{-1} of propidium iodide (PI). The suspension was subsequently passed through a 50-µm mesh nylon filter. For each sample, 3000–7000 nuclei were analyzed immediately following preparation using a Partec CCA (Munster, Germany) flow cytometer. Linear amplification was used for most of the samples; only for the analyses of calli of *G. kurroo*was logarithmic amplification applied because of the highly elevated endopolyploidy present in the cells of this plant material. PI fluorescence histograms were analyzed using DPAC V.2.2 software (Partec, GmbH, Münster, Germany). For callus samples, the mean *C*-value (mean ploidy; μ) was calculated (Lemontey et al. 2000). The nuclear DNA content of plants was calculated using the linear relationship between the ratio of the G_0/G_1 peak positions *Gentiana*/internal standard, on the fluorescence intensity histogram. Furthermore, the 1Cx-value (DNA content of a monoploid genome with chromosome number *x*) was calculated.

Chromosome Counting

Roots of *G. kurroo* and *G. tibetica* regenerants and control plants were used for evaluation of the number of chromosomes. Root tips were pretreated with 2 mM 8-hydroxyquinoline for 2 h at 21 °C and for the following 2 h at 4 °C, and then were fixed in an absolute ethanol: glacial acetic acid (3:1, v/v) for at least 24 h. After macerating for 50 min in 5 M HCl at 21 °C, root tips were stained in Schiff's reagent (Sigma-Aldrich) for 2 h in the dark and then rinsed three times with potassium metabisulfite solution (225 mM $K_2S_2O_5$, 5 mM HCl). Metaphase plates were prepared by gently squashing of root-tip meristems in a drop of 45 % (v/v) acetic acid on a microscopic slide. Observations were carried out using a Vanox AHBT3 microscope (Olympus) at 1000× magnification. Chromosomes were counted for at least 10 different, well-spread metaphase plates per each plant.

Stomatal guard cell length and frequency

The size and frequency of stomata were estimated for the abaxial epidermis of the youngest pair of fully expanded leaves of *G. kurroo* and *G. tibetica* plants grown in vitro. Observations were performed using a Vanox AHBT3 light microscope. The guard cell length of all stomata visible in 10 randomly selected fields of view per each leaf was measured using "analySIS® FIVE" software. Also, all stomata from 10 randomly selected microscopic fields of view per leaf were counted. The stomatal frequency was expressed as the number of stomata per 1 mm^2 area of the leaf blade.

Statistical Analysis

Each in vitro culture experiment was repeated three times. Three Petri dishes per each combination of culture conditions represented each replicate. Statistical analyses were performed with the help of Statistica ver. 6.0 (StatSoft Polska Sp. z o.o., Poland). One- or two-way analysis of variance was performed. Means were compared using Tukey's honestly significant difference (HSD) test, at the 0.05 level of probability. The correlation between mean size of protoplasts from different species and their yield was computed using Pearson product-moment correlation coefficient.

RESULTS

Protoplast Isolation and Culture of Five *Gentiana* Species

Protoplasts were successfully isolated from leaf mesophyll tissue of all five *Gentiana*species. All protoplasts were spherical in shape and mostly rich in chloroplasts (Fig. 1a). Their mean size ranged from 24.23 (*G. kurroo*) to 36.96 µm in diameter (*G. cruciata*) (Table 2). The size was negatively correlated with the yield of protoplasts (Pearson correlation coefficient $r = -0.91$), which varied from 3.46×10^5 protoplasts per 1 g of fresh leaf tissue of *G. septemfida* to 11.41×10^5 protoplasts from 1 g of *G. kurroo* leaf material (Table 2). The protoplast viability after isolation, as revealed by FDA staining (Fig. 1b) was similar for all species and reached 80–90 % (Table 2).

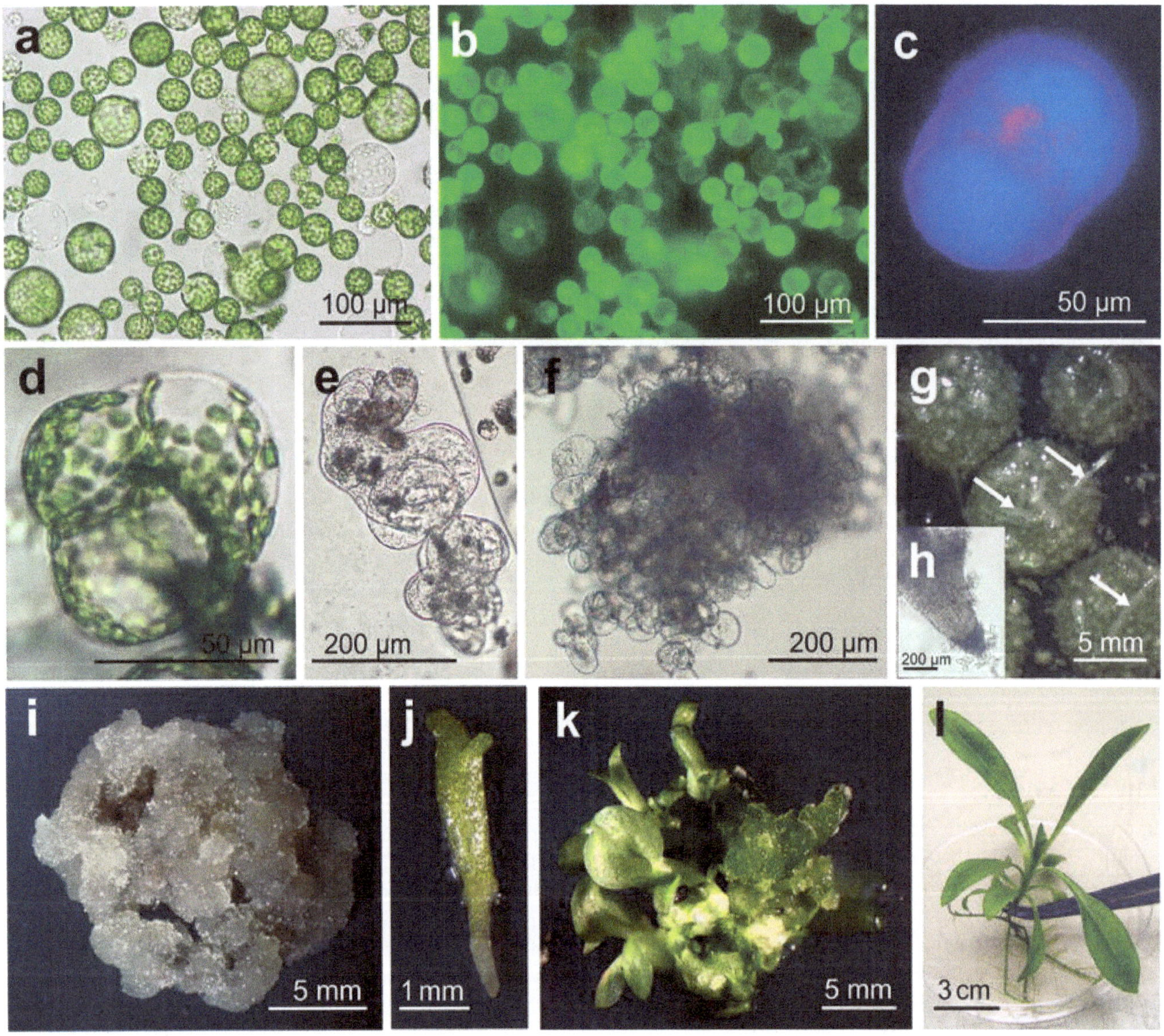

Figure. 1: Plant regeneration in green leaf mesophyll protoplast cultures of *Gentiana* species: **a**freshly isolated protoplasts of *G. tibetica* and **b** their viability revealed by FDA staining, **c**regeneration of the cell wall following 5 days of protoplast culture, and emphasized by treatment with Calcofluor White. Note the formation of a cell plate during first cell division, **d** three daughter cells formed as a result of the first and second division of protoplast-derived cell after 7 days of culture, **e** multicellular aggregate of *G. kurroo*formed after 5 and **f** 9 weeks of protoplast culture, respectively, **g** small minicalli of *G. tibetica* formed inside agarose beads with spontaneously regenerated roots (*arrows*), **h**magnification of spontaneously regenerated root, **i** callus tissue arising from single agarose bead, **j** somatic embryo of *G. tibetica* regenerated after 6 weeks of callus culture on PRM1 medium, **k** shoots of *G. kurroo* regenerated after 8 weeks of callus culture on PRM2 medium, **l** regenerated plant of *G. tibetica*

Table 2: Comparison of the mean size, yield and viability of freshly isolated protoplasts of five*Gentiana* species

Species	Protoplast diameter (µm)	Protoplast yield × 10^5 per 1 g of fresh weight	Protoplast viability (%)
G. cruciata	36.96 ± 7.05 *a*	5.36 ± 0.85 *bc*	90.66 ± 2.82 *a*
G. kurroo	24.23 ± 4.66 *c*	11.41 ± 2.25 *a*	87.85 ± 3.73 *a*
G. lutea	31.91 ± 8.01 *b*	5.06 ± 0.26 *bc*	89.66 ± 2.70 *a*
G. septemfida	35.67 ± 7.74 *a*	3.46 ± 1.25 *c*	84.91 ± 8.44 *a*
G. tibetica	31.98 ± 7.62 *b*	6.65 ± 1.48 *b*	79.25 ± 13.01 *a*

Values followed by the same letter are not significantly different at $P < 0.05$ (HSD test)

Within the first week of culture in agarose beads, protoplasts of all studied species, except for *G. lutea* completed the regeneration of the cell wall and started to divide (Fig. 1c–d). Protoplasts of *G. lutea* burst following 48 h of culture on both PCM1 and PCM2 medium.

For the remaining four species studied, culturing of protoplasts on PCM1 medium containing 0.1 mg l^{-1} TDZ resulted in higher plating efficiency than on PCM2 supplemented with 1 mg l^{-1} BAP (Table 3). Furthermore, the percentage of cell division noted at 26 °C was, on average, twice as much as at 21 °C (Table 3). Of all the species tested, *G. septemfida* was generally distinguished in that it demonstrated greatest plating efficiency at day 7 of culture.

Table 3: The percentage of dividing leaf mesophyll protoplasts-derived cells of four *Gentiana*species after 7 days of culture on two different media and at two different temperatures

Species	PCM1		PCM2	
	26 °C	21 °C	26 °C	21 °C
G. cruciata	4.83 ± 1.45 *abc*	2.00 ± 0.32 *cd*	3.45 ± 0.71 *abcd*	1.20 ± 0.34 *d*
G. kurroo	6.13 ± 1.60 *a*	3.46 ± 0.87 *abcd*	3.47 ± 0.51 *abcd*	2.31 ± 0.67 *cd*
G. septemfida	6.25 ± 1.74 *a*	2.98 ± 0.67 *bcd*	5.01 ± 1.92 *abc*	2.81 ± 0.94 *bcd*
G. tibetica	5.46 ± 0.73 *ab*	2.18 ± 0.68 *cd*	3.01 ± 0.58 *bcd*	1.72 ± 0.27 *d*

Values followed by the same letter are not significantly different at $P < 0.05$ (HSD test)

Despite the weekly addition of the fresh medium, cultures of *G. cruciata* and *G. septemfida* protoplasts turned brown and died within 2 weeks of culture. Protoplast-derived cells of *G. kurroo* and *G. tibetica* continued to divide and formed multicellular aggregates after 5–9 weeks on either PCM1 or PCM2 (Fig. 1e–f). After 10–12 weeks of culture, visible microcalli measuring of 1–2 mm in diameter were obtained. In addition, spontaneous rhizogenesis was observed at times in *G. tibetica* cultures maintained on both media (Fig. 1g–h).

Callus Culture and Plant Regeneration

Callus proliferation was achieved for all agarose beads transferred onto CPM media (Fig. 1i). The formation of callus tissue of both species, *G. kurroo* and *G. tibetica*, was faster when protoplasts were previously cultured on medium supplemented with TDZ (i.e. PCM1), than when cultured on the medium containing BAP (i.e. PCM2). The presence of TDZ in CPM1 medium promoted rapid growth of protoplast-derived callus as well (Online Resource 1). Effective callus proliferation of both *G. kurroo* and *G. tibetica* was also achieved on CPM3 medium supplemented with dicamba, NAA, BAP, and adenine sulfate.

The process of plant regeneration from protoplast-derived calli was mostly via somatic embryogenesis (Fig. 1j). For both species, PRM3 medium was the most effective in inducing somatic embryogenesis, but only from calli maintained previously on CPM1 and CPM3 media. The greatest number of *G. kurroo* somatic embryos per agarose bead (1.08) was recorded for those regenerated on PRM3 medium from callus cultured previously on CPM1 medium, whereas for *G. tibetica* the best embryogenic response (0.55 somatic embryos regenerated per agarose bead) was achieved for callus maintained on CPM3 medium. In a few *G. kurroo* callus cultures subjected to PRM2 medium, the regeneration of shoots was also observed (Fig. 1k, Online Resource 2).

Most of mature somatic embryos of both species easily developed into plantlets on half-strength MS medium with 2 % (w/v) sucrose. On the same medium, most of regenerated shoots of *G. kurroo* also rooted. Most of regenerated plants (Fig. 1l) grew vigorously on full strength MS medium.

Cell Ploidy of Protoplast-Derived Calli

Well over 90 % of the nuclei of leaves from seedlings of both *Gentiana* species investigated contained 2C DNA, only a few having 4C nuclei (Figs. 2a, 3a). Conversely, all *G. kurroo*protoplast-derived calli analyzed cytometrically exhibited mixoploidy, containing, beside 2C and 4C nuclei, some endopolyploid nuclei with DNA content above 4C

(Fig. 2). Nuclei with DNA content 2C, 4C, 6C, 8C, and 16C were detected, and in some calli a few nuclei with 32C were present. In three of five calli, 4C nuclei predominated, the remaining two having 8C nuclei. The mean *C*-value for calli of this species varied from 4.87 to 11.22. In the case of protoplast-derived calli of *G. tibetica*, the degree of mixoploidy was lower than for *G. kurroo*, and very few endopolyploid nuclei were detected (Fig. 3). However, surprisingly, in some calli, nuclei with DNA content lower than 2C (about 1.4C) occurred. Consequently, the mean *C*-value for *G. tibetica* callus cells varied from 1.48 to 2.51.

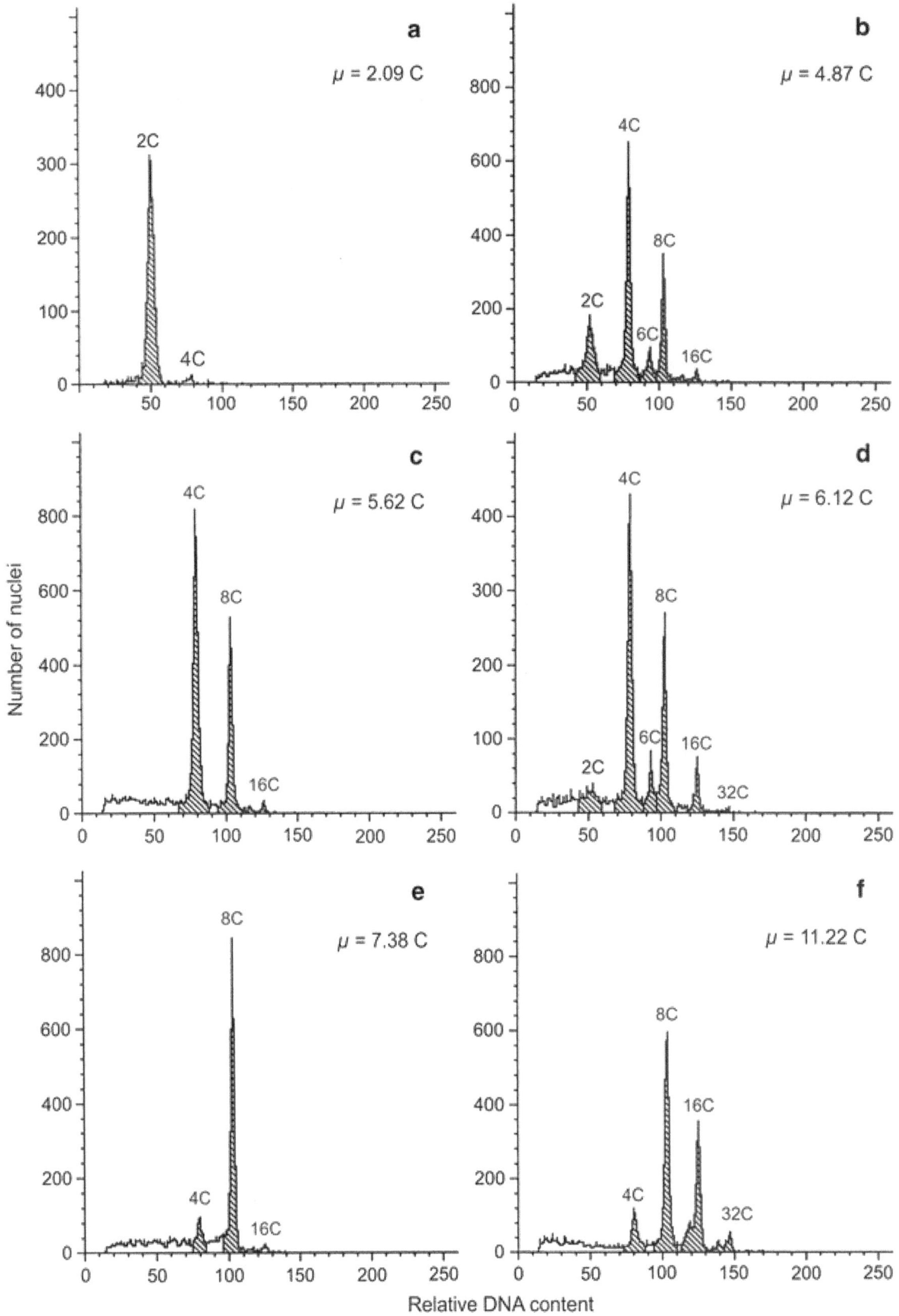

Figure. 2: Histograms of relative nuclear DNA content obtained after flow cytometric analysis of nuclei isolated from *G. kurroo* leaves and calli: **a** leaf of a seed-derived control plant, **b**–**f**protoplast-derived 4-month-old callus cultures. Note that cells with 2C DNA content are absent in the samples (**c**), (**e**) and (**f**). μ—mean *C*-value (mean ploidy)

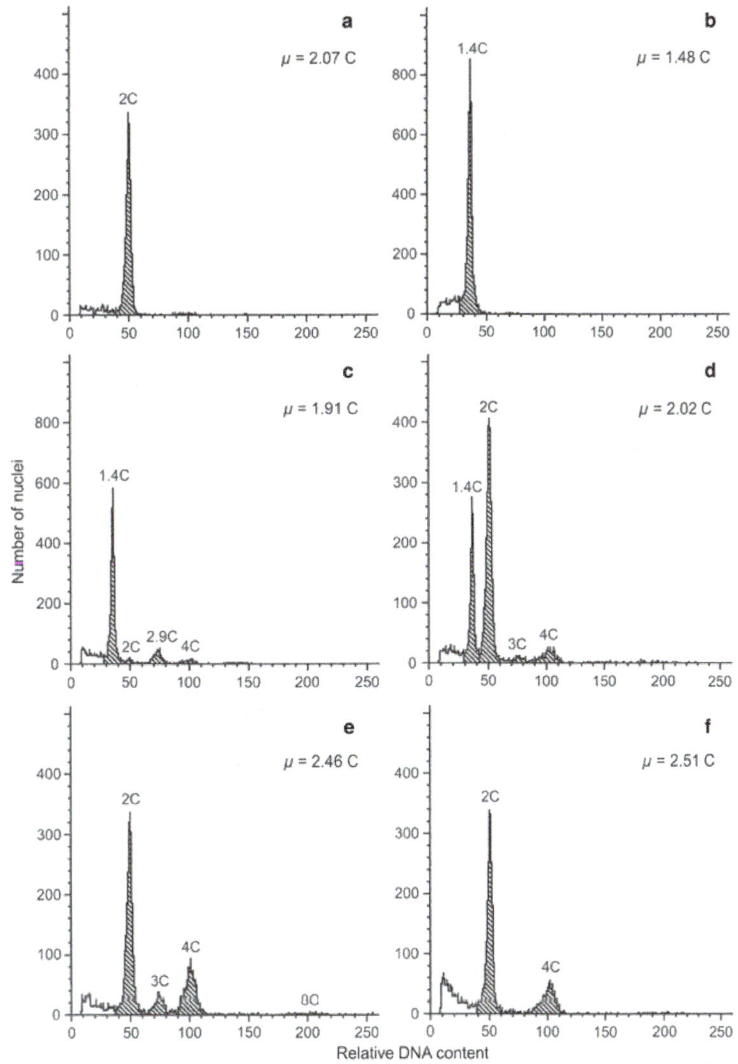

Figure. 3: Histograms of relative nuclear DNA content obtained after flow cytometric analysis of nuclei isolated from *G. tibetica* leaves and calli: **a** leaf of a seed-derived control plant, **b**–**f**protoplast-derived 4-month-old callus cultures. Note the presence of cells with relative nuclear DNA content below 2C in the samples (**b**), (**c**) and (**d**). μ—mean *C*-value (mean ploidy)

DNA Content and Chromosome Number of Regenerated Plants

Flow cytometry and chromosome counting revealed that control seed-derived plants of *G. kurroo* possessed 3.32 pg/2C DNA and 26 chromosomes (Fig. 4a; Table 4). Control plants of *G. tibetica* possessed 6.91 pg/2C DNA and 52 chromosomes (Fig. 4d; Table 4).

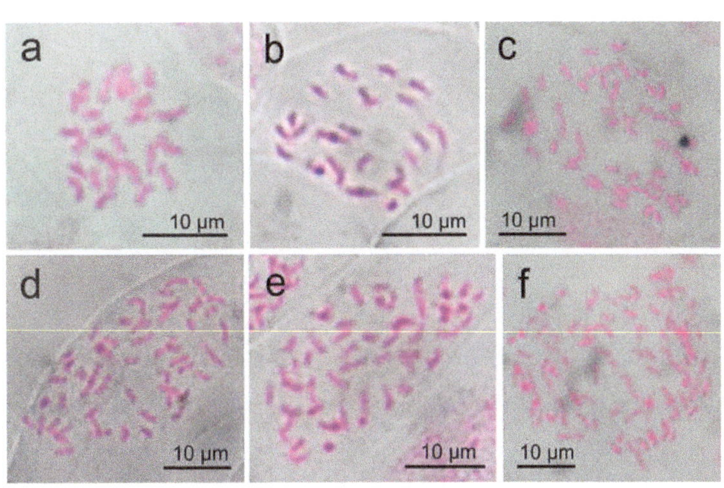

Figure. 4: Chromosome counting of *G. kurroo* and *G. tibetica* plants: **a** 26 mitotic metaphase chromosomes in a root-tip cell of a seed-derived parent plant of *G. kurroo* and **d** 52 chromosomes in a root-tip cell of a seed-derived parent plant of *G. tibetica*, **b** protoplast-derived regenerants of *G. kurroo* with 26 and **c** 52 chromosomes, **e** protoplast-derived regenerants of *G. tibetica* with 52 and **f** about 100 chromosomes

Table 4: Direct and indirect characteristics for ploidy estimation of control plants and regenerants from leaf mesophyll protoplasts of two *Gentiana* species studied

Species	Plant material	Number of plants (%)	Nuclear DNA content (pg/2C)	1Cx (pg)	Chromosome number	Stomata length (μm)	Number of stomata per mm²of leaf blade area	Ploidy
G. kurroo	Control	24 (100 %)	3.32 ± 0.07*b**	1.66	26	33.02 ± 4.58*b*	227.1 ± 66.7*a*	2*x*
	Regenerants	2 (10.0 %)	3.34 ± 0.15*b*	1.67	26	31.39 ± 2.83*c*	218.4 ± 41.3*a*	2*x*
		17 (85.0 %)	6.58 ± 0.14*a*	1.65	52	40.89 ± 3.45*a*	106.1 ± 23 *b*	4*x*
		1 (5.0 %)	8.75	1.46	n.t.	n.t.	n.t.	~6*x*
G. tibetica	Control	35 (100 %)	6.91 ± 0.20*b*	1.73	52	46.22 ± 5.6*b*	70.3 ± 11.9*a*	4*x*
	Regenerants	60 (85.7 %)	6.91 ± 0.36*b*	1.73	52	45.46 ± 3.83*b*	63.3 ± 15.8*a*	4*x*
		10 (14.3 %)	13.50 ± 0.95*a*	1.69	~100	66.86 ± 5.35*a*	34.4 ± 6.6 *b*	8*x*

n.t. not tested

* values for a certain species followed by the same letter are not significantly different at$P = 0.05$ (Tukey's test)

In 20 protoplast-derived plants of *G. kurroo*, 10.0 % possessed a genome and a chromosome number equal in size to that of the controls (Fig. 4b; Table 4), whereas in 85 %, nuclear DNA content and number of chromosomes had doubled (Fig. 4c; Table 4). One of *G. kurroo* regenerants was identified as a possible aneuploid having a 2C DNA content of 8.75 pg, below that expected for the hexaploid (assuming that 1Cx of this species is equal to about 1.7 pg, hexaploid DNA content would be about 10 pg). The chromosome number of this plant could not be assessed because of its inability to root and because it died rapidly in vitro. Of the 70 regenerants of *G. tibetica* also about 85 % had the same nuclear DNA content and number of chromosomes as control plants (Fig. 4e; Table 4), whereas the remaining protoplast-derived plants had their DNA content and twice as many chromosomes as the control (Fig. 4f; Table 4).

Plant Morphology and Stomatal Guard Cell Length and Frequency

Two plants regenerated from protoplasts of *G. kurroo*, which had the same nuclear DNA content and chromosome number as control plants, possessed a similar number of stomata per mm² of leaf blade area (Table 4). Regenerants with doubled DNA content and chromosome number, with the exception of leaf blades (Fig. 5b) which were wider than those of parent plants (Fig. 5a), showed no visible morphological differences under in vitro conditions. However, they had two-fold fewer stomata on the abaxial leaf surface than control plants (Table 4). Also, in comparison to the controls, the stomata were slightly smaller in regenerants with the same nuclear DNA content, but significantly larger in plants with doubled DNA content (Table 4). The stomata were not analyzed for an aneuploid *G. kurroo* regenerant, however severe leaf blade malformations such as swelling and crinkling were observed in this plant (Fig. 5c).

Figure. 5: Leaf morphology of *G. kurroo* and *G. tibetica* plants: **a** seed-derived parent plant of *G. kurroo*, **b** protoplast-derived regenerant of *G. kurroo* with doubled DNA content and chromosome number, **c** aneuploid protoplast-derived regenerant of *G. kurroo*, **d** seed-derived parent plant of *G. tibetica*, **e** protoplast-derived regenerant of *G. tibetica* with doubled DNA content and chromosome number

Similar relationships between ploidy level and the number and length of stomata were detected for *G. tibetica* protoplast-derived plants (Table 4). However, the regenerants with octoploid DNA content were dwarf and possessed thicker and wider leaves (Fig. 5e) than control plants and regenerants with tetraploid DNA content (Fig. 5d). They also developed shorter and thicker roots.

Discussion

Assessment of the morphogenic potential in protoplast culture, followed by the development of the protoplast-to-plant system, are the first steps in researching somatic hybridization or direct genetic transformation in species of interest. In this work, for the first time, leaf mesophyll protoplasts were isolated for five different gentian species of high medicinal and ornamental importance. Early response of protoplasts to culture conditions was compared for *G. cruciata*, *G. kurroo*, *G. septemfida* and *G. tibetica*. Regeneration potential and ploidy stability were evaluated for *G. kurroo* and *G. tibetica*.

The Influence of Culture Conditions on Protoplast-Derived Cell Viability and Divisions

Leaf mesophyll tissue of in vitro grown gentians is a suitable plant material for routine isolation of viable protoplasts. The physical and chemical conditions used for protoplast release allowed us to obtain yields of *G. kurroo* protoplasts similar to those of *G. decumbens* (Tomiczak et al. 2015) and *G. crassicaulis* (Meng et al. 1996), whereas the yields of protoplasts for other species was comparable to those that of *G. acaulis* (Jomori et al. 1995), *G. triflora*, and *G. triflora* × *G. scabra* (Nakano et al. 1995). A large number of factors related to plant material used, including plant genotype, age, and preculture conditions, influence the yield of released protoplast (Wright 1985; Grèzes et al. 1994; Hu et al. 2015). Our study indicates that yield of protoplasts can be also affected by protoplast size, thereby the cell size of source material.

Despite their relatively great viability directly after isolation, protoplasts of *G. lutea* did not survive the first 48 h in culture. Probably the concentration and/or the type of osmoticum, or the concentration of macro- and micronutrients in the protoplast culture medium were inadequate. Takahata et al. (1995) referred to the use of half-strength MS gellan gum-solidified medium supplemented with 1.5 % sucrose and 0.25 M mannitol for cell suspension-derived *G. lutea* protoplasts. However, to our best knowledge, no further reports on their regeneration into entire plants have ever been published. Furthermore, in our experience, leaves of *G. lutea* provide explants that respond poorly and have very low morphogenic potential (Fiuk and Rybczyński 2008b).

Influenced by the discovery of Nakano et al. (1995) and Takahata et al. (1995) that TDZ is a more effective cytokinin than BAP for culturing mesophyll protoplast of *G. triflora* and*G. triflora* × *G. scabra*, as well as by our previous work on *G. decumbens* (Tomiczak et al.2015), we compared the influence of TDZ and BAP on cell division efficiency in protoplast cultures of *G. cruciata*, *G. kurroo*, *G. septemfida*, and *G. tibetica*. TDZ, a substituted phenylurea compound, induces a diverse array of cultural responses ranging from callus formation to regeneration of shoots and somatic embryos (Murthy et al. 1998). For four gentian species studied in this work, a greater percentage of dividing cells was obtained when protoplasts were cultured on medium containing TDZ than when grown on medium supplemented with BAP.

Browning of the protoplast cultures of *G. cruciata* and *G. septemfida*, which was observed within 2 weeks from culture initiation, is thought to be caused by oxidation of phenolic compounds released from plant cells into the culture medium (Zhu et al. 1997). This phenomenon was also described by Takahata and Jomori (1989); Jomori et al. (1995); Kunitake et al. (1995) and Nakano et al. (1995), and indicates that it is one of the major problems in protoplast cultures of *Gentianaceae*. In the case of *G. cruciata* and *G. septemfida*, weekly medium refreshment was insufficient to prevent cell browning. Whether the addition of activated charcoal or polyvinylpyrrolidone (PVP) is beneficial for these species, as in *Eustoma grandiflorum* (Kunitake et al. 1995) or *Vitis vinifera* (Zhu et al. 1997), would be worth investigating. Another possibly valuable modification would be conditioned (Chen et al. 2004; Komai et al. 2006) or nurse type of protoplast culture (Zhou et al. 2005).

The influence of culture conditions on callus proliferation and plant regeneration

The positive effect of TDZ on proliferation of protoplast-derived callus of *G. kurroo* and *G. tibetica* was manifested in two ways. Firstly, a greater volume of callus tissue was obtained from agarose beads cultured previously on TDZ-supplemented PCM1 medium, than from beads cultured on PCM2 containing BAP. Very similar effects were observed during callus formation derived from mesophyll protoplasts of *G. decumbens* (Tomiczak et al. 2015). Secondly, it transpired that CPM1 medium is the most appropriate for callus proliferation of *G. kurroo* and is almost as good as CPM3 for *G. tibetica*. CPM3 medium containing 1.0 mg l^{-1} dicamba, 2.0 mg l^{-1} BAP, 0.1 mg l^{-1} NAA and 80 mg l^{-1} adenine sulfate is an universal medium typically used for the initiation and maintenance of cell suspension cultures of *G. tibetica*, *G. cruciata* and *G. pannonica* (Mikuła et al. 2002, 2005).

Publications involving the development of a protoplast-to-plant system for *Gentiana*mesophyll protoplasts available pre-2015 described only one way of plant regeneration i.e. shoot organogenesis (Takahata and Jomori 1989; Nakano et al. 1995). Regeneration of shoots was achieved on media containing high concentrations (5–10 mg l^{-1}) of cytokinins (TDZ or BAP) in combination with 0.1 mg l^{-1} NAA (Nakano et al. 1995) or 1.0 mg l^{-1} IAA (Takahata and Jomori 1989). Recently, we succeeded in regenerating plants from leaf mesophyll protoplasts of *G. decumbens* by indirect somatic embryogenesis (Tomiczak et al. 2015). PRM3 medium containing 1.0 mg l^{-1} kinetin, 0.5 mg l^{-1} GA_3, 80 mg l^{-1} adenine sulfate, which led to the formation of somatic embryos from protoplast-derived tissues of*G. decumbens*, was the same medium as was widely used for the induction of somatic embryogenesis from callus and cell suspensions of *G. pannonica*, *G. cruciata*, *G. tibetica*(Mikuła and Rybczyński 2001; Mikuła et al. 2002), and *G. kurroo* (Fiuk and Rybczyński2008a), as well as from protoplasts of *G. kurroo* cell suspensions (Fiuk and Rybczyński2007). The greatest number of somatic embryos regenerated from mesophyll protoplast-derived callus of *G. tibetica* and *G. kurroo* was also achieved by using this medium. Shoot organogenesis was also observed in *G. kurroo* cultures, but only on PRM2 medium supplemented with a high concentration (6.0 mg l^{-1}) of BAP. Interestingly, single somatic embryos were obtained on PRM1 medium, containing 8.0 mg l^{-1} TDZ. However, as reviewed by Murthy et al. (1998), both traditional cytokinin and auxin-type responses to TDZ in tissue culture medium have been reported. Induction of highly morphogenic callus followed by regeneration of shoots and somatic embryos has been documented for a wide variety of plant species and for different explant types (Tegeder et al. 1995; Gill and Ozias-Akins 1999; Zhang et al. 2001; Mithila et al. 2003; Jones et al. 2007; Thomas 2007).

Ploidy of Cells of Callus and Regenerated Plants

Young leaves of *Gentiana* seem to provide adequate explant material for initiation of cytogenetically stable cell cultures, as they do not contain endopolyploid cells. Most gentian leaf mesophyll cell nuclei become arrested in the G_1/G_0 phase of the cell cycle (2C DNA content), only a small fraction of G_2 nuclei were detected (Mikuła et al. 2008; Fiuk et al. 2010). However, the conditions for protoplast isolation and culture, as well as an indirect method of plant regeneration (through long-term callus phase), can induce somaclonal variation (known as protoclonal variation in this case), causing numerical and structural chromosome changes, and consequently leading to cell aneuploidy and/or polyploidy (Karp et al. 1982; Karp 1994). This variation is especially evident in protoplast cultures, where a high frequency of errors occurs during mitosis. It was confirmed in the present experiment that extensive variation in cell ploidy occurs in protoplast-derived callus. This variation was especially evident in *G. kurroo*; apart from endoreduplication, resulting in the presence of 8C and 16C cells and an increased proportion of 4C cells, other chromosome disturbances occur, precipitating the formation of aneuploid cells having a DNA content ranging from 5C to 6C. Also, in the tetraploid species *G. tibetica*, probably some errors occurred during early divisions, with the result that a considerable proportion of cells had a DNA content of less than 2C. Similarly, in protoplast cultures of potato, endoreduplication, polyploidy, and aneuploidy occurred very early during the initial stages of protoplast development and callus induction. An increase in the nuclear DNA content and chromosome number was observed as early as the first 2 weeks of protoplast culture. This process continued in the callus and finally, when regeneration occurred, led to polyploidy or aneuploidy in regenerated plants (Sree Ramulu et al. 1984, 1989). The proportion of polyploid/aneuploid regenerants to true-to-type plants can vary from single aneuploid and polyploid regenerated plants like it was noted for *Pelargonium* × *hortorum*(Nassour et al. 2003) or *Eustoma grandiflorum* (Lindsay et al. 1994), through almost half of the regenerants having a ploidy value different from that of the mother plants (*Fragaria* × *ananassa*; Nyman and Wallin 1992), to only polyploid plants being obtained from protoplast culture (*G. decumbens;* Tomiczak et al. 2015). In the present research, over 85 % of *G. tibetica* regenerants possessed a genome of the same size as that of control plants, and only 14 % were polyploid. There was no regeneration from the cells having a DNA content lower than 2C, as also observed with other species (Elmaghrabi and Ochatt2006). However, in *G. kurroo*, the majority of regenerated plants (85 %) contained double the amount of DNA as control plants. This was probably due to the high proportion of 4C cells and the low proportion, or even lack of 2C nuclei in the calli of this species (as shown in Fig. 2c, d, f). By contrast, Nontaswatsri and Fukai (2005) reported a greater ability to regenerate in diploid cells compared with tetraploids and octoploids, resulting in preferential regeneration of diploid plants from mixoploid calli of *Dianthus* hybrid. The greater totipotency in vitro of diploid cells was also confirmed for *Cucumis sativus*(Kubaláková et al. 1996), *Pisum sativum* (Ochatt et al. 2000), and *Medicago truncatula*(Elmaghrabi and Ochatt 2006). The presence of aneuploid cells (5–6C) in some calli of *G. kurroo* in the present study resulted in the regeneration of a single plant, which, however, showed abnormal morphology and did not survive. In our previous studies on *G. kurroo,*about 30–37 % polyploids (tetraploids and hexaploids) was regenerated from cell suspension-derived protoplasts (Fiuk and Rybczyński 2007), and 14 % tetraploids from hypocotyl-derived cell suspension (Fiuk and Rybczyński 2008a).

According to the literature, *G. kurroo* is a diploid species having 26 chromosomes (Kaur et al. 2009; Yuan et al. 1998), and *G. tibetica* is a tetraploid possessing 52 chromosomes (Mishiba et al. 2009; Yuan et al. 1998). In our study, the chromosome number of control plants of both species was consistent with previously reported data, and regenerants with double the amount of DNA also possessed twice the number of chromosomes. Furthermore, we observed the typical correlation that length of stomata increased and the number of stomata decreased with increasing ploidy, as was previously reported for *G. decumbens* (Tomiczak et al. 2015) and for many other genera and species (reviewed by Knight and Beaulieu 2008). On the other hand, the relationships between ploidy and morphology of gentians regenerated both in this and a previous study (Tomiczak et al.2015) comply with the general statement that there is an optimum for ploidy increase, beyond which the individuals become dwarf or less vigorous (Müntzing 1936). Gentian plants with a ploidy level close to 6*x* and 8*x* exhibited dwarf stature, thick and wrinkled leaves, as well as shorter and thicker roots than diploid and tetraploid plants. This can indicate that hexaploidy and octoploidy are beyond optimal values for *Gentiana* species.

CONCLUSIONS

In conclusion, we have compared the morphogenic potential of five *Gentiana* species in green leaf mesophyll protoplast culture and succeeded in the indirect plant regeneration of plants of *G. kurroo* and *G. tibetica*. Polyploidy detected amongst regenerants of both species had its origin in the various ploidy levels of protoplast-derived callus cells. The protoplast-to-plant system developed in this study may provide the prerequisite for creating novel genotypes of these valuable medicinal species by means of genetic manipulations and somatic cell hybridization.

ACKNOWLEDGMENTS

The authors wish to thank Aleksandra Chomacka for her technical assistance in flow cytometric analysis. This work was supported by the Polish Ministry of Science and Higher Education (Grant No. 3P04C 037 23).

REFERENCES

1. Behera MC, Raina R (2012) *Gentiana kurroo*—a critically endangered bitter herb. Int J Med Aromat Plants 2(1):22–29
2. Cai YF, Liu YL, Liu ZH, Zhang F, Xiang FN, Xia GM (2009) High frequency embryogenesis and regeneration of plants with high content of gentiopicroside from the Chinese medicinal plant *Gentiana straminea* Maxim. In Vitro Cell Dev Biol Plant 45:730–739
3. Chen L-P, Zhang M-F, Xiao Q-B, Wu J-G, Hirata Y (2004) Plant regeneration from hypocotyl protoplasts of red cabbage (*Brassica oleracea*) by using nurse cultures. Plant Cell Tiss Org Cult 77:133–138
4. Elmaghrabi A, Ochatt S (2006) Isoenzymes and flow cytometry for the assessment of true-to-typeness of calluses and cell suspensions of barrel medic prior to regeneration. Plant Cell Tiss Org Cult 85:31–43
5. Fiuk A, Rybczyński JJ (2007) The effect of several factors on somatic embryogenesis and plant regeneration in protoplast cultures of *Gentiana kurroo* (Royle). Plant Cell Tissue Org Cult 91:263–271
6. Fiuk A, Rybczyński JJ (2008a) Factors influencing efficiency of somatic embryogenesis of*Gentiana kurroo* (Royle) cell suspension. Plant Biotechnol Rep 2:33–39
7. Fiuk A, Rybczyński JJ (2008b) Genotype and plant growth regulator-dependent response of somatic embryogenesis from *Gentiana* leaf explants. In Vitro Cell Dev Biol Plant 44:90–99
8. Fiuk A, Bednarek PT, Rybczyński JJ (2010) Flow cytometry, HPLC-RP, and metAFLP analyses to assess genetic variability in somatic embryo-derived plantlets of *Gentiana pannonica* Scop. Plant Mol Biol Rep 28:413–420
9. Frearson EM, Power JB, Cocking EC (1973) The isolation, culture and regeneration of*Petunia* leaf protoplasts. Dev Biol 33:130–137
10. Gill R, Ozias-Akins P (1999) Thidiazuron-induced highly morphogenic callus and high frequency regeneration of fertile peanut (*Arachis hypogaea* L.) plants. In Vitro Cell Dev Biol Plant 35:445–450
11. Grèzes J, Thomas D, Thomasset B (1994) Factors influencing protoplast isolation from*Coffea arabica* cells. Plant Cell Tissue Org Cult 36:91–97
12. Hayta S, Akgun IH, Ganzera M, Bedir E, Gurel A (2011) Shoot proliferation and HPLC-determination of iridoid glycosides in clones of *Gentiana cruciata* L. Plant Cell Tissue Org Cult 107:175–180
13. Hosokawa K, Oikawa Y, Yamamura S (1998) Mass propagation of ornamental gentian in liquid medium. Plant Cell Rep 17:747–751
14. Hu X, Yin Y, He T (2015) Plant regeneration from protoplasts of *Gentiana macrophylla*Pall. using agar-pool culture. Plant Cell Tissue Org Cult 121:345–351
15. Jomori H, Takahata Y, Kaizuma N (1995) Plant regeneration from leaf-derived calli of gentians and their protoplast culture. Acta Hort 392:81–86
16. Jones MPA, Yi Z, Murch SJ, Saxena PK (2007) Thidiazuron-induced regeneration of*Echinacea purpurea* L.: micropropagation in solid and liquid culture systems. Plant Cell Rep 26:13–19
17. Karp A (1994) Origin, causes and uses of variation in plant tissue cultures. In: Vasil IK, Thorpe TA (eds) Plant, cell and tissue culture. Kluwer Academic Publishers, Dordrecht, pp 139–151
18. Karp A, Nelson RS, Thomas E, Bright SWJ (1982) Chromosome variation in protoplast-derived potato plants. Theor Appl Genet 63:265–272
19. Kaur R, Panwar N, Saxena B, Raina R, Bharadwaj SV (2009) Genetic stability in long-term micropropagated plants of *Gentiana kurroo*—an endangered medicinal plant. J New Seeds 10:236–244
20. Knight CA, Beaulieu JM (2008) Genome size scaling through phenotype space. Ann Bot 101:759–766
21. Köhlein F (1991) Gentians. Timber Press, Portland
22. Komai F, Morohashi H, Horita M (2006) Application of nurse culture for plant regeneration from protoplasts of *Lilium japonicum* Thunb. In Vitro Cell Dev Biol Plant 42:252–255
23. Kubaláková M, Doležel J, Lebeda A (1996) Ploidy instability of embryogenic cucumber (*Cucumis sativus* L.) callus culture. Biol Plant 38:475–480
24. Kunitake H, Nakashima T, Mori K, Tanaka M, Mii M (1995) Plant regeneration from mesophyll protoplasts of lisianthus (*Eustoma grandiflorum*) by adding activated charcoal into protoplast culture medium. Plant Cell Tissue Org Cult 43:59–65
25. Larkin PJ (1976) Purification and viability determinations of plant protoplasts. Planta 128:213–216
26. Lemontey C, Mousset-Déclas C, Munier-Jolain N, Boutin JP (2000) Maternal genotype influences pea seed size by controlling both mitotic activity during early embryogenesis and final endoreduplication level/cotyledon cell size in mature seed. J Exp Bot 51:167–175
27. Lindsay GC, Hopping ME, O'Brien IEW (1994) Detection of protoplast-derived DNA tetraploid Lisianthus (*Eustoma grandiflorum*) plants by leaf and flower characteristics and by flow cytometry. Plant Cell Tissue Org Cult 38:53–55
28. Marie D, Brown SC (1993) A cytometric exercise in plant DNA histograms, with 2C values for 70 species. Biol Cell 78:41–51
29. Meng Y, Gao Y, Jia J (1996) Plant regeneration from protoplasts isolated from callus of*Gentiana crassicaulis*. Plant Cell Rep 16:88–91
30. Mikuła A, Rybczyński JJ (2001) Somatic embryogenesis of *Gentiana* genus I. The effect of the preculture treatment and primary explant origin on somatic embryogenesis of*Gentiana cruciata* (L.), *G. pannonica* (Scop.), and *G. tibetica* (King). Acta Physiol Plant 23:15–25
31. Mikuła A, Skierski J, Rybczyński JJ (2002) Somatic embryogenesis of *Gentiana* genus III. Characterization of three-year-old embryogenic suspensions of *G. pannonica* originated from various seedling explants. Acta Physiol Plant 24:311–322

32. Mikuła A, Rybczyński JJ, Skierski J, Latkowska MJ, Fiuk A (2005) Somatic embryogenesis of *Gentiana* genus IV: Characterization of *Gentiana cruciata* and *Gentiana tibetica*embryogenic cell suspensions. In: Hvolsef-Eide AK, Preil W (eds) Liquid culture systems for in vitro plant propagation. Springer, Dordrecht, pp 345–358

33. Mikuła A, Olas M, Sliwinska E, Rybczyński JJ (2008) Cryopreservation by encapsulation of *Gentiana* spp cell suspensions maintains regrowth, embryogenic competence and DNA content. CryoLetters 29:409–418

34. Mishiba K, Nishihara M, Abe Y, Nakatsuka T, Kawamura H, Kodama K, Takesawa T, Abe J, Yamamura S (2006) Production of dwarf potted gentian using wild-type *Agrobacterium rhizogenes*. Plant Biotechnol 23:33–38

35. Mishiba K, Yamane K, Nakatsuka T, Nakano Y, Yamamura S, Abe J, Kawamura H, Takahata Y, Nishihara M (2009) Genetic relationships in the genus *Gentiana* based on chloroplast DNA sequence data and nuclear DNA content. Breed Sci 59:119–127

36. Mithila J, Hall JC, Victor JMR, Saxena PK (2003) Thidiazuron induces shoot organogenesis at low concentrations and somatic embryogenesis at high concentrations on leaf and petiole explants of African violet (*Saintpaulia ionantha* Wendl.). Plant Cell Rep 21:408–414

37. Momčilović I, Grubišić D, Nešković M (1997) Micropropagation of four *Gentiana* species (*G. lutea*, *G. cruciata*, *G. purpurea* and *G. acaulis*). Plant Cell Tissue Org Cult 49:141–144

38. Müntzing A (1936) The evolutionary significance of autopolyploidy. Hereditas 21:263–378

39. Murashige T, Skoog F (1962) A revised medium for rapid growth and bioassays with tobacco tissue cultures. Physiol Plant 15:473–497

40. Murthy BNS, Murch SJ, Saxena PK (1998) Thidiazuron: a potent regulator of in vitro plant morphogenesis. In Vitro Cell Dev Biol Plant 34:267–275

41. Nakano M, Hosokawa K, Oomiya T, Yamamura S (1995) Plant regeneration from protoplasts of *Gentiana* by embedding protoplasts in gellan gum. Plant Cell Tiss Org Cult 41:221–227

42. Nakatsuka T, Mishiba K, Kubota A, Abe Y, Yamamura S, Nakamura N, Tanaka Y, Nishihara M (2010) Genetic engineering of novel flower colour by suppression of anthocyanin modification genes in gentian. J Plant Physiol 167:231–237

43. Nassour M, Chassériaux G, Dorion N (2003) Optimization of protoplast-to-plant system for *Pelargonium* × *hortorum* 'Alain' and genetic stability of the regenerated plants. Plant Sci 165:121–128

44. Nontaswatsri C, Fukai S (2005) Regenerative callus of *Dianthus* 'Telstar Scarlet' showing mixoploidy produce diploid plants. Plant Cell Tiss Org Cult 83:351–355

45. Nyman M, Wallin A (1992) Improved culture technique for strawberry (*Fragaria* × *ananassa* Duch.) protoplasts and the determination of DNA content in protoplast derived plants. Plant Cell Tiss Org Cult 30:127–133

46. Ochatt SJ, Mousset-Declas C, Rancillac M (2000) Fertile pea plants regenerate from protoplasts when calluses have not undergone endoreduplication. Plant Sci 156:177–183

47. Qureshi RA, Gilani SA, Ashraf M (2007) Ethnobotanical studies with special reference to plant phenology at Sudhan Gali and Ganga Chotti Hills (District Bagh, A.K.). EJEAFChe 6(7):2207–2215

48. Singh A (2008) Phytochemicals of *Gentianaceae*: a review of pharmacological properties. Int J Pharm Sci Nanotechnol 1(1):33–36

49. Sliwinska E, Zielinska E, Jedrzejczyk I (2005) Are seeds suitable for flow cytometric estimation of plant genome? Cytometry 64A(2):72–79

50. Sree Ramulu K, Dijkhuis P, Roest S, Bokelmann GS, De Groot B (1984) Early occurence of genetic instability in protoplast cultures of potato. Plant Sci Lett 36:79–86

51. Sree Ramulu K, Dijkhuis P, Roest S (1989) Patterns of phenotypic and chromosome variation in plants derived from protoplast cultures of monohaploid, dihaploid and diploid genotypes and in somatic hybrids of potato. Plant Sci 60:101–110

52. Suzuki M, Akihama T, Ishikawa M (2005) Cryopreservation of encapsulated gentian axillary buds following 2 step-preculture with sucrose and desiccation. Plant Cell Tiss Org Cult 83:115–121

53. Takahata Y, Jomori H (1989) Plant regeneration from mesophyll protoplasts of gentiana (*Gentiana scabra* Bungei). Plant Tiss Cult Lett 6:19–21

54. Takahata Y, Jomori H, Miyano S, Kunitake H, Mii M (1995) Regeneration of plants from protoplasts of *Gentiana* species (Gentian). In: Bajaj YPS (ed) Biotechnology in agriculture and forestry, vol 34., Plant protoplasts and genetic engineering VISpringer-Verlag, Berlin, pp 57–62

55. Tanaka D, Niimo T, Isuzugawa K, Hikage T, Uemura M (2004) Cryopreservation of shoot apieces of in vitro grown gentian plants: comparison of vitrification and encapsulation-vitrification protocols. CryoLetters 25:167–176

56. Tegeder M, Gebhardt D, Scheider O, Pickardt T (1995) Thidiazuron-induced plant regeneration from protoplasts of *Vicia faba* cv. Mythos. Plant Cell Rep 15:164–169

57. Thomas TD (2007) Pretreatment in thidiazuron improves the in vitro shoot induction from leaves in *Curculigo orchioides* Gaertn., an endangered medicinal plant. Acta Physiol Plant 29:455–461

58. Tiwari RK, Trivedi M, Guang ZC, Guo G-Q, Zheng G-C (2007) Genetic transformation of*Gentiana macrophylla* with *Agrobacterium rhizogenes*: growth and production of secoiridoid glucoside gentiopicroside in transformed hairy root cultures. Plant Cell Rep 26:199–210

59. Tomiczak K, Mikuła A, Sliwinska A, Rybczynski JJ (2015) Autotetraploid plant regeneration by indirect

somatic embryogenesis from leaf mesophyll protoplasts of diploid*Gentiana decumbens* L.f. In Vitro Cell Dev Biol Plant 51:350–359

60. Wójcik A, Rybczyński JJ (2015) Electroporation and morphogenic potential of *Gentiana kurroo* (Royle) embryogenic cell suspension protoplasts. BioTechnologia 96:19–29
61. Wright DC (1985) Factors affecting isolation of protoplasts from leaves of grape (*Vitis vinifera*). Plant Cell Tiss Org Cult 4:95–100
62. Yuan Y-M, Küpfer P, Zeltner L (1998) Chromosomal evolution of *Gentiana* and *Jaeschkea*(Gentianaceae), with further documentation of chromosome data for 35 species from western China. Plant Syst Evol 210:231–247
63. Zhang C-L, Chen D-F, Elliott MC, Slater A (2001) Thidiazuron-induced organogenesis and somatic embryogenesis in sugar beet (*Beta vulgaris* L.). In Vitro Cell Dev Biol Plant 37:305–310
64. Zhao ZL, Dorje G, Wang ZT (2010) Identification of medicinal plants used as Tibetan Traditional Medicine Jie-Ji. J Ethnopharm 132:122–126
65. Zhou J, Wang B, Zhu L (2005) Conditioned culture for protoplasts isolated from chrysanthemum: an efficient approach. Colloids Surf B 45:113–119
66. Zhu YM, Hoshino Y, Nakano M, Takahashi E, Mii M (1997) Highly efficient system of plant regeneration from protoplasts of grapevine (*Vitis vinifera* L.) through somatic embryogenesis by using embryogenic callus and activated charcoal. Plant Sci 123:151–157

CHAPTER 14

THE IN VITRO PROPAGATION TECHNIQUES FOR PRODUCING BANANA USING SHOOT TIP CULTURES

Munguatosha Ngomuo[1], Emerald Mneney[2], Patrick A. Ndakidemi[1]

[1] School of Life Sciences, Nelson Mandela African Institute of Science and Technology, Arusha, Tanzania

[2] Mikocheni Agricultural Research Institute, Dar es Salaam, Tanzania

ABSTRACT

Banana is an important food crop and the second most important fruit crop. Despite the significant commercial value of the crop, the main production constrain is the availability of reliable and safe planting material. The planting materials obtained through conventional methods (suckers) do not meet the increasing demand for planting and they are of poor quality. Tissue culture is the approach which can solve these problems. Micro propagation of the crop is also faced with challenges which need to be addressed in order to improve its production. Some of the problems which hinder the success of the crop include oxidative browning of the wounded tissues and low number of shoots produce per explant. This review highlights the challenges encountered in tissue culture of banana and explores the in vitro propagation techniques by using shoot tip cultures of banana as the possibilities to overcome these problems.

INTRODUCTION

Banana (Musa accuminata L.) belongs to the Musaceae family. Banana provides millions of people throughout the tropics and subtropics with staple food and accounts for one of the most widely exported fruits in the world. Presently, banana is grown in around 150 countries across the world on an area of 4.84 million ha, producing 95.6 million tonnes [1]. It accounts for approximately 22% of the fresh fruit production and as the second most important fruit crop; Africa accounts for 35% of the total world production [2]. It is one of the oldest fruits known to mankind and is the most delicious fruit used as subsidiary food. It is used in different ways such as for table purpose as well as culinary fruit; its leaves and stems are chopped and used as cattle feed. Some species of banana yield fibre, which is used for making ropes. The tip of inflorescence is cooked as a vegetable in some places [3]. The plant is also used for decoration purpose in wedding, festivals and fairs. It is also used as raw material in industries for preparation of banana powder, chips, juices and beer. The juice of banana stem is used in making paper bond, tissue paper etc. Despite this important use of banana, its production has been decreasing since nineteen seventies. In north western parts of Tanzania in Kagera region, production decreased from 10 t to 4 t per Hectare due to declining soil fertility and emergence of pests and diseases [4]. This is very low production compared with banana production in other parts of the world like India where the production was on an average of 26.7 t per Hectare. Banana plants are usually propagated by vegetative means by using suckers which grow from lateral buds originating from corms, and suckers are used for production of individual plants. In some instances, complete or separated corms with one or

Citation: Munguatosha Ngomuo, Emerald Mneney, Patrick A. Ndakidemi, The in Vitro Propagation Techniques for Producing Banana Using Shoot Tip Cultures, http://dx.doi.org/10.4236/ajps.2014.511175.

several buds may be used. This process is very slow as the rate of multiplication of suckers through conventional vegetative means has been found to express several negative impacts which include transmission of diseases, low production and poor preservation of original plant genetic material [5]. The non-professional cultivation practices, pest epidemic and viral diseases affect the yield and quality of banana crop [6]. The problem of emerging diseases can be solved by propagating banana through tissue culture [7]. Tissue culture (TC) technology offers mass propagation and clean planting material. In vitro banana production technology is a superior technology over traditional method (Sucker-propagated) of banana production with respect to optimal yield, uniformity, disease-free planting material and true to type plants. Mass multiplication of tissue culture plants could be done in a short time. They are cheaper to transport than conventional suckers and the coupling with virus indexing allows for safe movement, and exchange and conservation of germplasm. In addition, bananas produced using the TC technology are reported to be more vigorous, higher yielding and produce better quality fruits than those produced by conventional means [8]. For commercialization, it is necessary that consistent supplies of good quality bananas are produced to meet the increasing demand. The high rate of multiplication of genetically uniform, pest and disease free planting material can be achieved through tissue culture. Tissue culture approaches would also allow for off-season production of fast growing plants compared with conventionally propagated ones [9]. The in vitro propagation technique for producing banana using shoot tip cultures is a necessary approach in dealing with the problems encountered in tissue culture. These techniques will ensure sustainable production of banana planting materials.

EFFECTS OF MEDIA COMPOSITION ON GROWTH AND DEVELOPMENT OF BANANA SHOOT TIPS

The ingredients of plant tissue culture media can be categorized as inorganic salts, organic compounds, complex natural preparations and inert supportive materials [10]. The successes with plant cell and organ cultures have been depended on use of appropriate nutrient media. By providing the necessary chemicals in good combinations and forms, it has been possible to establish cultures from virtually every plant part [10]. Inorganic macronutrient and micronutrient levels used in most plant tissue culture media are based on levels established in the plant tissue culture medium developed by [11] for tobacco tissue culture "MS medium" [12]. As such no single medium can be suggested as being entirely satisfactory for all types of plant tissues and organs [13]. Murashige and Skoog [11], is the most widely used plant culture medium [14]. Several media formulations have been reported for banana shoot tip culture with slight modifications of MS media [15]. Other popular media include B5 [16], SH (Schenk and Hildebrant [17], N6 [18] and Linsmaier and Skoog (LS) [19] media [20] [21]. The MS medium of Murashige and Skoog (1962) is a salt composition that supplies the needed macro and micronutrients. In order to achieve growth and differentiation of cells or tissues, concentrations of inorganic nutrients must be optimized such that the medium meets the requirements of the cells or tissues used [22]. The active factor in the medium is the ions of different types rather than the compounds. One type of ion can be contributed by more than one compound, example NO3− ions may be contributed by NH4NO3 as well as KNO3 [13]. Micro propagation of bananas in East Africa with the aim of improved or suitable growth media started recently [23]. However, the success in these efforts is limited. For example, Maerere et al. (2003) successfully managed to (develop a protocol for cv. Bukoba and Uganda) in Tanzania but the field performance of these varieties is not yet evaluated. Optimal growth and morphogenesis of tissues may vary from plant to plant based on their nutritional requirements. The source of explants from different parts of same plants may also have different requirements for satisfactory growth [21]. Therefore a need for developing a media with suitable composition for a specific banana cultivar is important in ensuring successful in vitro growth and development of each cultivar

THE ROLE OF VITAMINS IN BANANA TISSUE CULTURE

Although the basis of all nutrient media is a composition of essential nutrients [24], vitamins are required in trace amounts to serve the catalytic functions in enzyme systems [22]. Some plants can synthesize the essential vitamins for their growth. However when plants are grown in vitro, vitamins may act as limiting factors for cell growth and differentiation [25]. Other studies have confirmed that the vitamins thiamine and nicotinic acid affected cellular division in the pea root meristem [26] [27]. These vitamins worked in combination, i.e. both the presence of thiamine and nicotinic acid promoted the root growth. The four vitamins; myo-inositol, thiamine, nicotinic acid, and pyridoxine are ingredients of Murashige and Skoog (1962) medium and have been used in varying proportions for the culture of tissues of many plant species [28]. The requirements of cells for added vitamins vary according to the nature of the plant and the type of culture [29]. This makes it impossible to conclude that all the vitamins which have been used in a particular experiment were essential or they will work for another experiment. In banana, vitamins supplementation to the tissue culture media has been scarcely studied and often research workers tend to adopt a "belt and braces" attitude to minor media components, such as vitamins and add unusual supplements just to ensure that there is no missing factor which will limit the success of their experiment. Sometimes complex mixtures of as many as nine or ten vitamins have been employed [29] [30]. Therefore, there is a need to investigate the optimum amount of commonly used vitamins for each banana variety in tissue culture.

THE USE OF ANT OXIDANT AND ACTIVATED CHARCOAL IN REDUCING OXIDATIVE BROWNING IN BANANA TISSUE CULTURE

Banana contains constituents of phenolic enzymes principally polyphenoloxidase enzymes. The polyphenoloxidase enzymes serve as a very important phyto auxine in banana and help to defend the plant against infection from fungi, viruses and bacteria when injured [31]. The constituent of phenols in Musa spp. are principally doparnine, catechin, chlorogenic acid, cinnamic acid, hydroxylbenzoic, resorcinol, progallic acid, salicylic acid, ferulic acid, vanillin coumurin, P-coumaric acid and phenol [32]. Browning reactions and astringency of the fruit are caused by phenolic compounds, and are responsible for high mortality rate (lethal browning) in tissue culture [30]. This process is initiated by browning of the surface of plant tissues due to the oxidation of phenolic compounds resulting in the formation of quinines which are highly reactive and toxic to plant tissue [30] [33]. Apart from being an important group of secondary metabolites, phenolics may act as modulators of plant development by regulating indole acetic acid (IAA) catabolism [34]. They also play effective role in plant growth regulation, cell differentiation and organogenesis [35]. Their concentration is often affected by several internal and external factors example nutrients [36] [37]. Other factors include stress factors such as drought, water, radiation and pathogen infection from injured surfaces, which directly affects the concentrations of phenolics in plants [38]. Activated charcoal is used in control of lethal browning. Its use in micro propagation was first reported by Lu et al. (1990) and it was suggested that charcoal was responsible for adsorption and desorption which controlled the release of nutrients in the production of synthetic seeds [39]. Activated charcoal has a very fine network of pores with large inner surface area on which many substances can be adsorbed [40]. Activated charcoal (AC) is often used in tissue culture to improve cell growth and development. The main benefit is its adsorption of inhibitory substances in culture media. The phenolic oxidation or brown exudates accumulation can be significantly reduced by activated charcoal [41] [42]. However, AC can also adsorb vitamins, cytokinins and auxins, thus changing the ratios of medium components and subsequently influencing plant regeneration [43]. This makes the researchers using AC unaware of the actual quantity available to the plant tissues. Obtaining target levels of adequate free exogenous hormone may be difficult [42]. Control of lethal browning in tissue culture of banana has been reported by different studies. Onuoha et al. [31], reported the use of ant oxidant potassium citrate and citrate (K-C: C) in prevention of browning in plantain culture. As an antioxidant, potassium citrate-citrate reduced browning within two hours before culturing the tissues. Ascorbic acid is also an antioxidant used to control oxidation of phenols [44]-[46]. It is able to scavenge oxygen radicals produced when plant tissue is wounded, thus protecting the cells from the damage resulting from the injury. The detoxification of the free radicals by AA produced through oxidation of the phenolic compounds reduces the extent of browning [33]. In controlling of lethal browning in faba beans, [47] reported that the use of activated charcoal, ascorbic acid, cystene and silver nitrate had a significant effect on the number of shoots and the length of shoots regenerated per explants. [48] reported that addition of cysteine to the growth media reduced explants blackening in banana tissue culture. Understanding the processes contributing to the oxidation of phenols and how these can be minimized when initiating banana tissue culture is critical for successful in vitro culture, not only of banana but also of other crops. Among the banana varieties susceptible to tissue browning, elimination or reduction of this process is necessary requirement for successful culture establishment. Development of suitable and efficient treatment to minimize tissue browning of these varieties by focusing on suitable antioxidant, the optimum concentration and method of application during explants preparation is of paramount importance.

THE ROLE OF GROWTH REGULATORS (AUXINS AND CYTOKININ) ON GROWTH AND DEVELOPMENT OF BANANA SHOOT TIPS

Growth regulators play a key role for developing a specific mode of growth in the cultured cells or tissues, which may be due to accumulation of specific biochemical contents in them. The single or combination of different hormones in the medium causes maintenance of specific and balanced inorganic and organic contents in the growing tissue. This leads the cells or tissues to develop either into shoots/ or roots or even death [49]. In tissue culture, plant growth regulators are important media components in determining the development and developmental pathway of the plant cells. Growth regulators are used in different proportions to break dormancy and enhance shoot formation as the apical dormancy is under control of these growth regulators [50]. The cytokinins and auxins are of importance in in vitro culture as the later are concerned with root formation, the former is mainly required in the media for shoot formation and growth of buds [30]. These growth regulators are required in combination in the media as it is always the manipulation and variation of auxins and cytokinins levels that can successfully change the growth behavior of plant cultures [51]. Cytokinins such as benzyl aminopurine (BAP) and kinetin are known to reduce the apical meristem dominance and induce both auxiliary and adventitious shoot formation from meristematic explants in banana [52]. However, the application of higher BAP concentrations inhibits elongation of adventitious meristems and the conversion into complete plants [53]. Auxins and other growth regulator such as gibberellins play important roles in the growth and differentiation of cultured cells and tissues [54] [55]. Auxins such as Naphtalene acetic acid (NAA) have been reported to promote plant rooting in vitro [5] [14]. The use of plant growth regulators (PGRs) in plant nutrient media for in vitro culture depends on plant tissue growth stage and expected end product. Cytokinins play an important role in buds formation in vitro. However buds proliferation in vitro is influenced by apical dominance which is controlled by various growth regulators [56] [57]. Genotype of given plant species determines its buds proliferation in vitro. Apart from the influence of genotypes, shoot proliferation rate and elongation are influenced by cytokinin types and their concentration. Adenine-based cytokinins are used in several Musa spp. for in-vitro propagation [58]. N6-benzylaminopurine (BAP) is the most commonly preferred cytokinin [14]. The others are isopentyladenine (2-ip), zeatin and kinetin [59]. The concentration of exogenous cytokinin appears to be the main factor affecting multiplication. Many studies have reported the use of auxins and cytokinin in tissue culture. Gubbuk and Pekmzci [58], reported that moderate concentrations of cytokinins increased the shoot proliferation rate, but very high concentrations decreased multiplication and especially depressed shoot elongation [58]. Also they reported higher shoot proliferation and elongation with Thidiazuron (TDZ) than with BAP. However, BAP above 20 μM and TDZ over 2 μM decreased shoot elongation. The use of TDZ is known to inhibit shoot elongation. In a study conducted by Lee (2004), it was found that TDZ at 0.91 μM induced the largest number of shoots, but at higher concentration of TDZ (9.1 μM), elongation of shoots was inhibited and clumps of small globular buds appeared at the base of shoots [60]. In a study on effects of auxin/cytokinin combination on shoot proliferation on banana cultivars, [61], reported that incorporation of a strong auxin in the media suppressed the shoot proliferation rates of the banana cultivars. On media modified with low cytokinin/auxin ratios, for example 16.8/1.0 and 16.8/1.2 ZN/NAA combinations, the East African Highland banana (AAA-EA) cultivars showed single shoot development and callus induction due to apical dominance resulting from increased level of auxin concentration [61]. In another study, Buah et al. [62], demonstrated that differences existed in the relative strengths of different cytokinin types in inducing shoots. This differential ability of different hormones in inducing shoots in vitro was attributed to factors such as stability, mobility and the rate of conjugation and oxidation of hormones [62]. The concentration and combination of auxins and cytokinins in the nutrient mediums is an important factor which determines successful plant regeneration [22]. Thus for efficient in vitro propagation of banana the establishment of optimum combination of cytokinins and auxins and their interaction in a tissue culture medium for a specific cultivar is necessary.

THE EFFECT OF BUD SPLITTING ON SUPPRESSION OF APICAL DOMINANCE AND TO INDUCE MULTIPLE BUDS DEVELOPMENT

The demand for banana planting material is a major constraint to the expansion of banana production. The long time required for conventional planting material to mature and the prevalence of pests and diseases contributes to these problems. In the application of bud splitting technique, the banana plantlets of a younger physiological age (single leaf) are longitudinally split to increase the number of explants during tissue culture [5]. This is thought to promote buds proliferation by breaking the apical dominance [63]. It is established that lateral bud development in stems is under the control of the apical bud and that auxin application can sufficiently replace the stem apex in this function in as much as it has also been found that auxin stimulates C_2H_4 production in stems [64] [65]. The development of auxiliary meristems is inhibited by apical dominance [66]. These auxiliary meristems are commonly the source of buds formation in nature especially when the apical shoots are damaged or injured [67]. Poor ratooning of plantains results from the strong apical dominance exhibited by the main plant and the competition between suckers [68]. The apical dominance is controlled by growth substances released by the terminal bud, which inhibit growth of lateral buds [69]. Apical dominance is also under the control of several plant growth regulators [56] [57] [63]. Thus, sword suckers of plantains remain small until after flowering of the mother plant because their demand for assimilates (sinks) is too weak under the influence of apical dominance [70]. Bud splitting is not a common practice in banana tissue culture. Some reports on the role of apical dominance in shoot production from explants of banana are conflicting. For example, [71] confirmed that the arrest of apical dominance by removing the shoot tips was essential for the production of multiple shoot initials in cv. Cavendish. However, other studies reported that multiple initials were produced in presence of apical domes in cv. Robusta [72]. [73] reported that removal of apical dome was not essential for multiple shoots formation. On different approach to the same treatment i.e. spliting of the shoot tip; [72] reported a strongly increase in growth rate and shoot proliferation as compared with solid media on subculture of split shoot tips in liquid media. [74] found that temporary immersion of split shoot tips in liquid media for 20 minutes produced the highest multiplication rate after 24 hours [74]. On another study, [75] noticed that longitudinal cuts of buds induced a threefold increase in multiplication. The multiplication rate was also found to depend on the origin of sucker bud, where by lateral buds doubled in size within three weeks while the apical buds reached three times their size. In general, different cultivars shows variation in the degree of their shoot bud proliferation tendency and the type of multiple budding can be distinguished. This is because the multiple budding appears to be linked to genome configuration of a given cultivar [73] and apical dominance which is under control of PGRs. Thus appropriate concentration of PGRs and bud splitting technique in effort to break dormancy and induce multiple buds development is a necessary study to ensure large numbers of shoots are produced in vitro.

CONCLUSION

Banana is a staple food and it accounts for one of the most widely exported fruits in the world. Demand for planting material of banana is very high throughout the tropics and sub tropics. Despite this high demand, availability of safe and reliable planting material is a challenge facing small-scale and large-scale farmers. Propagation of banana through tissue culture is reliable solution to the problems facing farmers. The success in producing banana planting material through tissue culture is limited by the technical difficulties associated with our public laboratories among other things. From this background efficient in vitro propagation techniques are necessary to overcome these challenges. Through control of lethal brown, optimum concentration of auxins and cytokinin and arrest of apical dominance through bud splitting technique, these constrains would be solved and greatly contribute to the safe and reliable planting material of banana.

ACKNOWLEDGEMENTS

This study was funded by the Nelson Mandela African Institute of Science and Technology through research funds from Commission for Science and Technology (COSTECH) in Tanzania.

REFERENCES

1. Singh, H., Uma, S., Selvarajan, R. and Karihaloo, J. (2011) Micropropagation for production of quality banana planting material in Asia-Pacific. Asia-Pacific Consortium on Agricultural Biotechnology (APCoAB), New Delhi, Vol. 92.
2. Vuylsteke, D., Swennen, R. and De Langhe, E. (1990) Tissue Culture Technology for the Improvement of African Plantains. Sigatoka Leaf Spot Diseases of Bananas. RA Fullerton and RH Stover. INIBAP, Montpellier, 316-337.
3. Naduvinamani, R. (2007) Economics of Red Banana Production under Contract Farming in Karnataka. University of Agricultural Sciences, Bangalore.
4. Gallez, A., Runyoro, G., Mbehoma, C., Van den Houwe, I. and Swennen, R. (2004) Rapid Mass Propagation

and Diffusion of New Banana Varieties among Small-Scale Farmers in North Western Tanzania. African Crop Science Journal, 12, 7-17. http://dx.doi.org/10.4314/acsj.v12i1.27657

5. Hussein, N. (2012) Effects of Nutrient Media Constituents on Growth and Development of Banana (Musa spp.) Shoot Tips Cultured in Vitro. African Journal of Biotechnology, 11, 9001-9006.

6. Wambugu, F., Njuguna, M., Acharya, S. and Mackey, M. (2008) Socio-Economic Impact of Tissue Culture Banana (Musa Spp.) in Kenya through the Whole Value Chain Approach. International Conference on Banana and Plantain in Africa: Harnessing International Partnerships to Increase Research Impact, 879, 77-86.

7. Ali, A., Sajid, A., Naveed, N.H., Majid, A., Saleem, A., Khan, U.A., Jafery, F.I. and Naz, S. (2011) Initiation, Proliferation and Development of Micro-Propagation System for Mass Scale Production of Banana through Meristem Culture. African Journal of Biotechnology, 10, 15731-15738. http://dx.doi.org/10.5897/AJB11.2079

8. Hwan, S.C., Chen, C.L., Lin, J.C. and Lin, H.L. (1976) Cultivation of Banana Using Plantlets from Meristem Culture. HortScience, 19, 231-232.

9. Eckstein, K. and Robinson, J. (1995) Physiological Responses of Banana (Musa AAA; Cavendish Sub-Group) in the Subtropics. IV. Comparison between Tissue Culture and Conventional Planting Material during the First Months of Development. Journal of Horticultural Science, 70, 549-559.

10. Huang, L.C. and Murashige, T. (1977) Plant Tissue Culture Media: Major Constitutents, Their Preparation and Some Applications. Methods in Cell Science, 3, 539-548.

11. Murashige, T. and Skoog, F. (1962) A Revised Medium for Rapid Growth and Bio-Assays with Tobacco Tissue Cultures. Physiologia Plantarum, 15, 473-497. http://dx.doi.org/10.1111/j.1399-3054.1962.tb08052.x

12. Nassar, A.H. (2004) Effect of Some Copper Compounds on Rhizogenesis of Micropropagated Banana Shoots. International Journal of Agriculture & Biology, 6, 552-556.

13. Bhojwani, S.S. and Razdan, M.K. (1986) Plant Tissue Culture: Theory and Practice. Access Online via Elsevier.

14. Vuylsteke, D.R. (1989) Shoot-Tip Culture for the Propagation, Conservation and Exchange of Musa germplasm.

15. Browning, G., Ognjanov, V., Passey, A. and James, D. (1987) Multiple Shoot and Root Regeneration from Pear Embryo Cotyledon Explants in Vitro. Journal of Horticultural Science, 62, 305-311.

16. Gamborg, O.L., Miller, R.A. and Ojima, K. (1968) Nutrient Requirement of Suspension Cultures of Soybean Root Cells. Experimental Cell Research, 50, 151-158. http://dx.doi.org/10.1016/0014-4827(68)90403-5

17. Schenk, R.U. and Hildebrant, A.C. (1972) Medium and Techniques for Induction and Growth of Monocotyledonous and Dicotyledonous Plant Cell Cultures. Canadian Journal of Botany, 50, 199-204. http://dx.doi.org/10.1139/b72-026

18. Chu, C.C., Wang, C.C., Sun, C.S., Hsu, C., Yin, K.C., Chu, C.Y. and Bi, F.Y. (1975) Establishment of an Efficient Medium for Another Culture of Rice through Comparative Experiments on the Nitrogen Sources. Scientia Sinica, 18, 659-668.

19. Lismaier, E.M. and Skoog, F.F. (1975) Organic Growth Factor Requirements of Tobacco Tissue Cultures. Physiologia Plantarum, 18, 100-127. http://dx.doi.org/10.1111/j.1399-3054.1965.tb06874.x

20. Hussein, N. (2012) Effects of Nutrient Media Constituents on Growth and Development of Banana (Musa spp.) Shoot Tips Cultured in Vitro. African Journal of Biotechnology, 11, 9001-9006.

21. Saad, A.I. and Elshahed, A.M. (2012) Plant Tissue Culture Media.

22. North, J., Ndakidemi, P. and Laubscher, C. (2010) The Potential of Developing an in Vitro Method for Propagating Strelitziaceae. African Journal of Biotechnology, 9, 7583-7588.

23. Msogoya, T., Maerere, A., Kusolwa, P. and Nsemwa, L. (2006) Field Performance of Improved Banana Cv. Fhia 17 and Fhia 23 in the Eastern Zone of Tanzania. Journal of Agronomy, 5, 533-535. http://dx.doi.org/10.3923/ja.2006.533.535

24. Ramage, C.M. and Williams, R.R. (2002) Mineral Nutrition and Plant Morphogenesis. In Vitro Cellular & Developmental Biology—Plant, 38, 116-124. http://dx.doi.org/10.1079/IVP2001269

25. Torres, K.C. (ed.) (1989) Tissue Culture Techniques for Horticultural Crops. Chapman and Hall, New York, London.

26. Bonner, J. and Addicott, F. (1937) Cultivation in Vitro of Excised Pea Roots. Botanical Gazette, 99, 144-170.

27. Ohira, K., Ikeda, M. and Ojima, K. (1976) Thiamine Requirements of Various Plant Cells in Suspension Culture. Plant and Cell Physiology, 17, 583-590.

28. Murashige, T. (1974) Plant Propagation through Tissue Cultures. Annual Review of Plant Physiology, 25, 135-166. http://dx.doi.org/10.1146/annurev.pp.25.060174.001031

29. George, E.F., Hall, M.A. and De Klerk, G.J. (2008) The Components of Plant Tissue Culture Media II: Organic Additions, Osmotic and pH Effects, and Support Systems. In: George, E.F., Hall, M.A. and De Klerk, G.J., Eds., Plant Propagation by Tissue Culture, Springer, Berlin, 115-173.

30. North, J., Ndakidemi, P. and Laubscher, C. (2012) Effects of Antioxidants, Plant Growth Regulators and Wounding on Phenolic Compound Excretion during Micropropagation of Strelitzia Reginae. International Journal of Physical Sciences, 7, 638-646.

31. Onuoha, I.C., Eze, C.J. and Unamba, C.I. (2011) In Vitro Prevention of Browning in Plantain Culture. OnLine Journal of Biological Sciences, 11, 13-17.

32. Khalil, M., Moustafa, A. and Naguib, N. (2007) Growth, Phenolic Compounds and Antioxidant Activity of Some Medicinal Plants Grown under Organic Farming

Condition. World Journal of Agricultural Sciences, 3, 451-457.

33. Titov, S., Bhowmik, S.K., Mandal, A., Alam, M.S. and Uddin, S.N. (2006) Control of Phenolic Compound Secretion and Effect of Growth Regulators for Organ Formation from Musa spp. cv. Kanthali Floral Bud Explants. American Journal of Biochemistry and Biotechnology, 2, 97-104. http://dx.doi.org/10.3844/ajbbsp.2006.97.104

34. Ozyigit, I.I., Kahraman, M.V. and Ercan, O. (2007) Relation between Explant Age, Total Phenols and Regeneration Response in Tissue Cultured Cotton (Gossypium hirsutum L.). African Journal of Biotechnology, 6, 3-8.

35. Mato, M., Rúa, M. and Ferro, E. (1988) Changes in Levels of Peroxidases and Phenolics during Root Formation in Vitis Cultured in Vitro. Physiologia Plantarum, 72, 84-88. http://dx.doi.org/10.1111/j.1399-3054.1988.tb06626.x

36. Lux-Endrich, A., Treutter, D. and Feutch, W. (2000) Influence of Nutrients and Carbohydrate Supply on the Phenol Composition of Apple Shoot Cultures. Plant Cell, Tissue and Organ Culture, 60, 15-21.

37. Zapprometov, M. (1989) The Formation of Phenolic Compounds in Plant Cell and Tissue Cultures and Possibility of Its Regulation. In: Maramorosch, K. and Sato, G.H., Eds., Advances in Cell Culture, Academic Press, Waltham, 240- 245.

38. Kefeli, V.I., Kalevitch, M.V. and Borsari, B. (2003) Phenolic Cycle in Plants and Environment. Journal of Molecular Cell Biology, 2, 13-18.

39. Ganapathi, T.R. and Higgs, N. (1999) Transformation and Regeration of the Banana Cultivar Rasthali (AAB). Proceedings of the International Symposium on the Molecular and Cellular Biology of Bananas, Ithaca, New York, 22-25 March 1999, 13.

40. Pan, M.J. and van Staden, J. (1998) The Use of Charcoal in in Vitro Culture—A Review. Plant Growth Regulation, 26, 155-163. http://dx.doi.org/10.1023/A:1006119015972

41. Fridborg, G., Pedersén, M., Landstörm, L.E. and Erikson, T. (1978) The Effect of Activated Charcoal on Tissue Cultures: Adsorption of Metabolites Inhibiting Morphogenesis. Physiologia Plantarum, 43, 104-106. http://dx.doi.org/10.1111/j.1399-3054.1978.tb01575.x

42. Thomas, T.D. (2008) The Role of Activated Charcoal in Plant Tissue Culture. Biotechnology Advances, 26, 618-631. http://dx.doi.org/10.1016/j.biotechadv.2008.08.003

43. Teng, W.L. and Ngai, Y.W. (1999) Regeneration of Oxalis triangularis spp. Triangularis from Suspension Cells Cultured in Three Different Systems (Solid, Liquid-Flask and Bioreactor Cultures). Plant Cell Reports, 18, 701-706. http://dx.doi.org/10.1007/s002990050646

44. Bhardwaj, L. and Ramawat, K. (1993) Effect of Anti-Oxidants and Adsorbents on Tissue Browning Associated Metabolism in Cocculus Pendulus Callus Cultures. Indian Journal of Experimental Biology, 31, 715-718.

45. Chawla, H. (2002) Introduction to Plant Biotechnology. Science Publishers.

46. Abeyaratneb, W. and Lathiff, M. (2002) In-Vitro Propagation of "Rathambala" (Musa AAA) and the Occurrence of Phenotypic Variations in the Pseudostem. Annals of the Sri Lanka Department of Agriculture (LKA), 4, 191-197.

47. Abdelwahd, R., Hakam, N., Labhilili, M. and Udupa, S.M. (2008) Use of an Adsorbent and Antioxidants to Reduce the Effects of Leached Phenolics in in Vitro Plantlet Regeneration of Faba Bean. African Journal of Biotechnology, 7, 997- 1002.

48. Strosse, H., Houwe, I., Panis, B., Jain, S. and Swennen, R. (2004) Banana Cell and Tissue Culture—Review. Banana Improvement: Cellular, Molecular Biology, and Induced Mutations. Proceedings of a Meeting, Leuven, 24-28 September 2001, 1-12.

49. Dahot, M.U. (2007) Morpho-Physiological Aspects of Micro-Propagating Banana under Different Hormonal Conditions. Asian Journal of Plant Sciences, 6, 496-501. http://dx.doi.org/10.3923/ajps.2007.496.501

50. Madhulatha, P., Anbalagan, M., Jayachandran, S. and Sakthivel, N. (2004) Influence of Liquid Pulse Treatment with Growth Regulators on in Vitro Propagation of Banana (Musa spp. AAA). Plant Cell, Tissue and Organ Culture, 76, 189-192. http://dx.doi.org/10.1023/B:TICU.0000007291.31439.6c

51. Gonzales, R.A. (1994) Plant Cell Culture: A Practical Approach. Oxford University Press, Oxford.

52. Khalid, N. (2011) Effect of Benzylaminopurine (BAP) Pulsing on in Vitro Shoot Multiplication of Musa acuminata (Banana) cv. Berangan. African Journal of Biotechnology, 10, 2446-2450

53. Buising, C.M., Shoemaker, R.C. and Benbow, R.M. (1994) Early Events of Multiple Bud Formation and Shoot Development in Soybean Embryonic Axes Treated with the Cytokinin, 6-Benzylaminopurine. American Journal of Botany, 81, 1435-1448. http://dx.doi.org/10.2307/2445317

54. Alexandrova, K., Denchev, P. and Conger, B. (1996) Micropropagation of Switchgrass by Node Culture. Crop Science, 36, 1709-1711. http://dx.doi.org/10.2135/cropsci1996.0011183X003600060049x

55. Bohidar, S., Thirunavoukkarasu, M. and Rao, T. (2008) Effect of Plant Growth Regulators On In Vitro Micropropagation of 'Garden Rue' (Ruta graveolens L.). International Journal of Integrative Biology, 3, 36-43.

56. Wickson, M. and Thimann, K.V. (1958) The Antagonism of Auxin and Kinetin in Apical Dominance. Physiologia Plantarum, 11, 62-74. http://dx.doi.org/10.1111/j.1399-3054.1958.tb08426.x

57. Cline, M.G. (1994) The Role of Hormones in Apical Dominance. New Approaches to an Old Problem in Plant Development. Physiologia Plantarum, 90, 230-237. http://dx.doi.org/10.1111/j.1399-3054.1994.tb02216.x

58. Gübbük, H. and Pekmezci, M. (2004) In Vitro Propagation of Some New Banana Types (Musa spp.). Turkish Journal of Agriculture and Forestry, 28, 355-361.

59. De Langhe, E. and Vuylstek, D. (1985) Feasibility of in Vitro Propagation of Banana and Plantains. Tropical Agriculture (Trinidad), 62, 323-328.

60. Shirani, S., Mahdavi, F. and Maziah, M. (2009) Morphological Abnormality among Regenerated Shoots of Banana and Plantain (Musa spp.) after in Vitro Multiplication with TDZ and BAP from Excised Shoottips. African Journal of Biotechnology, 8, 5755-5751.

61. Arinaitwe, G., Rubaihayo, P. and Magambo, M. (1999) Effects of Auxin/Cytokinin Combinations on Shoot Proliferation in Banana Cultivars. African Crop Science Journal, 7, 605-611. http://dx.doi.org/10.4314/acsj.v7i4.27755

62. Buah, J., Danso, E., Taah, K., Abole, E., Bediako, E., Asiedu, J. and Baidoo, R. (2010) The Effects of Different Concentrations Cytokinins on the in Vitro Multiplication of Plantain (Musa sp.). Biotechnology, 9, 343-347. http://dx.doi.org/10.3923/biotech.2010.343.347

63. Woolley, D.J. and Wareing, P.F. (1972) The Interaction between Growth Promoters in Apical Dominance. New phytologist, 71, 781-793. http://dx.doi.org/10.1111/j.1469-8137.1972.tb01957.x

64. Ma, S.S., Shii, C.T. and Wang, S.O. (1978) Regeneration of Banana Plants from Shoot Meristem Tips and Inflorescence Sections in Vitro. 20th International Horticultural Congress, Sydney, August 1978, 15-23.

65. Kumar, D. and Wareing, P.F. (1972) Factors Controlling Stolon Development in the Potato Plant. New phytologist, 71, 639-648. http://dx.doi.org/10.1111/j.1469-8137.1972.tb01274.x

66. Razdan, M.K. (2003) Introduction to Plant Tissue Culture. Science Publishers.

67. Burrows, G. (1989) Developmental Anatomy of Axillary Meristems of Araucaria Cunninghamii Released from Apical Dominance Following Shoot Apex Decapitation in Vitro and in Vivo. Botanical Gazette, 150, 369-377. http://dx.doi.org/10.1086/337782

68. Swennen, R., Wilson, G. and De Langhe, E. (1984) Preliminary Investigation of the Effects of Gibberellic Acid (GA3) on Sucker Development in Plantain (Musa cv. AAB) under Field Conditions. Tropical Agriculture, 61, 253-256.

69. De Langhe, E., Swennen, R. and Wilson, G. (1983) Aspects hormonaux du rejetonnage des bananiers plantains. Fruits. 38, 318-325.

70. Ortiz, R. and Vuylsteke, D.R. (1994) Genetics of Apical Dominance in Plantain (Musa spp., AAB Group) and Improvement of Suckering Behavior. Journal of the American Society for Horticultural Science, 119, 1050-1053.

71. Ma, S.S. and Shii, C.T. (1972) In Vitro Formation of Adventitious Buds in Banana Shoot Apex Following Decapitation. Journal of the Chinese Society for Horticultural Science, 18, 135-142.

72. Swamy, R.D., Rao, N. and Chacko, E.K. (1983) Tissue-Culture Propagation of Banana. Scientia Horticulturae, 18, 247-252. http://dx.doi.org/10.1016/0304-4238(83)90028-6

73. Khatri, A., Khan, I.A., Siddiqui, S.H., Ahmad, M. and Siddiqui, K. (1997) In Vitro Culture of Indigenous and Exotic Banana Clones for Maximising Multiplication. Pakistan Journal of Botany, 29, 143-150.

74. Alvard, D., Cote, F. and Teisson, C. (1993) Comparison of Methods of Liquid Medium Culture for Banana Micropropagation. Plant Cell, Tissue and Organ Culture, 32, 55-60. http://dx.doi.org/10.1007/BF00040116

75. Mateille, T. and Foncelle, B. (1988) Micropropagation of Musa AAA cv. Poyo in the Ivory Coast. Tropical Agriculture (Trinidad), 65, 325-328.

CHAPTER 15

THERAPEUTIC AGENTS FROM TISSUE CULTURES OF MEDICINAL PLANTS

Ciddi Veeresham and Praveena Chitti

University college of Pharmaceutical Sciences, Kakatiya University, Warangal, Andhra Pradesh, India

ABSTRACT

Plants are being the important continuous source of pharmacologically active compounds, with many blockbuster drugs being derived directly or indirectly from plants. Despite the current occupation with synthetic chemistry as a vehicle to discover and manufacture drugs, the contribution of plants to disease treatment and prevention is still enormous. However several challenges have been associated with supply of biologically active pharmaceuticals from natural sources. Alternative avenues for plant products have gained prominence during the past few years and plant biotechnology has a major role to play in plant based industries. Recently the production of secondary metabolites using plant cells has been the subject of extended research. Plant cell culture can be obtained from any plant species. In such culture, each cell has all genes necessary for all the functions of a plant including secondary metabolism. Different strategies can also be applied for the improvement of secondary metabolite production. A recent development to overcome the difficulties arising with cell suspension cultures is the genetic transformation of plants with Agrobacterium rhizogenes. Hairy roots have been found to be suitable for the production of secondary metabolites because of their stable and high productivity in hormone-free culture conditions. Plant transformed technology has now reached a platform of commercial reality. Here in this article we outline results of production of few bioactive secondary metabolites from in vitro cultures established in our laboratory.

INTRODUCTION

Nature is an infinite resource for drug development and created almost an inexhaustible array of molecular entities. It is not possible to obtain a precise figure for the total number of species existing on earth. Pimm et al. [1] made an effort to express biodiversity in number, estimating the total number of species to be 10-100 million and plants existing range from 2,50,000 to 7,50,000. Only 5-10% of these species however have been acknowledged through scientific evaluation to have therapeutic value [2]. Plant derived drugs are being widely used not only in developing countries but also in the most advanced countries. It has been reported by WHO that about 80% of the world's population rely on medicinal plants for their primary health care. Until recently plants are being the important source of novel pharmacologically active compounds, with many blockbuster drugs derived directly or indirectly from plants. Despite the current occupation with synthetic chemistry as a vehicle to discover and manufacture drugs, the contribution of plants to disease treatment and prevention is still enormous [3,4]. According to recent report utility of natural products as

Citation: Ciddi Veeresham and Praveena ChittiTherapeutic Agents from Tissue Cultures of Medicinal Plants, http://dx.doi.org/10.4172/2329-6836.1000118.

sources of novel structures is still alive and well. Up to 50% the approved drugs during the last 30years are from either directly or indirectly from natural products. In the area of cancer, over the time frame from around the 1940s to date, of the 175 small molecules 85 actually being either natural products or directly derived there from [5]. Although many of the drugs are made by synthetic chemistry, most of the core structures or scaffolds for synthetic chemicals are based upon natural products. Not all natural products can be fully synthesized and many natural products have very complex structures that are too difficult and expensive to synthesize on an industrial scale. Because of their complex structures and in spite of extensive efforts to develop partial or total chemical synthesis, isolation of such plant derived compounds from their natural source remains the only viable option, with very few exceptions. And most of the medicinal plants are not cultivated; rather they are collected from wild. In the past, quantities needed to meet demand were relatively low; however, increasing commercial demand is fast outpacing supply. Production and isolation of the chemicals by conventional techniques face several problems leading the plants to become endangered and resulting in loss of biodiversity [6]. Alternative avenues for plant products have gained prominence during the past few years and the production of secondary metabolites using plant cells has been the subject of extended research. Hence forth plant biotechnology has a major role to play in plant based industries (Table 1).

PLANT TISSUE CULTURE

Plant in vitro techniques, in which plant cells, tissues and organs are cultivated under aseptic conditions totally independent of geographical and climatic factors, offers alternatives for producing important metabolites, and several technologies based on them have been developed

Table 1: Bioactive secondary metabolites from plant tissue cultures

Secondary metabolite	Plant	Culture
Baccatin–III, Deacetyl baccatin-III	*Taxus wallichiana, T. canadensis*	callus, cell suspension
Camptothecin	*Nothapodytes foetida*	callus, cell suspension
Castanospermine	*Castanospermum australe*	callus, cell suspension
Forskolin	*Coleus forskohlii*	callus, cell suspension, hairy root
Nitidine	*Toddalia asiatica*	callus, cell suspension
Hypercin, Pseudohypercin	*Hypericum perforatum*	shoot cultures
Ajmalicine and Serpentine	*Catharanthus roseus*	hairy root

Plant tissue culture technology could be a potential alternative approach for bio production of phytoconstituents of therapeutic value and might be attractive under certain conditions if, for example, the source plant is difficult to cultivate, has a long cultivation period or has a low metabolite yield; if chemical synthesis has not been achieved or if it is technically problematic. In vitro plant cell culture includes mainly four main approaches, namely callus, suspension, immobilized cells and differentiated cultures. In brief callus culture involves growing a disorganized aggregate of cells from plant explants by culturing on a semi-solid support which contains nutrients and any hormones required to promote growth of the cells. Suspension cultures result when callus is suspended in liquid growth medium and growing cells as dispersed cell culture. With their relatively fast growth - doubling times and their ease of manipulation, suspension cultures are widely employed in the study of secondary metabolite production by plant cells. The advantages of this approach are obvious, as biomass production is rapid than that of whole plant, nutritional and environmental requirements can be easily controlled allowing the production of pharmaceutical throughout the year if necessary, nutrient uptake is enhanced by submerged culture conditions which stimulate the multiplication rate and higher yield of bioactive compounds. The major problems of using suspension cultures are instability of the cell lines and low product yield that in some cases is virtually zero. It is generally acknowledged that lack of basic information about biosynthetic pathways and mechanisms underlying the production of secondary metabolites is the bottleneck that needs to be addressed [7,8]. Immobilized cell cultures result when free cells are entrapped in alginate, agarose or other similar matrices, or else are allowed to grow into a porous support such as reticulated polyurethane foam. Particularly in the case of cells immobilized in foams, where cell-to-cell contact is maximized and where the process is technically simple, such systems combine many of the advantages of both cell suspensions and callus. The final system in general use, an alternative method of producing biologically active compounds and one which is growing in importance, is the use of differentiated cultures such as shoot cultures, root cultures, transformed cultures because their behaviour is more predictable than that of undifferentiated cell suspension cultures. By manipulating the external hormone balance or by influencing the internal hormonal status of non-transformed differentiated cultures can be initiated. The structural integrity of organ cultures is maintained by manipulating the external hormonal balance. Such morphological stability is in turn associated with improved genetic and metabolic stability compared with undifferentiated cultures. Such organ cultures produce yields of secondary metabolites comparable to intact plants [9-11].

Alternatively, genetically transformed hairy root cultures have been extensively researched for the synthesis of root derived chemicals. The recent advances and developments in plant genetics and recombinant DNA technology have helped to improve and boost research into secondary metabolite biosynthesis. Genetic transformation would be a powerful tool for enhancing the productivity of novel secondary metabolites of limited use. Hairy roots, transformed with Agrobacterium rhizogenes have been found to be suitable for the production of secondary metabolites because of their stable and high productivity in hormone-free culture conditions. In certain instances, particularly transformed roots can grow almost as rapidly as the best suspension cultures whilst still maintaining high secondary metabolite productivity through genetic and biochemical stability. A number of plant species including many medicinal plants have been successfully transformed with Agrobacterium rhizogenes. Plant transformed technology has now reached a platform of commercial reality [12-15]. However, most often trails with plant cell culture fail to produce the desired products. In such cases, strategies to improve the production must be considered. Elicitor treatment, precursor feeding, carbon source, permeabilization, light, pH, temperature etc [16-19].

RESEARCH WORK CARRIED OUT IN AUTHOR'S LAB

Callus and Cell Suspension Cultures

Plant tissue culture studies carried were carried out on three species of T. axus, T. wallichiana, T. baccata and T. canadensis (Taxaceae) with an objective to enhance the bio production of pharmaceutically important phytoconstituents under the influence of different biotic and abiotic elicitors. Different explants such shoot buds, young leaves, roots as explants and three different nutrient media (MS, WPM, B5) were used for establishment of callus cultures. Young leaves of T. Wallichiana, T. baccata and shoot buds of T. canadensis were found to give best response of callus induction on Gamborg's B5 medium supplemented with Dicamba (2 mg/l), BA(0.2 mg/l), glutamate(292 mg/l) and sucrose (1%w/v). HPLC analysis of 15 weeks old callus cultures of T. wallichiana revealed the presence of 0.005% deacetyl baccatin-III on dry weight basis. Suspension cultures of T. wallichiana and T. canadensis after 3 passages from growth medium to production medium (1/4 B5, sucrose 5%w/v, 2,4-D 2 mg/l, zeatin 0.5 mg/l) incidentally shown the maximum concentration of baccatin–III and deacetyl baccatin-III. Abiotic elicitors arachidonic acid and methyl jasmonte significantly improved the production of baccatin–III and deacetyl baccatin-III over the control cultures of T. wallichiana and cinnanic acid enhanced the production of baccatin–III and deacetyl baccatin-III in cell suspension cultures of T. canadensis. Calcium chloride in low concentration was found to be effective in enhancing the yield of these taxanes from T. canadensi suspension cultures. Biotic elicitors employed in the study did not exhibit any activity on bio production of baccatin–III and deacetyl baccatin-III, on the other hand found to partially retard the bio production [20,21]. Camptothecin (CPT) and its analogues 10-hydroxy CPT, 9-methoxy CPT (MCPT) and semi synthetic derivatives topotecan, irinotecan are promising anticancer drugs with unique mechanism of action. Nothapodytes foetida (Iccacinaceae) is a potential source of CPT and its analogues than any other botanical sources of camptothecins. Though it is a potential source of camptothecins the concentration of these compounds is very low and needs large quantities of plant material. In addition biosynthesis is significantly influenced by physiological shifts caused by changes in environment leading to highly fluctuating productivity. Natural propagation is very slow and lack of information on vegetative propagation necessitated the attention for multiplication of N. foetida and alternative methods of production of CPT and its analogues. MS medium supplemented with Thiadiazuron (0.1 mg/l) gave good initiation of shoots from nodal explants of in vitro raised seedlings. Maintenance and elongation of shoots was accomplished on low cytokinin level i.e., Thiadiazuron (0.05 mg/l). MS medium supplemented with NAA (0.1 mg/l), and BA (0.05 mg/l) was found suitable for rooting.

MS medium supplemented with Picloram (2 mg/l), BA (1.0 mg/l) GA3 (1.0 mg/l) gave good callus induction frequency of 95% from leaf explants. CPT and MCPT were detected in callus and media. Cell suspension cultures initiated on same phytohormonal combination accumulated high levels of CPT and MCPT than callus cultures. However in both the case the yield of these compounds is lower than the intact plant. Addition of methy jasmonate and copper sulphate on day-7 to cell cultures enhanced the yields of CPT and MCPT. Supplementation of cell cultures with phenyl alanine, sodium acetate and tryprtophan on day-0 also enhanced the bio production which could be due their incorporation in biosynthetic pathway of CPT and analogues as precursors [22-24]. Coleus forskohlii, (Lamiaceae/Labiatae) is an important ancient root drug in Indian Ayurvedic Medicine contains an important secondary metabolite labdane diterpenoid forskolin which occurs exclusively in this plant. Forskolin has several biological and pharmacological activities that have been linked to its role as an activator of adenylate cyclase. Forskolin content of the roots obtained from natural habitats showed a range of variation from 0.04 to 0.44% dry cell weight. The total synthesis of forskolin reported involves 30 steps. The sole source of forskolin is the roots of wild or cultivated C. forskohlii plants. Biotechnological approaches especially plant tissue culture are found to have potential as a supplement to the traditional agriculture in the industrial production of bioactive plant metabolites. MS medium with NAA (0.5 mg/l), 2, 4-D (0.25 mg/l), BA (1 mg/l) and sucrose (3% w/v) was found to be suitable for callus induction from young leaves and the established callus cultures shown the presence of forskolin. The cell cultures initiated in MS growth medium supplemented with NAA (0.5 mg/l), BA (1.0 mg/l), 2,4-D (0.5 mg/l), casein hydrolysate (600 mg/l) and sucrose (3% w/v) were also found to produce forskolin. To increase the yield elicitors and precursors were added to cell suspension cultures. Elicitor's salicylic acid, methyl jasmonate had a profound influence on forskolin production [25]. Castanospermum australe (Leguminosae), the only species of genus of Castanospermum found in Australia. Castanospermine, a polyhydroxy indolizidine alkaloid is the important constituent of C. australe. Semisynthetic derivatives of Castanospermine, Bu cast and Celgosivir are in clinical trials for anti-HIV and hepatitis infections respectively. In vitro bio production potential of C. australe and studies to enhance the production of its secondary metabolites in cell culture was not extensively studied. Callus cultures developed from leaf explants on MS medium supplemented with 2, 4-D(2 mg/l), NAA(2 mg/l) and Kn(0.5 mg/l) and were found to have biosynthetic potential to produce castanospermine and related alkaloids. Suspension cultures initiated were also bio productive. Intracellular amount of castanospermine was found be about 6 times more than the extracellular accumulation. Methy jasmonate, copper sulphate, cinnamic acid, calcium chloride employed as abiotic elicitors enhanced production of castanospermine however methy jasmonate, copper sulphate had a more significant effect with 6.6 and 6.2 fold increment in castanospermine levels than control cultures. Biotic elicitor Verticillium dahilae shown 2.7 fold improvement over control cultures. Addition of L-lysine, sodium acetae and L-pipecolic acid also found to enhanced the production of castanospermine [26].

Toddalia asiatica (Rutaceae), well known for its antimalarial activity traditionally, have been reported to have anticancer and anti-HIV activities. These activities are mainly attributed to benzophenanthridine alkaloid nitidine. The growing interest on search for plant derived anticancer and anti–HIV agents has given an intuition to work on this plant. The yield of the nitidine (alkaloid) in intact plant is too low and plant tissue culture has been used as an alternative method for increasing the production of secondary metabolites. MS medium supplemented with NAA (2 mg/L) and Kinetin (1 mg/L) was found to be more suitable for induction of callus culture. This callus cultures showed 10 fold production of the alkaloid than that of intact plant. Suspension cultures established from callus cultures have also shown biopotential as much as of callus cultures. Copper sulphate, salicylic acid and methyl jasmonate were had significant effect in increasing the yield of nitidine as elicitors. Positive precursor feeding response as a result of the treatment of α-ketoglutaric acid and tyrosine on bio production of nitidine from cell cultures also observed. Cells immobilized by ionotropic gelation of alginate using sodium alginate and calcium chloride. The entrapped cells in beads were transferred to MS medium with NAA (2 mg/l), Kinetin (1 mg/l) and sucrose (3% w/v) and subjected to elicitation and precursor feeding. There was 6 fold increase in the production of nitidine over the control cell cultures upon addition of salicylic acid and ~ five folds by copper sulphate over control cell cultures but was less than that produced in cell cultures. Addition of tyrosine and α-ketoglutaric acid on further enhanced the production of nitidine compared to control cell cultures [27].

Differentiated / Shoot Cultures

Hypericum perforatum (Hypericaceae), commonly known as St. John's Wort, is an important medicinal plant and is one of the top selling herbs in U.S. It is an herbaceous perennial plant belonging to Hypericaceae family. It produces a wide range of chemical constituents like dianthrones, phloroglucinol derivatives, xanthones and flavonoids. Hypericins and hyperforins are the main constituents responsible for the therapeutic activity in the treatment of depression, AIDS and cancer. The hypericins are present in very low amounts, their content varies greatly in wild plants and the synthetic derivatives are biologically inactive. As a result, cell and tissue cultures of H. perforatum have been studied by various researchers for the production of significant quantities of therapeutically valuable hypericins. Callus cultures were established from hypocotyl segments of in vitro germinated seedlings on MS media containing 2, 4-D (1 mg/l) and KN (0.5 mg/l). But analysis of callus cultures has not shown the presence of Hypericins. Even the cell suspension cultures initiated from callus cultures did not produce hypercin or its analogues. Elicitor and precursor feeding studies could not influence the biosynthetic ability of cell cultures. This necessitated the induction of differentiated cultures [28]. Shoot cultures were initiated from callus cultures accumulated significantly higher levels of hypercin and pseudohypercin. Effect of carbon source/sugar on growth and secondary metabolite production was studied and it has been found that sucrose promoted cell proliferation and production of secondary metabolite compared to other sugars (fructose, maltose and glucose) employed in the investigation. Various biotic and abiotic elicitors were screened for their effect on shoot cultures and chitosan and methyl jasmonate were found to be effective elicitors. Precursors, organic compounds and growth inhibitors were also employed to evaluate their effect on biosynthesis of secondary metabolites. These treatments significantly improved the production of hypercins [29].

Transformed Cultures

Vinblastine and vincristine are widely used in the chemotherapy of cancer while ajmalicine is an anti-hypertensive agent. These anticancer alkaloids are present in very low amount in intact plant Catharanthus roseus (Apocynaceae) and organic synthesis of chemically complex molecules is not economical. Although semi-synthesis using monomeric precursors is possible but the plant produces a very little amount of catharanthine as compared to that of vindoline. As a result, cell and tissue cultures of C. roseus have been studied for many years for the production of therapeutically interesting alkaloids.

Hairy roots were obtained from leaf explants following infection with A4 strain. Ajmalicine, serpentine and catharanthine were identified but neither vindoline nor vinblastine could be detected in hairy root cultures. With an aim to increase the production of indole alkaloids, hairy root cultures were treated with abiotic and biotic elicitors. Heavy metals ions copper and silver were found to stimulate the production as well as their release into the medium. Highest concentrations of Cu (1000μM) produced more release than lower concentrations (500 μM and 100μM). Salt stress by Sodium chloride (250 μM) and osmotic shock by mannitol did not affect adversely growth of hairy roots on the other hand it stimulated the production and release of these alkaloids. Significant improvement in ajmalicine and serpentine production was observed with signal transduction molecules like arachidonic acid and linoleic acid. Generally, hairy root culture C. roseus have been found to accumulate alkaloids within the cells. The artificial release of metabolites not only facilitates downstream processing but also provides opportunities for increasing the productivity by combining with other strategies like in situ removal. Hence hairy root were subjected to permeabilization with compounds with compounds like chitosan, dimethylsulfoxide and Tween 20. But surprisingly these chemical compounds did not induce the release of alkaloids into the culture medium. However with higher concentrations of Tween 20 (1 and 2%v/v), traces of ajamalicine and serpentine could be detected in culture medium. Nevertheless, all permeabilizing agents stimulated significantly the alkaloid production, increasing the root contents. As permeabilizing agents did not induce the release of the alkaloids from hairy root cultures when used individually, it was thought that worthwhile to see whether the combined addition of these agents changes the accumulation pattern of alkaloids. Combined addition of permeabilizing agents produced a dramatic effect, inducing the release of indole alkaloids into the culture medium. The biosynthesis of IPP and hence of alkaloids in C. roseus hairy root cultures was studied by employing various metabolic inhibitors. Significant reduction in levels of indole alkaloids was observed with lovastatin and fosidomycin treatment compared to control cultures [30,31]. Hairy root cultures induced from leaf explants of Coleus forskohlii were investigated for the bio production ability. It was found that forskolin production was parallel with the growth indicating that forskolin is growth associated product, accumulated to maximum levels during the exponential growth phase in parallel with the biomass. Precursors such as L-phenylalanine, α-ketoglutaric acid and elicitors such as salicylic acid, copper sulphate and methyl jasmonate significantly improved the production of forskolin in hairy root cultures [32].

CONCLUSION

Plants have been an important source of medicine from immemorial time. The use of plant tissue cultures for the biotechnological production of phytoconstituents has a number of attractions as mentioned above. Several reports are demonstrating the production of wide array of secondary metabolites of plants and some being produced on commercial scale. Tissue culture protocols have been developed for several plants but there are many other species, which are over exploited in pharmaceutical industries and other fields and do need conservation. Tissue culture technique is useful not only for production of phytoconstituents but also for multiplying and conserving the species, which are difficult to regenerate by conventional methods and save them from extinction. Still the word "potential" is used in connection with tissue culture technology. Hence there is much scope to develop and bring it to commercial scale for the benefit of human welfare and to preserve biodiversity

REFERENCES

1. Pimm SL, Russell GJ, Gittleman JL, Brooks TM (1995) The future of biodiversity. Science 269: 347-350.
2. Verpoorte R (1998) Exploration of nature's chemodiversity: the role of secondary metabolites as lead drugs for drug development. Drug development Today
3. 3: 232-238. 3. Raskin I, Ribnicky DM, Komarnytsky S, Ilic N, Poulev A, et al. (2002) Plants and human health care in the twenty –first century. Trends in Biotechnology 20: 522 531.
4. Newman DJ, Cragg GM, Snader KM (2000) The influence of natural products upon drug discovery. Natural Products Reports 17: 215-234.
5. Newman DJ, Cragg GM (2012) Natural products as sources of new drugs over the 30 years from 1981 to 2010. Journal of Natural Products 75: 311−335.
6. Kolewe ME, Gaurav V, Roberts SC (2008) Pharmaceutically active natural product synthesis and supply via plant cell culture technology. Molecular Pharmaceutics 5: 243-256.
7. Tripathi L, Tripathi JN (2003) Role of biotechnology in medicinal plants. Tropical Journal of Pharmaceutical Research 2: 243-253.
8. Vanisree M, Lee CY, Lo SF, Satish MN, Lin CY, et al. (2004) Studies on the production of some important secondary metabolites from medicinal plants by plant tissue cultures. Botanical Bulletin of Academia Sinica 45: 1-22.
9. Parr AJ (1989) The production of secondary metabolites by plant cell cultures. Journal of Biotechnology 10: 1–26.
10. Heble MR (1985) Multiple shoot cultures: A viable alternative in vitro system for the production of known and new biologically active plant constituents. Primary and Secondary Metabolism of Plant Cell Cultures. Springer-Verlag, Berlin, 281-289.
11. Lindsey K, Yeoman MM (1983) Novel experimental systems for studying the production of secondary metabolites by plant tissue cultures. Plant Biotechnology. SEB Seminar Series 18, Cambridge University Press, Cambridge, 39-66.
12. Stephanie G, Jocelyne TG, Pratap Kumar P, Marc R, Pascal G. (2006) Hairy root research: recent scenario and exciting prospects. Current Opinion in Plant Biology 9: 341–346.
13. Srivastava S, Srivastava AK (2007) Hairy Root Culture for Mass-Production of High-Value Secondary Metabolites. Critical Reviews in Biotechnology 27: 29-43.
14. Georgiev MI, Pavlov AI, Bley T (2007) Hairy root type plant in vitro systems as sources of bioactive substances. Applied Microbiology Biotechnology 74: 1175-1185.
15. Galera SG, Pelacho AM, Gene A, Capell T, Christou P (2007) The genetic manipulation of medicinal and aromatic plants. Plant Cell Reports 26: 1689-1715.
16. Mantell SH, Smith H (1983) Cultural factors that influence secondary metabolite accumulations in plant cell and tissue cultures. Plant Biotechnology, SEB Seminar Series 18. Cambridge University Press, Cambridge, 75-108.
17. Hamill JD, Rhodes MJC (1993) Manipulating secondary metabolism in culture. Biosynthesis and manipulation of plant products. Plant Biotechnology Series178-200.
18. Endress R (1994) Plant cells as producers of secondary compounds. In: Plant cell biotechnology. Springer – Verlag Berlin Heiderg, Germany, 121-255.
19. Veeresham C (2004) Medicinal Plant Biotechnology. Medicinal plant biotechnology. 506.
20. Veeresham C, Mamatha R, Prasad Babu Ch, Srisilam K, Chandrakanth K (2003) Production of taxol and its analogues from cell cultures of Taxus wallichiana. Pharmaceutical Biology: 41: 426-430.
21. Prasad Babu Ch, Mamatha R, Chandrakanth K, Veeresham C (2001) Elicitation of Taxus wallichiana (Himalayan yew) cell cultures for the production of taxanes. Indian drugs 38: 502 505.
22. Sudaravealan R, Brahma Redddy DR, Veeresham C (2003) Microprogation of Nothpodoytes foetida using tissue culture. Indian Journal of Natural Products 18: 18-21.
23. Sundaravelan R, Brahma Reddy DR Veeresham C (2004) Production of camptotecines from callus Cultures of Nothapodytes foetida. Indian Journal of Biotechnology 3: 452-453.
24. Sundervelan R, Brahma Reddy DR, Veeresham C (2002) Effect of elicitors on the production of camptothecine and its analogues in cell cultures of Nothapodytes foetida. Ethiopian Pharmaceutical Journal 20: 39-46.
25. Srinivas Reddy Ch, Praveena Ch, Veeresham C (2011) Approaches to enhance the production of Forskolin in cell cultures of Coleus forskohlii Briq. International Journal of Pharmaceutical Sciences. 3: 1434-1442.
26. Brahma Reddy DR, Srinivas Reddy Ch, Veeresham C (2004) Production of castanospermine from cell cultures of Castanospermum austral. 56th Indian Pharmaceutical congress (IPC) Kolkata, India.

27. Rajkumar M, Harish Chandra R,Veeresham C (2010) Production of Nitidine from callus cultures of Toddalia asiatica. International Journal of Pharmaceutical Sciences and Nanotechnology 2: 1028-1033.

28. Rao UMV, Gaviraj EN, Veeresham C (2009) Studies on synthesis of Hypericins by undifferentiated cultures of Hypericum perforatum L. Ethiopian Pharmaceutical Journal 27: 25-32.

29. Rao UMV, Gaviraj EN, Veeresham C (2010) Effect of ethephon and ancymidol on the production of hypericins in shoot cultures of Hypericum perforatum L. Indian dugs 47: 77- 79.

30. Gaviraj EN, Veeresham C (2011) Enhanced alkaloid production by elicitation in hairy root cultures of Catharanthus roseus var Nimal. Indian drugs 48: 20-25.

31. Gaviraj EN, Veeresham C (2005) Chemical-induced production and release of alkaloids from hairy root cultures of hairy root cultures of Catharanthus roseus var Nirmal. Ethiopian Pharmaceutical Journal 23:31-38.

32. Srinivas Reddy C, Praveena Ch, Veeresham C (2012) Strategies to improve the production of Forskolin from hairy root cultures of Coleus forskohlii Briq. International Journal of Pharmaceutical Sciences and Nanotechnology 5: 1720-1726.

CHAPTER 16

ELIMINATION OF GRAPEVINE BOIS NOIR PHYTOPLASMA BY TISSUE CULTURE COUPLED OR NOT WITH HEAT THERAPY OR HOT WATER TREATMENT

L Chalak[1,2], A Elbitar[1], N Mourad[1], C Mortada[1] and E Choueiri[1]

[1]Lebanese Agricultural Research Institute, Tal Amara, P.O. Box 287, Zahlé, Lebanon

[2]Lebanese University, Faculty of Agricultural Sciences, Dekwaneh, Beirut, Lebanon

INTRODUCTION

Phytoplasmas are wall-less intracellular bacteria restricted to sieve tubes and transmitted by leafhoppers in which they multiply [1]. They affect a wide range of plants causing severe diseases, generally known as yellows [2,3]. In grapevines (Vitisvinifera L.), Flavescence dorée (FD) caused by an elm yellows-type phytoplasma belonging to the 16Sr-V subgroups C and D [4] and Bois noir (BN) elicited by a stolbur-type phytoplasma belonging to 16SrXII-A group [5] are the most common yellows in Europe [6].

BN, one of the most widespread yellows in the European grape industry [7], occurs also in several Middle Eastern countries, like Israel [8], Lebanon [9] and Syria [10]. BN is particularly serious in Lebanon, especially in the Bekaa Valley, whose damages are the cause of increasing concern for the growers. The control of FD is based on chemical treatments against the vector (Scaphoideustitanus Ball), a strategy that is not effective for BN containment because of the marked differences in its epidemiology [7]. Propagation of infected plants by nurseries widely contributes to the introduction of the disease to viticulture area previously free from it [11]. Therefore, steps are undertaken towards the development of protocols for the production of phytoplasma-free propagating material.

In vivo hot water treatment has been proposed since 1966 [12] for curing dormant woody plant material from phytoplasma and, more recently, as an efficient way to contain spreading of grapevine yellows [13-16]. However, it has been reported that hot water treatment applied to the dormant buds prior to grafting can interfere with the vitality of woody propagation material leading to a weak recovery of grafted vines [17-19].

Whereas tissue culture techniques, alone or coupled with heat treatment, have been successfully employed for eliminating viruses from a wide range of hosts [20-25] their potentialities for phytoplasma elimination have been much less explored. Notable examples are the successful use of shoot tip grafting for the elimination of the phloemrestricted huanglongbing liberibacter from citrus [26] and the use of apical meristems and embryogenic callus for knocking out sugarcane yellows phytoplasma from sugarcane [27]. More recently, Candidatus Phytoplasma phoenicium was successfully eliminated from different almond varieties using stem cuttings and shoot tip cultures associated or not with thermotherapy [28]. The main advantage of tissue culture techniques is the possibility of large scale propagation of the sanitized stocks that could enter certification schemes.

Because, as mentioned, vineyards severely touched by BN, particularly of cv. Chardonnay, are widespread in some Middle Eastern countries, the control of this disease through the development of a protocol to produce phytoplasma-

Citation: Chalak L, Elbitar A, Mourad N, Mortada C, Choueiri E (2013) Elimination of Grapevine Bois Noir Phytoplasma by Tissue Culture Coupled or not With Heat Therapy or Hot Water Treatment. Adv Crop Sci Tech 1:107. doi: 10.4172/2329-8863.1000107.

free propagating material has become an issue of primary importance. In the present paper, we report on the efficiency of procedures using tissue culture techniques associated or not with heat or hot water treatment for producing phytoplasma-free vines of cv. Chardonnay.

MATERIALS AND METHODS

In summer 2011, shoots were collected in vineyards in West Bekaa (Lebanon) from cv. Chardonnay vines showing yellows symptoms (i.e. leaf rolling, yellowing of the leaves and incomplete wood ripening) that were affected by BN based on PCR assays, RFLP analysis and rDNA sequence [9]. Young shoots from the current season's growth (June) were surface sterilized with a 30% solution of Na-hypochloride for 20 min, rinsed 3 times in sterile distilled water, and blotted on filter paper. Single-node stem cuttings (1 cm in length) and shoot tips (0.4-0.5 mm in diameter) were excised in a laminar flow cabinet, and used as explants. The growth medium was MS [29] supplemented with 15 mg.l-1 ascorbic acid and 15 mg.l-1 citric acid, 0.5 mg.l-1 BAP (6-benzylaminopurine) and 0.01 mg.l-1 NAA (naphthalene acetic acid), solidified with 0.8% BactoDifco agar, and autoclaved at 118°C for 20 min. All cultures were placed in a growth cabinet with a photoperiod of 16 h of artificial light and 8 h of darkness at 25 ± 2°C, except for cultures exposed to thermotherapy.

Six treatments were tested: (i) culture of stem cuttings with or without thermotherapy at 38 ± 1°C for 40 days from culture initiation; (ii) culture of stem cuttings after a hot water bath at 50°C for 15 and 30 min prior to culture initiation; (iii) culture of shoot tips with and without thermotherapy, where shoots were exposed or not to 38 ± 1°C for 40 days prior to shoot tip collection. For each treatment, 100 explants were used (20 explants×5 replications) vs. 30 explants for the negative control (10 explants×3 replications) as indicated in Table 1.

Newly developed shoots derived from the different treatments underwent subculturing for a period of 30 days before being tested for the first time by nested PCR. Then, explants were multiplied for two other successive subcultures to undergo a second PCR testing before being rooted on MS medium supplemented with 1 mg.l-1 NAA. One positive and one negative controls taken from infected and healthy grapevine samples and assayed for the presence of BN were also used.

Total nucleic acids were extracted from portions of whole shoots using the CTAB (cethyl-trimethyl-ammmonium bromide) extraction protocol described by Maixner et al. [3]. Phytoplasma universal PCR primers P1/P7 based on the sequencing of the 16S-23S gene [30] were used for amplification of ribosomal DNA, followed by nested PCR using R16F2n/R16R2 [31]. The amplification of a 1800 bp band for the first PCR and of a 1200 bpamplicon for the nested PCR were expected. PCR products were analyzed by agarose gel electrophoresis, stained with ethidium bromide and visualized under a UV transilluminator. The percentage of explants regenerating new shoots and phytoplasmafree shoots was determined. Mean differences among treatmentswere evaluated by Duncan test (General Linear Models Procedure, SAS Institute, Cary, USA).

RESULTS

Viability and Regeneration Ability

Figure 1 shows the successive stages of culture evolution for the different explants of cv. Chardonnay while table 1 reports the number of shoots regenerated from the explants after been submitted to the various sanitation treatments. According to the treatments, a certain number of explants did not survive due to an oxidation problem that developed 2-3 days after cultures initiation. Oxidation was particularly evident in stem cuttings and shoot tips that were subjected to heat therapy or hot water bath which led to the loss of a significant number of explants ranked between 22 and 50%.

The survival rate of explants observed 20 days after the initiation of culture varied with the type of explant and treatment applied between 50 and 100%. The best results were obtained with stem cuttings (Figure 1A and 1B), but their treatment with hot water significantly reduced the survival from 100 to 50%. The average number of regenerated shoots per explant observed 30 days after the initiation of culture varied between 1.8 and 2.3. Of all the treatments, was the stem cuttings treated with hot water that gave the lowest number of regenerated plantlets with an average of 1.8 new shoots per explant.

With time, these differences among treatments became lighter and the proliferation ability of the explants issued from heat treatment increased progressively during the successive subcultures (Figure 1C), reaching 2.2 to 2.5 new shoots per explant. These results were comparable to those obtained from control (healthy) explants. Moreover, shoots regenerated from infected material developed normally and similarly to the healthy ones. About a week after the transfer of new shoots to the rooting medium, whitish callus began to appear at the base of the stems and first roots appeared from day 15 (Figure 1D). The rate of rooting recorded after 40 days varied between 84 and 90% without any significant differences among the treatments from which the shoots were issued (not shown).

Effect on Sanitation

Table 2 reports the proportion of phytoplasma-free shoots as determined by nested PCR. The expected DNA fragments of 1800 bp and 1200 bp, respectively, were clearly amplified for the positive control samples (Figure 2) confirming their infection with BN while all healthy samples tested were PCR negative. Here it is worthy noted the reliability of results obtained by nested PCR compared to the direct one since, on the whole shoots tested, only three samples found negative in direct PCR (Lanes 15 and 17) were positive with nested PCR, while all the samples found positive in the first test remained positive by nested PCR.

Regarding the various treatments, the nested PCR operated at the end of the first subculture showed sanitation rates varying between 36 and 100%. All tested shoots regenerated from the stem cuttings treated in hot water bath at 50°C for 15 or 30 min were phytoplasma-free, whereas this value dropped to 76% for the shoots regenerated from stem cuttings exposed to heat treatment. As to shoots regenerated from shoot tips, heat treatment increased the sanitation rate from 36% to 76%.

Our sanitation rates are indeed similar or even higher than those previously reported for the elimination of BN from different grapevine genotypes by in vivo treatment with hot water at 50°C for 30 and 45 min of dormant buds prior to grafting [14,15,18,19,32]. By converse, the in vitro hot water treatment of vegetating material proved to be more effective than previously reported in terms of viability and growth ability. A hot water bath for the duration of 15 min only, which proved to be efficient in our study, has apparently not been tested on vines before.

The second nested PCR conducted at the end of the third subculture (after 120 days of culture) confirmed the absence of phytoplasmas in the shoots that were negative in the first test. Among 35 shoots regenerated from the different treatments and found phytoplasma-positive in the first test, only one originating from shoot tip culture after four months of culture proved to be phytoplasma-free during the second PCR (not shown).

Discussion

The survival rate of explants used in the different sanitation treatments was influenced by the type of explants and of the treatment (moderately hot air or hot water). Oxidation was the major drawback, resulting in the loss of explants, especially in the case of heat therapy applied to both shoot tips and stem cuttings. Nevertheless, the explant survival rates remained broadly acceptable to make worthwhile the used sanitation techniques, especially since the ability to proliferate was progressively normalized during successive subcultures. It is worth noting that previous studies involving thermotherapy treatment reported its negative impact on the vitality and growth of woody propagation material which led to a weak recovery of the vines either in the nursery or at later stages [17-19].

The various protocols of in vitro culture tested in this study for the sanitation of cv. Chardonnay from BN were all effective, with rates varying between 36 and 100%. In general, however, the BN agent appeared to be more difficult to eradicate than phytoplasma infecting other fruits species. Indeed, Candidatus Phytoplasma Phoenicium was completely knocked out from two almond varieties using both stem cutting cultures coupled with thermotherapy and shoot tip culture with or without thermotherapy [28].

On the other hand, our study showed the potential of successive subcultures for getting rid of phytoplasma from infected vines without any particular treatment. Indeed, the elimination of both FD and BN through the simple culture of nodal sections has previously been reported with percentages varying between 50 and 100%, after 9 months of subculturing [33].

Even though our results are encouraging, they may be confirmed by histological and ultra-structural studies. Furthermore, it remains to be assessed whether the sanitized shoots will remain symptomless after the hardening stage. Among the different techniques assayed in this study, stem cutting culture coupled with heat treatment seemed to be the most practical and easier way for regenerating phytoplasma-free shoots, with a 78% explants survival rate. Nevertheless, and despite the lower explants survival rate at the beginning, the hot water treatment might be the most efficient sanitation technique as it yielded 100% BN-free shoots. Thus, a protocol based on tissue culture followed by micropropagation of sanitized explants, as shown also by other authors [25,28,34] seems to be a good candidate to become a routine technique for producing phytoplasma-free vines within certification programmes.

ACKNOWLEDGEMENTS

The authors thank Prof. Giovanni Paolo Martelli and Dr. IvanaGribaudo for their critical review of the manuscript. The kind assistance of Ms. A. Kadri and Mr. A. Chehade at LARI is gratefully acknowledged.

REFERENCES

1. Marwitz R (1990) Diversity of yellows disease agents in plant infections. ZentralblattfürBakteriologie 20:431-434.
2. Dosba F, Lansac M, Mazy K, Garnier M, Eyquard JP (1991) Incidence of different diseases associated with mycoplasma-like organisms in different species of Prunus. ActaHorticulturae 283:311-320.
3. MaixnerM, Ahrens U, Seemuller E (1995) Detection of the German grapevine yellows (VergibungsKransKheit) MLO in grapevine, alternative hosts and a vector, by a specific PCR procedure. European Journal of Plant Pathology 101: 241-250.
4. Boudon-Padieu E (2003) The situation of grapevine yellows and current research directions: distribution, diversity, vectors, diffusion and control. Extended Abstracts of the14th Meeting of the International Council for the study of Viruses and Virus-like Diseases of the Grapevine (ICVG).
5. Lee IM, Giundersen DE, Davis RE, Bartosyk IM (1998) Revised classification scheme of phytoplasma based on RFLP analyses of 16sRNA and ribosomal protein gene sequences. International Journal of Systematic Bacteriology 48:1153-1169.
6. Martelli GP, Boudon-Padieu E(2006) Directory of infectious diseases of grapevines. Options Mediterranéennes B 55:157-194.
7. Maixner M (2006) Grapevine yellows – current developments and unsolved questions. 15th Meeting of theInternational Council for the study of Viruses and Virus-like Diseases of the Grapevine (ICVG).
8. Orenstein S, Zahavi T, Weintraub P (2001)Distribution of phytoplasma in grapevines in the Golan Heights, Isarael, and development of a universal primer. Vitis 40:219-223.
9. Choueiri E, Jreijiri F, El Zammar S, Verdin E, Salar P, et al. (2002) First report of grapevine "Bois Noir" Disease and a New phytoplasma infecting Solanaceous Plants in Lebanon. Plant Disease 86: 697.
10. Contaldo N, Soufi Z, Bertaccini A (2011) Preliminary identification of phytoplasmas associated with grapevine yellows in Syria.Bulletin of Insectology 64: S217-S218.
11. Mannini F (2007) Hot water treatment and field coverage of mother plant vineyards to prevent propagation material

from phytoplasma infections. Bulletin of Insectology 60: 311-312.

12. Caudwell A (1966) L'inhibitionin vivodu virus de la "flavescence dorée" par la chaleur. Etudes de virologie. Annales des Epiphyties 17:61-66.

13. Caudwell A, Larrue J, Valat C, Grenan S (1990) Hot water treatments against Flavescencedorée of grapevine on dormant wood. Proceedings of the 10th Meeting of theInternational Council for the study of Viruses and Virus-like Diseases of the Grapevine (ICVG).

14. Bianco PA, Scattini G, Casati P, Fortusini A (2000)Thermotherapy of grapevine cuttings for Flavescencedorée eradication. Extended Abstracts of the 13thmeeting of the International Council for the study of Viruses and Virus-like Diseases of the Grapevine (ICVG).

15. Tassart-Subirats V, Clair D, Grenan S, Boudon-Padieu E, Larrue J (2003) Hot water treatment: curing efficiency for phytoplasma infection and effect on plant multiplication material. Extended Abstracts of the14thMeeting of theInternational Council for the study of Viruses and Virus-like Diseases of the Grapevine (ICVG).

16. Mannini F, Marzachi C (2007) Termoterapia in acqua contro I fitoplasmi della vite.L'InformatorioAgrario63: 62-65.

17. Mannini F,GribaudoI (2008) Strategies to prevent or eliminate phytoplasma from grapevine propagation material. Petria 18:241-243.

18. Mannini F, Argamante N, Gambino G, Mollo A (2009) Phytoplasma diffusion through grapevine propagation material and hot water treatment. Extended abstracts of the 16thMeeting of theInternational Council for the study of Viruses and Virus-like Diseases of the Grapevine (ICVG).

19. Bianco PA, Zorloni A, Parisi N, Casati P, Colombo A, et al. (2010)Sanitation of grapevine yellows affected cultivars of Lombardia region by hot water treatment. Petria 20:635-802.

20. Nicoli M (1985) La régénération des agrumes en Corse par la technique du micro-greffage des méristèmes in vitro.Fruits 40: 113-136.

21. Sanchez GE, Slack SA, Dodds JH (1991) Response of selectedSolanumspecies to virus eradication therapy. American Potato Journal 68: 299-315

22. Cupidi A, Barba M (1993) Ottimizzazione del microinnestoin vitroper il risanamento della vite. Vignevini 20:43-46.

23. Loebenstein G, Berger PH, Brunt AA, Lawsan RG (2001) Virus and Virus-like Diseases of Potatoes and Seed-potatoes. Kluwer Academic Publishers, Dodrescht.

24. Chalak L, Elbitar A, Masaad W, Choueiri E (2004) Assainissement de la pomme de terre infectée par le virus PVYNTNpar culture de méristèmes.Lebanese Science Journal 1: 37-44.

25. Chalak L, Elbitar A, Chehade T, El Zammar S, Jreijiri J, et al. (2007)Assainissement de variétés dePrunusinfectées par Prunus Necrotic Ring spot Virus. Lebanese Science Journal 8: 175-181.

26. Navarro L, Civerolo EL, Juarez J, Garnsey SM (1991) Improving therapy methods for citrus germplasm exchange. Extended Abstracts of the 11thConference of the International Organization of Citrus Virologists (IOCV).

27. Parmessur Y, Aljanabi S, Saumtally S, Dookun-Saumtally A (2002) Sugarcane yellow leaf virus and sugarcane yellow phytoplasma: Elimination by tissue culture. Plant Pathology 51:561-566.

28. ChalakL, ElbitarA, RizkR, ChoueiriE, SalarP, etal.(2005) AttemptstoeliminateCandidatusphytoplasmaphoenicium from infected Lebanese almond varieties by tissue culture techniques combined or not with thermotherapy. Eur J Plant Pathol 1: 85-89.

29. Murashige T, Skoog F (1962) A revised medium for rapid growth and bioassay with tobacco tissue cultures. PhysiologiaPlantarum 15:473-97.

30. Smart CD, Schneider B, Blomquist CL, Guerra LJ, Harrison NA, et al.(1996) Phytoplasma-specific PCR primers based on sequences of the 16S-23S rRNA spacer region. Applied and Environmental Microbiology 62:2988-2993.

31. Gundersen DE, Lee IM (1996) Ultrasensitive detection of phytoplasmas by nested-PCR assays using two universal primers pairs. PhytopathologiaMediterranea 35: 144-151.

32. Bertaccini A, Borgo M, Bertotto L, Bonetti A, Botti S, et al. (2001) Termoterapia e chemioterapia per eliminare i fitoplasmi da materiali di moltiplicazione della vite. L'Informatorio Agrario 42: 137-144.

33. Gribaudo I, Ruffa P, Cuozzo D, Gambino G, Marzachi C (2007) Attempts to eliminate phytoplasmas from grapevine clones by tissue culture techniques. Bulletin of Insectology 60: 315-316.

34. Augé R, Beauchesne G, Boccon-Gibod J, Decourtye L, Digat B, et al. (1989)La culturein vitroet ses applications horticoles. Ed Tec & Doc-Lavoisier, Paris.

CHAPTER 17

USE OF TISSUE CULTURE TECHNIQUES FOR PRODUCING VIRUS-FREE PLANT IN GARLIC AND THEIR IDENTIFICATION THROUGH REAL-TIME PCR

Hatıra Taşkın,[1] Gökhan Baktemur,[2] Mehmet Kurul,[3] and Saadet Büyükalaca[4]

[1]Department of Plant Production and Technologies, Faculty of Agricultural Sciences and Technologies, University of Niğde, 51240 Niğde, Turkey

[2]Department of Biology, Faculty of Arts and Science, University of Osmaniye Korkut Ata, 80000 Osmaniye, Turkey

[3]Roche Diagnostics, 34394 Istanbul, Turkey

[4]Department of Horticulture, Faculty of Agriculture, University of Çukurova, 01330 Adana, Turkey

ABSTRACT

This study was performed for comparison of meristem culture technique with shoot tip culture technique for obtaining virus-free plant, comparison of micropropagation success of two different nutrient media, and determination of effectiveness of real-time PCR assay for the detection of viruses. Two different garlic species (Allium sativum and Allium tuncelianum) and two different nutrient media were used in this experiment. Results showed that Medium 2 was more successful compared to Medium 1 for both A. tuncelianum and A. sativum (Kastamonu garlic clone). In vitro plants obtained via meristem and shoot tip cultures were tested for determination of onion yellow dwarf virus (OYDV) and leek yellow stripe virus (LYSV) through real-time PCR assay. In garlic plants propagated via meristem culture, we could not detect any virus. OYDV and LYSV viruses were detected in plants obtained via shoot tip culture. OYDV virus was observed in amount of 80% and 73% of tested plants for A. tuncelianum and A. sativum, respectively. LYSV virus was found in amount of 67% of tested plants of A. tuncelianum and in amount of 87% of tested plants of A. sativum in this study.

INTRODUCTION

Total garlic production of Turkey was 105, 363 tons [1], and 18–20% of this production is provided from Taşköprü district of Kastamonu province. Kastamonu garlic clone (A. sativum) is suitable for winter consumption, can be stored for a long time, and is suitable for processing due to dry matter content [2]. Allium tuncelianum, called Tunceli garlic, is only common in Tunceli province of Turkey, especially around Munzur Mountains in Ovacık district, and it is known as an endemic to this region [3, 4]. Unlike other garlics with multiple-cloved bulb, A. tuncelianum has single-cloved bulbs and has small formations like a small bulbs, and it can also produce fertile flowers and seeds. The most important problem of garlic production in Turkey is infected areas due to pests and diseases carried via contaminated seedling. Garlic production areas and seedling production areas are not separated in Turkey. Pests and diseases can be carried via contaminated material in garlic propagated vegetatively because of sexual sterility. The most important ones of these diseases and pests are nematode (Ditylenchus dipsaci), white rot disease (Sclerotium cepivorum), and viruses that cause a loss of between 30% and 100% of production and the most common viruses are onion yellow dwarf virus (OYDV) and leek yellow stripe virus (LYSV). Fidan et al. [5] reported that these two viruses together affect the plant negatively and result in a yield loss

Citation: Hatıra Taşkın, Gökhan Baktemur, Mehmet Kurul, and Saadet Büyükalaca, "Use of Tissue Culture Techniques for Producing Virus-Free Plant in Garlic and Their Identification through Real-Time PCR," The Scientific World Journal, vol. 2013, Article ID 781282, 5 pages, 2013. doi:10.1155/2013/781282

of up to 78%. There is no effective chemical control method against viruses directly. For this reason, the most common method for obtaining virus-free plant is meristem culture technique. Different researchers [6, 7] have also reported that shoot tip remained insufficient for obtaining the virus-free plants. Biological (mechanical inoculation, indexing), serological (enzyme-linked immunosorbent assay—ELISA), and molecular (polymerase chain reaction—PCR) methods are used for detection of viruses in plants and identification of these viruses. Real-time 4 PCR assay, which is one of the PCR types, is able to provide quantitative assessment. One of the main advantages of the real-time PCR is the very low chances of infection and detection of viruses with high efficiency. Detection of viruses through real-time PCR method has been performed successfully by many researchers such as Roberts et al. [8], Korimbocus et al. [9], Balaji et al. [10], Lunello et al. [11], Yılmaz [12], Mehle et al. [13, 14], and Fidan et al. [5].

The objectives of this study were (i) comparison of meristem culture technique with shoot tip culture technique to obtain virus-free plants, (ii) comparison of micropropagation success of two different nutrient media and two different garlic species (A. tuncelianum and A. sativum), (ii) to see the effectiveness of real-time PCR assay on detection of viruses.

MATERIAL AND METHODS

This study was conducted at the Department of Horticulture, University of Çukurova Turkey, and Adana Veterinary Control Institute, Turkey between the years 2009 and 2010. Allium sativum (Kastamonu garlic clone) and Allium tuncelianum garlic species determined to be infected with viruses were used as plant material. A. sativum and A. tuncelianum garlic samples were provided by Kastamonu-Taşköprü and Tunceli-Ovacık regions of Turkey, respectively.

Meristem and Shoot Tip Culture Studies

A. sativum and A. tuncelianum garlic species were propagated via meristem and shoot tip cultures. Cloves of garlic were waited in 25% sodium hypochlorite for 20 minutes for surface sterilization and washed 4-5 times with sterile water. Meristem and shoot tips from sterile cloves were extracted by sterile forceps and scalpels under a stereobinocular microscope in laminar flow and placed into glass tubes containing MS nutrient medium [15]. All cultures were incubated in the growth chamber at 25°C, under 3000 lux light and 8 hours dark and 16 h light photoperiod conditions. Germinated meristem and shoot tips were transferred to two different nutrient media (Medium 1: MS medium containing 0.5 mg L^{-1} 2-IP, 0.2 mg L^{-1} NAA, and 30 g L^{-1}sucrose; Medium 2: MS containing 2 mg L^{-1} BA, 0.5 mg L^{-1} IBA, and 30 g L^{-1} sucrose) for micropropagation. After 6 weeks, number of shoots per plant were recorded, and the shoots were subcultured. The experiment was designed in a completely randomized experimental design with three replications and thirty explants included per replication.

Variance analysis was conducted to evaluate the results, and Tukey test was used for controlling the significance of the differences.

Real-Time PCR Studies

Sixty in vitro plants (15 of them from meristem culture of A. tuncelianum, 15 of them from shoot tip culture ofA. tuncelianum, 15 of them from meristem culture of A. sativum, and 15 of them from shoot tip culture of A. sativum) were used for real-time PCR studies. High Pure Viral Nucleid Acid Kit was used for RNA extraction of garlic samples. 200 µL working solution and 50 µL proteinase K were added to 1.5 mL nuclease-free microcentrifuge tube containing 100 mg plant samples and pulverized. After the mixture was mixed, it was incubated for 10 min at 72°C. Then, 100 µL binding buffer was added and mixed. Mixture was transferred into the upper reservoir of combination of high pure filter tube and the collection tube and spun for 1 min at 8,000 g in a centrifuge. Collection tube was removed and filter tube was placed into a new collection tube. Then 500 µL inhibitor removal buffer was added and spun for 1 min at 8,000 g in a centrifuge. Collection tube was changed with a new one. Then 500 µL wash buffer was added and spun for 1 min at 10,000 g in a centrifuge. The collecting tube was changed, and the washing step was repeated. Filter tube was placed into new sterile 1.5 mL nuclease-free microcentrifuge tube, and 75 µL elution buffer to elute the viral nucleic acid followed by spinning for 1 min at 13,000 g in a centrifuge. Filter tubes placed in the Eppendorf tubes were removed, and purified viral nucleic acid was extracted.

Obtained pure viral nucleic acids were converted to cDNA with reverse transcription PCR (RT-PCR) and incubated at −15/−20°C in a freezer. Transcriptor High Fidelity cDNA Synthesis Kit was used for converting RNA to cDNA. This process was completed in two steps. 2 µL random primer, 1 µL PCR-grade water, and 8.4 µL RNA per sample were used to prepare PCR mixture in the first step. This mixture was incubated at 65°C for 10 min. Prepared mixture in the second step consisted of 4 µL transcriptor reverse transcriptase reaction tampon, 0.5 µL protector RNAse inhibitor, 2 µL deoxynucleotide mix, 1 µL DTT, and 1.1 µL transcriptor reverse transcriptase enzyme. These two mixtures were mixed and incubated at 45–55°C for 10–30 min and then at 85°C for 5 min. Thereafter, cDNA synthesis was completed and stored at −20°C. Real-time PCR mix-conducted in a 20 µL volume containing 1xSYBR Green master mix (Roche applied science), 5 nM forward-reverse primers, and PCR grade water, using the following program: 10 min at 95°C for activation of DNA polymerase, 45 cycles of 10 s at 95°C, 30 s at 60°C, 1 s at 72°C, 30 min at 40°C and 30 s at 40°C for cooling in Roche LightCycler 2.0 Real-time PCR. Evaluation was performed by qualitative detection program of this thermocycler. Mixture containing 5 µL isolated DNA from samples, 2 µL SYBER Green master (it consisted, fast start Taq DNA polymerase, reaction buffer, $MgCl_2$, dNTP mix, and SYBER Green-Roche), 5 µL forward and reverse primers, 2.4 mM Mg, and PCR-grade water (total volume 20 µL) was prepared for the detection in real-time PCR using 600 s at 95°C for denaturation of DNA, 45 cycles of 10 s at 95°C, 30 s at 60°C, 2 min at 72°C, and 30 min at 40°C for cooling.

SYBER Green was used for detection of OYDV and LYSV viruses. The primers used in this study (Table 1) for OYDV and LYSV viruses were performed using primers registered in genbank [11] as forward OYDV primer (F-OYDV), 1099–1122 positions of sequences registered as AJ292223 in GenBank

Database, as reverse OYDV primer (R-OYDV) 1306–1325 positions of AJ292223, as forward LYSV primer (F-LYSV) 789–808 positions of sequences registered as AY007693, and as reverse LYSV primer (R-LYSV) 839–854 positions of AY007693 [11].

Table 1: Primers used in the study [11]

Primer		bp
Forward-LYSV	5-TCTCGCACGGTATGCATTTG-3	20
Reverse-LYSV	5-GCCTCGCGCGCTCTAA-3	16
Forward-OYDV	5-AGTGATGCAGCTGAAGCATACATT-3	24
Reverse-OYDV	5-ACGTTACCATCCAGGCCAAA-3	20

RESULTS

Ninety shoot tips and ninety meristems were used for A. tuncelianum in this experiment. Meristems transformed to plants in eight weeks. While 85 of meristems could grow without any problems, three of them could not germinate, and as infection problem was observed in two meristems. Cultured shoot tips were turned into plants in 3-4 weeks. Eighty-four shoot tips germinated healthy; only 1 shoot tip could not grow and 5 of ninety shoot tips were found to be infected. Plants obtained meristems of A. tuncelianum were transferred to two different nutrient media for micropropagation. Number of shoots were recorded in both first micropropagation and subculture were higher in Medium 2 compared to Medium 1, and the differences between the two media were found to be significant statistically. In the first propagation, 14.10 shoots/plant and 4.63 shoots/plant were obtained from Medium 2 and Medium 1, respectively. Similar results were provided from subculture. 13.27 shoots/plant in Medium 2 and 3.93 shoots/plant in Medium 1 were determined. Shoot tip culture results showed that Medium 2 was better compared to Medium 1. 11.37 shoots/plant and 2.41 shoots/plant were obtained from Medium 2 and Medium 1 in the first propagation, respectively. 8.93 shoots/plant and 2.36 shoots/plant were observed in Medium 2 and Medium 1 in subculture, respectively. Considered in terms of explants, meristem explants were found to be more successful compared to shoot tip explants in both garlic species and nutrient media.

The experiment for A. sativum was designated with 90 meristems and 90 shoot tips. While 89 meristems could transform into plant in meristem culture (the remaining one meristem was probably damaged), 87 shoot tips could germinate in shoot tip culture (the remaining 3 plants were infected). Meristem and shoot tip culture results of A. sativum showed that Medium 2 was more successful compared to Medium 1. Number of shoots/plant obtained from different tissue culture techniques and different garlic species are given in Table 2. As seen in Table 2, 11.67 shoots/plant and 5.41 shoots/plant were counted in Medium 2 and Medium 1 in the first micropropagation, respectively. In subculture, 11.61 shoots/plant and 5.56 shoots/plant were obtained on Medium 2 and Medium 1, respectively. These results compared with results of A. tuncelianum; number of shoots of A. tuncelianum in Medium 2 were higher than of A. sativum in both nutrient media. In Medium 1, A. sativum gave slightly higher results compared to A. tuncelianum. In shoot tip culture of A. sativum, Medium 2 and Medium 1 gave 9.84 shoots/plant and 2.93 shoots/plant for the first micropropagation, respectively. 10.17 shoots/plant from Medium 2 and 3.61 shoots/plant from Medium 1 were obtained in subculture.

Table 2: Number of shoots/plant obtained from different tissue culture techniques on different garlic species

Explant Type	First micropropagation (shoots/plant)			Sub-culture (shoots/plant)		
	Medium 1	Medium 2	LSD (5%)	Medium 1	Medium 2	LSD (5%)
Meristem culture (AT)	4.63^b	14.10^a	0.86	3.93^b	13.27^a	1.86
Shoot tip culture (AT)	2.41^b	11.37^a	1.73	2.36^b	8.93^a	2.16
Meristem culture (AS)	5.41^b	11.67^a	1.21	5.56^b	11.61^a	2.58
Shoot tip culture (AS)	2.98^b	9.84^a	1.73	3.61^b	10.17^a	2.16

AT: *A. tuncelianum*; AS: *A. sativum*.

a,b Means followed by different letters in the same column are significantly different regarding Tukey's multiple range test at $P < 0.05$.

Sixty garlic plants propagated via meristem and shoot tip cultures were tested in terms of OYDV and LYSV viruses through real-time PCR method. Specific primer sequences and specific DNA fragments replicated with these sequences were detected with SYBER Green marking system. OYDV and LYSV viruses were not observed in plants obtained from meristem culture of both A. tuncelianum and A. sativum. Viruses were detected in shoot tip culture. OYDV virus was determined in 12 plants of A. tuncelianum and 11 plants of A. sativum. Ten plants of A. tuncelianum and 13 plants of A. sativum were found to be contaminated with LYSV virus. While OYDV virus was observed in the amount of 80% of tested plants and the amount of 73% of tested plants for A. tuncelianum and A. sativum, respectively, LYSV virus was found in the amount of 67% of tested plants of A. tuncelianum and the amount of 87% of tested plants of A. sativum in this study. Amplification curves of meristem and shoot tip cultures are given in Figures 1 and 2.

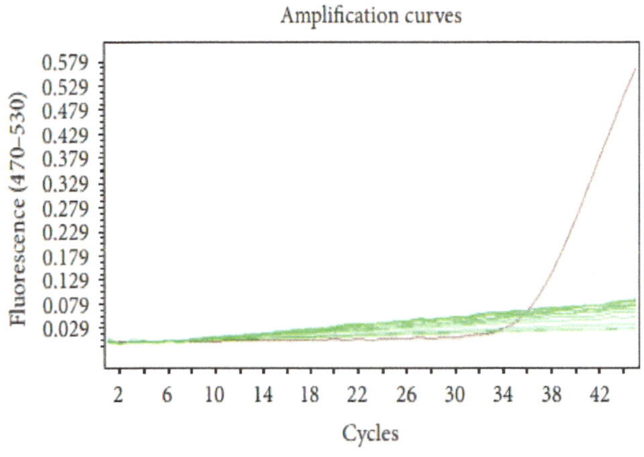

Figure 1: Application curves for meristem culture.

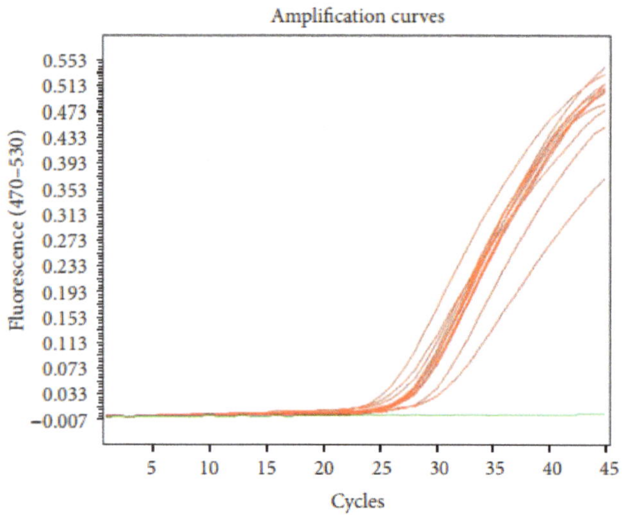

Figure 2: Application curves for shoot tip culture.

DISCUSSION

Results of this research showed clearly that meristem culture technique was more effective compared to shoot tip culture technique in obtaining virus-free plant. All plantlets propagated by meristem culture technique were found completely clear in terms of OYDV and LYSV. Whereas virus problem could not be solved by shoot tip culture technique, viruses were detected in the majority of plantlets obtained by this technique. In meristem culture, only apical dome and a few leaf primordia are isolated and placed to nutrient media [16]. Tips of growing shoots (2 cm or less than 2 cm) are used in shoot tip culture. According to some researchers, there is a competition between cell proliferation, and the formation of the virus particles in meristem region of plant. Nucleic acid production capacity in meristematic tissue during cell division is used for cell division and this situation prevents the reproduction of virus. According to other researchers, transportation of viruses to the meristem region of the plant is prevented due to lack of transport system in meristem [17].

ELISA and real-time PCR techniques are used commonly in diagnostics of plant viruses. Real-time PCR method is more sensitive and specific than other techniques. Therefore, recently researchers prefer to use this technique alone or as comparison with ELISA. Many researchers have used real-time PCR method to identify viruses in different plants: for Plum Pox virus in plum by Varga and James [18], for OYDV and LYSV viruses in garlic by Lunello et al. [11], for zucchini yellow mosaic virus (ZYMV) in squash by Çalışkan [19], and for OYDV and SLV viruses in onion and garlic by Fidan et al. [5]. Real-time PCR method was found to be more sensitive than other methods in diagnosis of potato virus Y (PVY) by Mehle et al. [14]. Therefore, real-time PCR method was used in this study to detect viruses in garlic plants.

Also two different nutrient media were tested in terms of micropropagation success. Positive effect of BA and IBA on micropropagation of garlic was determined by Baktır [20]. NAA and 2-IP were used for in vitropropagation of garlic by Bhojwani [6]. Therefore, effects of these hormones on micropropagation of garlic were tested through using two different nutrient media in this study. Medium 2 containing $2\,mg\,L^{-1}$ BA and $0.5\,mg\,L^{-1}$ IBA was found to be more successful compared to Medium 1 containing $0.5\,mg\,L^{-1}$ 2-IP and $0.2\,mg\,L^{-1}$NAA in this study.

One of the most important encountered problems in the production of garlic in Turkey is viruses. However, there is not any effective chemical application available against virus control. Meristem culture technique is widely used for obtaining virus-free plants. However, extracting the meristem regions of the plant is difficult, time-consuming and requires expertise. Therefore after achieving certain number of meristem, plants obtained from these meristems are multiplied using tissue culture techniques for commercial production. In this study, micropropagation capacity of A. tuncelianum, an endemic species of Turkey, and Kastamonu garlic clone of A. sativum were evaluated. The results obtained from the present study are very important for the scientist trying to develop virus-free plants and could also have significance of practical application which will in turn affect positively to increase yield with better quality of crop for the local farmers.

ACKNOWLEDGMENTS

The authors would like to thank the Republic of Turkey Ministry of Food, Agriculture and Livestock-General Directorate of Agricultural Research and Policy (TAGEM) Research Projects Coordinating Office (TAGEM/11/AR-GE/20), Roche Company, Çukurova University and Adana Veterinary Control Institute for supporting this research.

REFERENCES

1. Anonymous, http.www.fao.org, 2009.
2. N. Türkeş, Researches on breeding of Turkey garlic varieties through selection [Ph.D. thesis], University of Ankara, Ankara, Turkey, 1978.
3. M. Koyuncu and A. Güvenç, "Endemic Allium species of Turkey," in The Scientific and Technological Research Council of Turkey (TÜBİTAK) Project (Project No: TBAG-1089), Ankara, Turkey, 1994.
4. E. Yazar, Propagation of Tunceli garlic Allium tuncelianum (Kollman, N. Özhatay, D. Matthew, Ş. Şiraneci) through in vitro root and shoot tip culture [M.S. thesis], University of Ankara, Ankara, Turkey, 2006.
5. H. Fidan, S. Baloğlu, G. Koç, and N. Birişik, "New virus diseases for Turkey detected in onion and garlic: onion yellow dwarf virus and shallot latent virus," in Turkey›s National 3rd Plant Protection Congress, p. 131, Van, Turkey, July 2009.
6. S. S. Bhojwani, "In vitro propagation of garlic by shoot proliferation," Scientia Horticulturae, vol. 13, no. 1, pp. 47–52, 1980. ·
7. D. Walkey, M. J. W. webb, C. J. bolland, and A. Miller, "Production of virus-free garlic (Allium sativumL.) and shallot (A. ascalonicum L.) by meristem-tip culture," The Journal of Horticultural Science and Biotechnology, vol. 62, pp. 211–220, 1987.
8. C. A. Roberts, R. G. Dietzgen, L. A. Heelan, and D. J. MaclLean, "Real-time RT-PCR fluorescent detection of tomato spotted wilt virus," Journal of Virological Methods, vol. 88, no. 1, pp. 1–8, 2000. · ·

9. J. Korimbocus, D. Coates, I. Barker, and N. Boonham, "Improved detection of Sugarcane yellow leaf virus using a real-time fluorescent (TaqMan) RT-PCR assay," Journal of Virological Methods, vol. 103, no. 2, pp. 109–120, 2002. · ·
10. B. Balaji, D. B. Bucholtz, and J. M. Anderson, "Barley yellow dwarf and cereal yellow dwarf virus quantification by real-time polymerase chain reaction in resistant and susceptible plants,"Phytopathology, vol. 93, no. 11, pp. 1386–1392, 2003. ·
11. P. Lunello, C. Mansilla, V. Conci, and F. Ponz, "Ultra-sensitive detection of two garlic potyviruses using a real-time fluorescent (Taqman) RT-PCR assay," Journal of Virological Methods, vol. 118, no. 1, pp. 15–21, 2004. · ·
12. S. Yılmaz, Obtaining and rapid identification of Pseudomonas corrugata and Pseudomonas mediterranea isolates, causal agents of tomato pith necrosis disease in western Mediterranean region by multiplex-PCR and Real-time PCR [M.S. thesis], University of Akdeniz, Antalya, Turkey, 2005.
13. N. Mehle, J. Boben, and M. Ravnikar, "Real-time PCR in diagnosis of plant viruses," in 3rd international qPCR symposium: industrial exhibition and TATAA application workshop, V. Pfaffl and W. Michael, Eds., pp. 26–30, Freising-Weihenstephan,, Germany, March 2007, qPCR, 2007 event : proceedings. Freising: Technical Univesity Munich. Str. 54, 2007.
14. N. Mehle, M. Kovač, N. Petrovič et al., "Spread of potato virus Y^{NTN} in potato cultivars (Solanum tuberosum L.) with different levels of sensitivity," Physiological and Molecular Plant Pathology, vol. 64, no. 6, pp. 293–300, 2004. · ·
15. T. Murashige and F. Skoog, "A revised medium for rapid growth and bioassays with tobacco cultures,"Physiologia Plantarum, vol. 15, no. 3, pp. 473–497, 1962. ·
16. N. Espinoza, R. Estrada, P. Tovar, J. Bryan, and J. H. Dodds, Tissue Culture Micropropagation, Conservation and Export of Potato Germplasm, International Potato Center (CIP), Lima-Peru, Specialized Technology, 1984.
17. R. Hatipoğlu, Plant Biotechnology, Çukurova University Press, 1997.
18. Varga and D. James, "Real-time RT-PCR and SYBR Green I melting curve analysis for the identification of Plum pox virus strains C, EA, and W: effect of amplicon size, melt rate, and dye translocation," Journal of Virological Methods, vol. 132, no. 1-2, pp. 146–153, 2006. · ·
19. F. Çalışkan, Identification of Zucchini Yellow Mosaic Virus (ZYMV) isolate and search of possible using of plant activators in ZYMV control [Ph.D. thesis], University of Çukurova, Adana, Turkey, 2007.
20. İ. Baktır, "Propagation of Tunceli garlic (Allium tuncelianum) via in vitro techniques," in Turkey›s National GAP 4th Agriculture Congress, pp. 206–208, Şanlıurfa-Turkey, September 2005.

INDEX

P

Q

R

S

T

U

Z